Texts in Applied Mathematics 52

Texts in Applied Mathematics

(continued after index)

Mark H. Holmes

Introduction to Numerical Methods in Differential Equations

Mark H. Holmes
Academic Science of the Material Science and Engineering
Rensselaer Polytechnic Institute
Troy, NY 12180
holmes@rpi.edu

Series Editors
J.E. Marsden
Control and Dynamical Systems, 107–81
California Institute of Technology
Pasadena, CA 91125
USA
marsden@cds.caltech.edu

L. Sirovich
Division of Applied Mathematics
Brown University
Providence, RI 02912
USA
chico@camelot.mssm.edu

S.S. Antman
Department of Mathematics
and
Institute for Physical Science
  and Technology
University of Maryland
College Park, MD 20742-4015
USA
ssa@math.umd.edu

Mathematics Subject Classification (2000): 65L05, 65L06, 65L07, 65L12, 65M12, 65M70,
                                          65N12, 65N22, 65N35, 65N40, 68U05, 74S20

Library of Congress Control Number: 2006927786

ISBN-10: 0-387-30891-1
ISBN-13: 978-0387-30891-3

Printed on acid-free paper.

9 8 7 6 5 4 3 2 1

springer.com

To my parents

Preface

The title gives a reasonable first-order approximation to what this book is about. To explain why, let's start with the expression "differential equations." These are essential in science and engineering, because the laws of nature typically result in equations relating spatial and temporal changes in one or more variables. To develop an understanding of what is involved in finding solutions, the book begins with problems involving derivatives for only one independent variable, and these give rise to ordinary differential equations. Specifically, the first chapter considers initial value problems (time derivatives), and the second concentrates on boundary value problems (space derivatives). In the succeeding four chapters problems involving both time and space derivatives, partial differential equations, are investigated.

This brings us to the next expression in the title: "numerical methods." This is a book about how to transform differential equations into problems that can be solved using a computer. The fact is that computers are only able to solve discrete problems and generally do this using finite-precision arithmetic. What this means is that in deriving and then using a numerical algorithm the correctness of the discrete approximation must be considered, as must the consequences of round-off error in using floating-point arithmetic to calculate the answer. One of the interesting aspects of the subject is that what appears to be an obviously correct numerical method can result in complete failure. Consequently, although the book concentrates on the derivation and use of numerical methods, the theoretical underpinnings are also presented and used in the development.

This brings us to the remaining principal word in the title: "introduction." This has several meanings for this book, and one is that the material is directed to those who are first learning the subject. Typically this includes upper-division undergraduates and beginning graduate students. The objective is to learn the fundamental ideas of what is involved in deriving a numerical method, including the role of truncation error, and the importance of stability. It is also essential that you actually use the methods to solve problems. In other words, you run code and see for yourself just how successful, or unsuccessful, the method is for solving the problem. In conjunction with this it is essential that those who do computations develop the ability to effectively communicate the results to others. The only way to learn this is to do it. Consequently, homework assignments that involve an appreciable amount of computing are important to learning the material in this book. To help with this, a library of sample code for the topics covered is available at www.holmes.rpi.edu. Speaking of which, many of the problems considered in the book result in solutions that are time-dependent. To help visualize the dynamical nature of the solution, movies are provided for some of the example problems. These are identified in the book with an (M) in the caption of the associated figure.

Another meaning for "introduction" as concerns this textbook is that the subject of each chapter can easily produce one or more volumes in its own right. The intent here is to provide an introduction to the subject, and that means certain topics are either not discussed or they are presented in an abbreviated form. All told, the material included should fill a semester course. For those who might want a more in-depth presentation on a specific topic, references are provided throughout the text.

The prerequisites for this text include an introductory undergraduate course in differential equations and a basic course in numerical computing. The latter would include using LU to solve matrix equations, polynomial interpolation, and numerical differentiation and integration. Some degree of computing capability is also required to use the methods that are derived. Although no specific language or program is required to read this book, the codes provided at www.holmes.rpi.edu use mostly MATLAB, and the movies provided require QuickTime.

I would like to express my gratitude to the many students who took my course in numerical methods for differential equations at Rensselaer. They helped me immeasurably in understanding the subject and provided much-needed encouragement to write this book. It is also a pleasure to acknowledge the suggestions of Yuri Lvov, who read an early version of the manuscript.

Troy, New York   *Mark H. Holmes*
January, 2006

Contents

1

Initial Value Problems

1.1 Introduction

Even from casual observation it is apparent that most physical phenomena vary both in space and time. For example, the temperature of the atmosphere changes continuously at any given location and it varies significantly from point to point over the surface of the Earth. A consequence of this is that mathematical models of the real world almost inevitably involve both time and space derivatives. The objective of this book is to examine how to solve such problems using a computer; but to begin, we first consider more simplified situations. In this chapter we study problems involving only time derivatives and then in the next chapter we examine spatial problems. The remaining chapters then examine what happens when both time and space derivatives are present together in the problem.

A general form of the type of problem we consider is

$$\mathbf{y}'(t) = \mathbf{f}(t, \mathbf{y}), \quad \text{for } 0 < t, \tag{1.1}$$

where the initial condition is $\mathbf{y}(0) = \mathbf{a}$. The differential equation along with the initial condition form what is known as an initial value problem (IVP). It is assumed throughout the chapter that the IVP is well posed (i.e., there is a unique solution that is a smooth function of time). By smooth it is meant that $\mathbf{y}(t)$ and its various derivatives are defined and continuous.

It is the sad fact that most real-world problems are so complicated that there is no hope of finding an analytical solution. An example is shown in Figure 1.1. To study molecular machinery such as nanogears it is necessary to solve a system involving thousands of equations with a very complicated nonlinear function $\mathbf{f}$ in (1.1). The consequence of this is that numerical solutions are required. This brings us to the objective of this chapter, which is to develop an understanding of how to derive finite difference approximations for solving initial value problems (IVPs). In anticipation of this we identify a few IVPs that are used as test problems in this chapter.

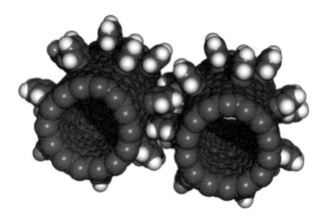

Figure 1.1. (M) These nanogears are composed of carbon (the grey spheres) and hydrogen (the white spheres) atoms. The rotation of the tubes, and the resulting meshing of the gear teeth, was carried out by solving a large system of IVPs (Han et al. [1997])

1.1.1 Examples of IVPs

Radioactive Decay

According to the law of radioactive decay, the mass of a radioactive substance decays at a rate that is proportional to the amount present. To express this in mathematical terms, let $y(t)$ designate the amount present at time t. In this case the decay law can be expressed as

$$\frac{dy}{dt} = -ry, \quad \text{for } 0 < t. \tag{1.2}$$

If we start out with an amount α of the substance then the corresponding initial condition is

$$y(0) = \alpha. \tag{1.3}$$

In the decay law (1.2), r is the proportionally constant and it is assumed to be positive. Because the largest derivative in the problem is first order, this is an example of a first-order IVP for $y(t)$. It is also linear, homogeneous, and has constant coefficients. Using an integrating factor, or separation of variables, one finds that the solution is

$$y(t) = \alpha e^{-rt}. \tag{1.4}$$

Consequently, the solution starts at α and decays exponentially to zero as time increases.

To put a slightly different spin on this, recall that $y = Y$ is an equilibrium, or steady-state, solution if it is constant and satisfies the differential equation.

Also, a steady-state Y is stable if any solution that starts near Y stays near it. If, in addition, initial conditions starting near Y actually result in the solution converging to Y as $t \to \infty$, then $y = Y$ is said to be asymptotically stable. With the solution in (1.4) we conclude that $y = 0$ is an asymptotically stable equilibrium solution for (1.2).

Logistic Equation

In the study of populations limited by competition for food one obtains the logistic equation, which is

$$\frac{dy}{dt} = \lambda y(1 - y), \quad \text{for } 0 < t, \tag{1.5}$$

where

$$y(0) = \alpha. \tag{1.6}$$

It is assumed that λ and α are positive. As with radioactive decay, this IVP involves a first-order equation for $y(t)$. However, because of the y^2 term, this equation is nonlinear. It is possible to find the solution using separation of variables, and the result is

$$y(t) = \frac{\alpha}{\alpha + (1 - \alpha)e^{-\lambda t}}. \tag{1.7}$$

Now, the equilibrium solutions for this equation are $y = 1$ and $y = 0$. Because $\lambda > 0$, the solution approaches $y = 1$ as t increases. Consequently, $y = 1$ is an asymptotically stable equilibrium solution, whereas $y = 0$ is not.

Newton's Second Law

The reason for the prominence of differential equations in science and engineering is that they are the foundation for the laws of nature. The most well known of these laws is Newton's second, which states that $F = ma$. Letting $y(t)$ designate position then this law takes the form

$$m\frac{d^2 y}{dt^2} = F(t, y, y'), \quad \text{for } 0 < t. \tag{1.8}$$

The above equation allows for the possibility that the force F varies in time as well as depends on position and velocity. Assuming that the initial position and velocity are specified, then the initial conditions for this problem take the form

$$y(0) = \alpha \text{ and } y'(0) = \beta. \tag{1.9}$$

This IVP is second order and it is nonlinear if the force depends nonlinearly on either y or y'.

It is possible to write the problem as a first-order system by introducing the variables

$$y_1 = y, \qquad (1.10)$$
$$y_2 = y'. \qquad (1.11)$$

Differentiating each of these equations, and using the original differential equation (1.8), we obtain the following

$$y_1' = y_2,$$
$$y_2' = \frac{1}{m} F(t, y_1, y_2).$$

By introducing the vector $\mathbf{y}(t)$, defined as

$$\mathbf{y} = \begin{pmatrix} y_1 \\ y_2 \end{pmatrix}, \qquad (1.12)$$

the IVP can be written as

$$\mathbf{y}'(t) = \mathbf{f}(t, \mathbf{y}), \quad \text{for } 0 < t, \qquad (1.13)$$

where the initial conditions (1.9) take the form

$$\mathbf{y}(0) = \begin{pmatrix} \alpha \\ \beta \end{pmatrix}. \qquad (1.14)$$

The function $\mathbf{f}(t, \mathbf{y})$ appearing in (1.13) is

$$\mathbf{f}(t, \mathbf{y}) = \begin{pmatrix} y_2 \\ \frac{1}{m} F(t, y_1, y_2) \end{pmatrix}. \qquad (1.15)$$

What is significant is that the change of variables has transformed the second-order problem for $y(t)$ into a first-order IVP for $\mathbf{y}(t)$. Like the original, (1.13) is nonlinear if F depends nonlinearly on either y_1 or y_2.

As an illustration of this transformation, for a linear mass–spring–dashpot system the force consists of a contribution from a spring $(-ky)$ and from a dashpot $(-cy')$. The resulting equation of motion is

$$my'' = -cy' - ky. \qquad (1.16)$$

This can be written as a first-order system as in (1.13), where

$$\mathbf{f}(t, \mathbf{y}) = \begin{pmatrix} y_2 \\ \frac{1}{m}(-cy_2 - ky_1) \end{pmatrix}. \qquad (1.17)$$

1.2 Methods Obtained from Numerical Differentiation

The task we now undertake is to approximate the differential equation, and its accompanying initial condition, with a problem we can solve using a computer. To explain how this is done we consider the problem of solving

$$\frac{dy}{dt} = f(t,y), \quad \text{for } 0 < t, \tag{1.18}$$

where

$$y(0) = \alpha. \tag{1.19}$$

The function $f(t,y)$ in assumed to be given. For example, with radioactive decay $f(t,y) = -ry$ and for the logistic problem $f(t,y) = \lambda y(1-y)$. The question is, can we accurately compute the solution directly from the problem without first finding an analytical solution? As it turns out, most realistic mathematical models of physical and biological systems cannot be solved by hand, so having the ability to find accurate numerical solutions directly from the original equations is an invaluable tool.

1.2.1 The Five Steps

To explain how we will construct a numerical algorithm that can be used to solve (1.18) it should be noted that the variables in this problem, t and y, are continuous. Our objective is to replace these with discrete variables so that the resulting problem is algebraic and therefore solvable using standard numerical methods. Great care must be taken in making this replacement, because the computed solution must accurately approximate the solution of the original IVP. The approach we take proceeds in a sequence of five steps, and these steps will serve as a template used throughout this book.

One point to make before beginning is that the computer cannot run forever. Therefore, we must specify just how large a time interval will be used in computing the solution. It is assumed in what follows that the interval is $0 \le t \le T$.

STEP 1. We first introduce the time points at which we will compute the solution. These points are labeled sequentially as $t_0, t_1, t_2, \ldots, t_M$ and a schematic drawing indicating their location along the time axis is shown in Figure 1.2. We confine our attention to a uniform grid with step size k, so, the formula for the time points is

$$t_j = jk, \quad \text{for } j = 0, 1, 2, \ldots, M. \tag{1.20}$$

Because the time interval is $0 \le t \le T$ we require $t_M = T$. Therefore, k and M are connected through the equation

$$k = \frac{T}{M}. \tag{1.21}$$

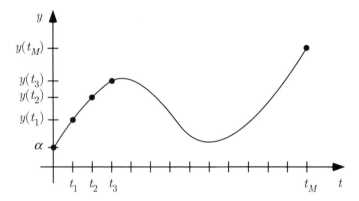

Figure 1.2. Grid system used to derive a finite difference approximation of the initial value problem. The points are equally spaced and $t_M = T$.

STEP 2. Evaluate the differential equation at the time point $t = t_j$ to obtain

$$y'(t_j) = f(t_j, y(t_j)). \qquad (1.22)$$

STEP 3. Replace the derivative term in STEP 2 with a finite difference formula using the values of y at one or more of the grid points in a neighborhood of t_j. This is where things get a bit interesting, because numerous choices can be made, a few of which are listed in Table 1.1. Different choices result in different numerical procedures, and as it turns out, not all choices will work. To start we take the first entry listed in Table 1.1, which means we use the following expression for the first derivative:

$$y'(t_j) = \frac{y(t_{j+1}) - y(t_j)}{k} + \tau_j, \qquad (1.23)$$

where

$$\tau_j = -\frac{k}{2}y''(\eta_j) \qquad (1.24)$$

and η_j is a point between t_j and t_{j+1}. Introducing this into (1.22) we obtain

$$\frac{y(t_{j+1}) - y(t_j)}{k} + \tau_j = f(t_j, y(t_j)), \qquad (1.25)$$

or equivalently,

$$y(t_{j+1}) - y(t_j) + k\tau_j = kf(t_j, y(t_j)). \qquad (1.26)$$

A couple of pithy comments are in order here. First, the difference formula in (1.23) uses a value of t ahead of the current position. For this reason it is referred to as a forward difference formula for the first derivative. Second, the term τ_j, as it appears in (1.25), represents how well we have approximated

Type	Difference Formula	Truncation Term
Forward	$f'(x_i) = \dfrac{f(x_{i+1})-f(x_i)}{h} + \tau_i$	$\tau_i = -\dfrac{h}{2}f''(\eta_i)$
Backward	$f'(x_i) = \dfrac{f(x_i)-f(x_{i-1})}{h} + \tau_i$	$\tau_i = \dfrac{h}{2}f''(\eta_i)$
Centered	$f'(x_i) = \dfrac{f(x_{i+1})-f(x_{i-1})}{2h} + \tau_i$	$\tau_i = -\dfrac{h^2}{6}f'''(\eta_i)$
One-sided	$f'(x_i) = \dfrac{-f(x_{i+2})+4f(x_{i+1})-3f(x_i)}{2h} + \tau_i$	$\tau_i = \dfrac{h^2}{3}f'''(\eta_i)$
One-sided	$f'(x_i) = \dfrac{3f(x_i)-4f(x_{i-1})+f(x_{i-2})}{2h} + \tau_i$	$\tau_i = \dfrac{h^2}{3}f'''(\eta_i)$
Centered	$f''(x_i) = \dfrac{f(x_{i+1})-2f(x_i)+f(x_{i-1})}{h^2} + \tau_i$	$\tau_i = -\dfrac{h^2}{12}f''''(\eta_i)$

Table 1.1. Numerical differentiation formulas. The points $x_1, x_2, x_3, \ldots$ are equally spaced with step size $h = x_{i+1} - x_i$. The point η_i is located between the left- and rightmost points used in the formula.

the original problem. For this reason it is the truncation error for the method, and from (1.24) it is seen that it is $O(k)$. It is essential that whatever approximations we use, the truncation error goes to zero as k goes to zero. This means that, at least in theory, we can approximate the original problem as accurately as we wish by making the time step k small enough. It is said in this case that the approximation is consistent. Unfortunately, as we demonstrate shortly, consistency is not enough to guarantee an accurate numerical solution.

STEP 4. Drop the truncation error. This is the step where we go from an exact problem to one that is, hopefully, an accurate approximation of the original. After dropping τ_j in (1.26) the resulting equation is

$$y_{j+1} - y_j = kf(t_j, y_j), \qquad (1.27)$$

or equivalently,

$$y_{j+1} = y_j + kf(t_j, y_j), \quad \text{for } j = 0, 1, 2, \ldots, M - 1. \qquad (1.28)$$

From the initial condition (1.19) we have that the starting value is

$$y_0 = \alpha. \qquad (1.29)$$

The finite difference equation (1.28) is known as the Euler method for solving (1.18). It is a recursive algorithm in which one starts with $j = 0$ and then uses

(1.28) to determine the solution at $j = 1$, then $j = 2$, then $j = 3$, etc. Because (1.28) gives the unknown y_{j+1} explicitly in terms of known quantities, it is an explicit method.

Example
Let's see how well Euler's method does with the logistic equation (1.5). Specifically, suppose the IVP is

$$\frac{dy}{dt} = 10y(1 - y), \quad \text{for } 0 < t, \tag{1.30}$$

where

$$y(0) = 0.01. \tag{1.31}$$

We will use the Euler method to calculate the solution for $0 \le t \le 1$. In this case, using (1.21), k and M are connected through the equation

$$k = \frac{1}{M}. \tag{1.32}$$

For this example, the finite difference equation in (1.28) takes the form

$$y_{j+1} = y_j + 10ky_j(1 - y_j), \quad \text{for } j = 0, 1, 2, \ldots, M - 1. \tag{1.33}$$

Taking $M = 6$, so $k = \frac{1}{6}$, the first few steps using the Euler method are shown in Table 1.2. For a more graphical picture of the situation, the exact solution, given in (1.7), and computed solutions are also shown in Figure 1.3 using successively smaller values of the time step k or, equivalently, larger values of M. It is seen that the numerical solution with $M = 4$ is not so good, but the situation improves considerably as more time points are used. In fact, it would appear that if we keep increasing the number of time points that the numerical solution converges to the exact solution. Does this actually happen? Answering this question brings us to the very important concept of error.

Error

As illustrated in Table 1.2, at each time point we have three different solutions, and they are

$$y(t_j) \equiv \text{exact solution of the IVP at } t = t_j; \tag{1.34}$$
$$y_j \equiv \text{exact solution of finite difference equation at } t = t_j; \tag{1.35}$$
$$\overline{y}_j \equiv \text{solution of difference equation at } t = t_j \text{ calculated}$$
$$\text{by the computer.} \tag{1.36}$$

We are interested in the difference between the exact solution of the IVP and the values we actually end up computing using our algorithm. Therefore, we are interested in the error $e_j = |y(t_j) - \overline{y}_j|$. The question we are going to ask

Index	Time t_j	Exact Solution $y(t_j)$ of Logistic Equation	Exact Solution y_j of Euler Equation	Computed Solution $\bar{y}_j$ of Euler Equation	$y(t_j) - y_j$	$y_j - \bar{y}_j$
$j = 0$	0	$\frac{1}{100}$	$\frac{1}{100}$	0.01	0	0
$j = 1$	$\frac{1}{6}$	$\left(1+99e^{-5/3}\right)^{-1}$	$\frac{53}{2000}$	2.6500e-02	2.43e-02	-3.47e-18
$j = 2$	$\frac{1}{3}$	$\left(1+99e^{-10/3}\right)^{-1}$	$\frac{55597}{800000}$	6.9496e-02	1.51e-01	-1.39e-17
$j = 3$	$\frac{1}{2}$	$\left(1+99e^{-5}\right)^{-1}$	$\frac{68073133591}{384000000000}$	1.7727e-01	4.23e-01	-2.78e-17
$\cdots$	$\cdots$	$\cdots$	$\cdots$	$\cdots$	$\cdots$	$\cdots$

Table 1.2. The first few time steps in solving the logistic equation (1.30) using the Euler method (1.33). The letter e in the values given in the last three columns is floating-point output that designates the exponent. So, for example, 2.6500e-02 = 2.6500 × 10^{-2}.

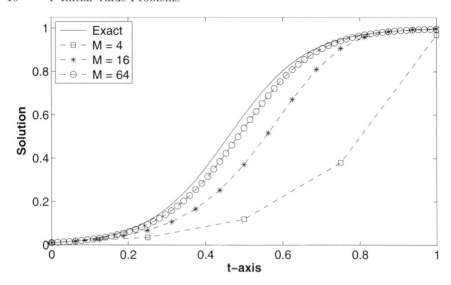

Figure 1.3. Solution of the logistic equation (1.30) using the Euler method (1.33) for three values of M. Also shown is the exact solution. The symbols are the computed values, and the dashed lines are drawn by the plotting program simply to connect the values.

is, if we increase M in (1.33) will $e_M = |y(T) - \bar{y}_M|$ converge to zero or at least decrease down to the level of the round-off? We want the answer to this question to be yes and, moreover, that it is true no matter what choice we make for $t = T$. If this holds then the method is convergent.

To help make it more apparent what is contributing to the error we rewrite it as follows

$$e_M = |y(T) - y_M + y_M - \bar{y}_M|. \tag{1.37}$$

From this the error can be considered as coming from the following two sources:

$y(T) - y_M$: This is the difference, at $t = T$, between the exact solution of the IVP and the exact solution of the problem we use as its approximation. As occurs in Table 1.2, this should be the major contributor to the error until k is small enough that this difference gets down to approximately that of the round-off.

$y_M - \bar{y}_M$: This is the error at $t = T$ that originates from round-off when one uses floating-point calculations to compute the solution of the difference equation. The last column of Table 1.2 gives the values of this error at the first few time points. Getting values of 10^{-15} or smaller, as occur in this calculation, is about as good as can be expected using double precision.

For the logistic equation example considered earlier, the error $e_M = |y(T) - \bar{y}_M|$ from the Euler method is plotted in Figure 1.4 as a function of the number of time points used to reach $T = 1$. It is seen that the error decreases linearly

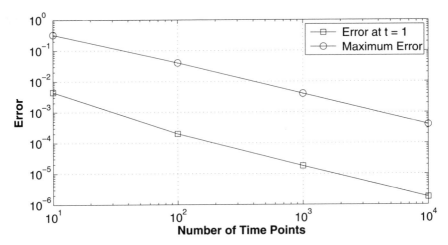

Figure 1.4. The difference between the exact and computed solutions, as a function of the number of time steps, M, used in solving the logistic equation (1.30) with the Euler method (1.33). Shown is the error $|y(T) - \bar{y}_M|$ at $t = 1$ as well as the maximum error as determined using (1.38).

in the log-log plot in such a way that increasing M by a factor of 10 decreases the error by the same factor. In other words, the error decreases as k^n, with $n = 1$. It is not a coincidence that this is the same order as for the truncation error (1.24). At first glance, because the term that is neglected in (1.26) is $k\tau_j = O(k^2)$, one might expect that the error in Figure 1.4 would decrease as k^2. However, $k\tau_j$ is the error we generate at each time step. To get to T we take $M = 1/k$ time steps so the accumulated error we generate in getting to T is reduced by a factor of k. Therefore, with a convergent method the order of the truncation error determines the order of the error.

We are using the error at $t = T$ to help determine how the approximation improves as the number of time steps increases. In many applications, however, one is interested in how well the numerical solution approximates the solution throughout the entire interval $0 \le t \le T$. For this it is more appropriate to consider using a vector norm to define the error. For example, using the maximum norm the error function takes the form

$$e_\infty = \max_{j=0,1,\dots,M} |y(t_j) - \bar{y}_j|. \tag{1.38}$$

To indicate how this differs from the error e_M used earlier, (1.38) is plotted in Figure 1.4 for the logistic equation example. As expected, e_∞ is larger than e_M but its dependence on M is still $O(k)$.

Example
The earlier example (1.30) is typical of what occurs in most applications. Namely, using the laws of physics or some other principles one obtains one or

more differential equations to solve, and the numerical method is constructed directly from them. It is informative to see whether the steps can be reversed. Specifically, suppose we start with (1.28) and ask whether it is based on a consistent approximation of (1.18). This is determined by plugging the exact solution into (1.28) and seeing how close it comes to satisfying this finite difference equation. In preparation for this we use Taylor's theorem to obtain

$$
\begin{aligned}
y(t_{j+1}) &= y(t_j + k) \\
&= y(t_j) + ky'(t_j) + \frac{1}{2}k^2 y''(t_j) + \dots \\
&= y(t_j) + kf(t_j, y(t_j)) + \frac{1}{2}k^2 y''(t_j) + \dots.
\end{aligned} \tag{1.39}
$$

Now, substituting $y(t_j)$ for y_j in (1.28) yields

$$
y(t_{j+1}) \stackrel{?}{=} y(t_j) + kf(t_j, y(t_j)). \tag{1.40}
$$

A question mark is put above the equal sign here, because we are investigating whether $y(t)$ satisfies (1.28) or, more precisely, how close it comes to satisfying this equation. With (1.39) the question of whether (1.40) is satisfied can be written as

$$
y(t_j) + kf(t_j, y(t_j)) + \frac{1}{2}k^2 y''(t_j) + \dots \stackrel{?}{=} y(t_j) + kf(t_j, y(t_j)),
$$

or equivalently,

$$
\frac{1}{2}k^2 y''(t_j) + \dots \stackrel{?}{=} 0. \tag{1.41}
$$

The conclusion from this last step is that $y(t_j)$ misses satisfying (1.28) by a term that is $O(k^2)$, and from this it follows that the truncation error is $O(k)$. Because the truncation error goes to zero with k it follows that the method is consistent. Of course we already knew this, but the above calculation shows that if necessary, it is possible to determine this directly from the finite difference equation.

Stability

STEP 5. It is not unreasonable to think that as long as the problem is approximated consistently, then the numerical solution will converge to the exact solution as the time step is refined. Unfortunately, as demonstrated shortly using the leapfrog method, consistency is not enough. To explain what is missing, the approximation that produced Euler's method means that even though y_0 is known exactly, the method computes a value for y_1 that differs a bit from the exact value $y(t_1)$. Moreover, this difference affects the values of all the y_j's that come afterwards. It is essential that the method not magnify these differences as t_j increases, and this is true irrespective of the time step at which the error is produced. This is the idea underlying the concept of stability. There are various ways to express this condition and we will use one of the stronger forms, something known as A-stability. This is determined by using the method to solve the radioactive decay equation

$$\frac{dy}{dt} = -ry, \tag{1.42}$$

where

$$y(0) = \alpha. \tag{1.43}$$

As before, it is assumed that r is positive. For this equation the Euler method (1.28) reduces to

$$y_{j+1} = (1 - rk)y_j. \tag{1.44}$$

The solution of this that satisfies the initial condition is

$$y_j = \alpha(1 - rk)^j. \tag{1.45}$$

In comparison, the exact solution to the IVP is $y(t) = \alpha e^{-rt}$, and this function approaches zero as t increases. It is required at the very least that the numerical solution of this problem not grow, and this is the basis for the following definition.

Definition 1.1. *If the method, when applied to (1.42), produces a bounded solution irrespective of the (positive) value of r and k, then the method is said to be A-stable. If boundedness occurs only when k is small then the method is conditionally A-stable. Otherwise, the method is unstable.*

The Euler solution in (1.45) remains bounded as j increases only as long as $|1 - rk| \le 1$. This occurs if the step size is chosen to satisfy the condition $k \le 2/r$. Therefore, the Euler method is conditionally A-stable. It is worth looking at what happens in the unstable case. If we take a step size that does not satisfy the stability condition, say $k = 3/r$, then $y_j = \alpha(-2)^j$. The solution in this case oscillates with an amplitude that increases as t_j increases. This is similar to what was seen when the Tacoma bridge collapsed, where relatively small oscillations grew and eventually became so large the bridge came apart. Whenever such growing oscillatory behavior appears in a numerical solution one should seriously consider whether one has been unfortunate enough to have picked a step size that falls in the instability region.

One last point to make here is that one of the central questions arising when using any numerical method to solve a differential equation concerns what properties of the problem the numerical method is able to preserve. For example, if energy is conserved in the original problem then it is natural to ask whether the numerical method does the same. As we will see, preserving particular properties, such as energy conservation or the monotonicity of the solution, can have profound consequences on how well the method works. It is within this context that the requirement of A-stability is introduced.

The radioactive decay problem possesses an asymptotically stable equilibrium solution $y = 0$. A-stability is nothing more than the requirement that $y = 0$ be at least a stable equilibrium solution for the method (i.e., any solution starting near $y = 0$ will remain near this solution as time increases). As we found earlier, the stability interval for Euler's method is $k \leq 2/r$. On the interior of this interval, so $k < 2/r$, the equilibrium solution $y = 0$ is asymptotically stable because $y_j \to 0$ as $t_j \to \infty$. This observation accounts for why you will occasionally see a requirement of strict A-stability, where boundedness is replaced with the requirement that $y_j \to 0$ as $t_j \to \infty$. The reason is that a strictly A-stable method preserves asymptotic stability. Our conclusion in this case would be that Euler is strictly A-stable when strict inequality holds, namely, $k < 2/r$.

End Notes

One might wonder why the radioactive decay problem is used as the arbiter for deciding whether a method is A-stable. To explain how this happens suppose the differential equation is a bit simpler than the one in (1.18) and has the form $y' = f(y)$. Also, suppose $y = Y$ is an asymptotically stable equilibrium solution. This means that the constant Y is a solution of the equation and any initial condition $y(0) = \alpha$ chosen close to Y will result in the solution of the IVP converging to Y as $t \to \infty$. Now, to determine how the solution behaves near $y = Y$ let $v(t) = y(t) - Y$. Substituting this into the equation, one gets that $v' = f(Y + v)$. With α close to Y we have that $v(t)$ starts out relatively small. Consequently, using Taylor's theorem, $f(Y + v) = f(Y) + vf'(Y) + O(v^2) \approx f(Y) + vf'(Y)$. Because $f(Y) = 0$ we conclude that $v' = -rv$, where $r = -f'(Y)$. Therefore the solution near $y = Y$ is governed by the radioactive decay equation, and that is why it is used to determine A-stability. For the earlier example using the logistic equation, $f = \lambda y(1 - y)$ and $r = \lambda(-1 + 2Y)$. Because $Y = 1$, so $r = \lambda$, then the stability requirement when using the Euler method to solve the logistic equation is $k \leq 2/\lambda$. Taking $\lambda = 10$, then the stability condition is $k \leq \frac{1}{5}$. This helps explain the rather poor showing of the $M = 4$ curve in Figure 1.3.

Another observation to make is that even though the discussion has centered on problems involving one differential equation, the ideas are easily extended to systems of equations. For example, assuming that the equation is $\mathbf{y}' = \mathbf{f}(t, \mathbf{y})$, then the vector version of the forward difference used in (1.23) is

$$\mathbf{y}'(t_j) = \frac{\mathbf{y}(t_{j+1}) - \mathbf{y}(t_j)}{k} + O(k). \tag{1.46}$$

Substituting this into the differential equation and dropping the truncation error produces the vector form of the Euler method given in Table 1.3. This formula can be obtained directly from the single-variable version in (1.28) by simply converting the appropriate variables to vectors. The same is true

for most of the other methods considered in this chapter, the exception arising with Runge–Kutta methods and this is discussed later. A limitation of
approximating vector derivatives in this way is that every equation is approximated the same way. For example, in (1.46) each component of the vector
$\mathbf{y}'(t_j)$ is approximated using the Euler formula. There are situations in which
it is better to use different approximations on different components, and an
example of this is explored in Section 1.6.

One last comment to make concerns A-stability for systems. The general-
ization of (1.42) for systems is the equation $\mathbf{y}' = -\mathbf{A}\mathbf{y}$, where $\mathbf{A}$ is a matrix
with constant coefficients. Similar to what occurred earlier, this matrix can be
thought of as derived from a local approximation of the Jacobian matrix $\frac{\partial f_i}{\partial y_j}$.
The requirement for A-stability remains the same, namely that the method
produces bounded solutions for any $\mathbf{A}$ that results in $\mathbf{y} = \mathbf{0}$ being an asymptotically stable equilibrium solution of the original problem. To illustrate what
is involved, if $\mathbf{A}$ is diagonalizable with eigenvalues $\lambda_1, \lambda_2, \ldots, \lambda_n$ then $\mathbf{y} = \mathbf{0}$
is an asymptotically stable equilibrium solution if $\mathrm{Re}(\lambda_i) > 0, \forall i$. Now, Euler
applied to $\mathbf{y}' = -\mathbf{A}\mathbf{y}$ yields the finite difference equation $\mathbf{y}_{j+1} = (\mathbf{I} - k\mathbf{A})\mathbf{y}_j$,
where $\mathbf{I}$ is the identity matrix. Using the diagonalizablity of $\mathbf{A}$ it is possible
to reduce this to scalar equations of the form $z_{j+1} = (1 - kr)z_j$, where r is
an eigenvalue of $\mathbf{A}$. Consequently, the problem has been reduced to (1.44),
except that r is now complex-valued with $\mathrm{Re}(r) > 0$. The conclusion is therefore the same, namely that Euler is conditionally A-stable. This example also
demonstrates that the scalar equation in (1.42) serves as an adequate test
problem for A-stability, and it is the one we use throughout this chapter.
Those interested in a more extensive development of A-stability for systems
should consult the text by Deuflhard et al. [2002].

1.2.2 Additional Difference Methods

The steps used to derive the Euler method can be employed to obtain a host
of other finite difference approximations. The point in the derivation that
separates one method from another is STEP 3, where one makes a choice
for the difference formula. Most of the formulas used in this book are listed
in Table 1.1. It is interesting to see what sort of numerical methods can be
derived using these expressions, and a few of the possibilities are discussed
below.

Backward Euler

If one uses the backward difference formula in Table 1.1, then in place of
(1.23), we get

$$y'(t_j) = \frac{y(t_j) - y(t_{j-1})}{k} + \tau_j, \tag{1.47}$$

where

	Methods for solving the differential equation $$\frac{d}{dt}\mathbf{y}(t) = \mathbf{f}(t,\mathbf{y})$$		
Method	Difference Formula	τ_j	Properties
Euler	$\mathbf{y}_{j+1} = \mathbf{y}_j + k\mathbf{f}_j$	$O(k)$	Explicit; Conditionally A-stable
Backward Euler	$\mathbf{y}_{j+1} = \mathbf{y}_j + k\mathbf{f}_{j+1}$	$O(k)$	Implicit; A-stable
Trapezoidal	$\mathbf{y}_{j+1} = \mathbf{y}_j + \frac{k}{2}(\mathbf{f}_j + \mathbf{f}_{j+1})$	$O(k^2)$	Implicit; A-stable
Heun (RK2)	$\mathbf{y}_{j+1} = \mathbf{y}_j + \frac{1}{2}(\mathbf{k}_1 + \mathbf{k}_2)$ where $\mathbf{k}_1 = k\mathbf{f}_j$ $\mathbf{k}_2 = k\mathbf{f}(t_{j+1}, \mathbf{y}_j + \mathbf{k}_1)$	$O(k^2)$	Explicit; Conditionally A-stable
Classical Runge–Kutta (RK4)	$\mathbf{y}_{j+1} = \mathbf{y}_j + \frac{1}{6}(\mathbf{k}_1 + 2\mathbf{k}_2 + 2\mathbf{k}_3 + \mathbf{k}_4)$ where $\mathbf{k}_1 = k\mathbf{f}_j$ $\mathbf{k}_2 = k\mathbf{f}(t_j + \frac{k}{2}, \mathbf{y}_j + \frac{1}{2}\mathbf{k}_1)$ $\mathbf{k}_3 = k\mathbf{f}(t_j + \frac{k}{2}, \mathbf{y}_j + \frac{1}{2}\mathbf{k}_2)$ $\mathbf{k}_4 = k\mathbf{f}(t_{j+1}, \mathbf{y}_j + \mathbf{k}_3)$	$O(k^4)$	Explicit; Conditionally A-stable

Table 1.3. Finite difference methods for solving an IVP. The points $t_1, t_2, t_3, \ldots$ are equally spaced with step size $k = t_{j+1} - t_j$. Also, $\mathbf{f}_j = \mathbf{f}(t_j, \mathbf{y}_j)$ and τ_j is the truncation error for the method.

$$\tau_j = \frac{k}{2}y''(\eta_j). \tag{1.48}$$

Introducing this into (1.26), we obtain

$$y(t_j) - y(t_{j-1}) + k\tau_j = kf(t_j, y(t_j)). \tag{1.49}$$

Dropping the truncation error τ_j, the resulting finite difference approximation is

$$y_j = y_{j-1} + kf(t_j, y_j), \quad \text{for } j = 1, 2, \ldots, M. \tag{1.50}$$

From the initial condition (1.19) we have that the starting value is

$$y_0 = \alpha. \tag{1.51}$$

The difference equation in (1.50) is the backward Euler method. It has the same order of truncation error as the Euler method. However, because of

Figure 1.5. The animation of deformable objects using physically based modeling involves solving differential equations in which A-stability is an essential property of the numerical scheme. The example shown here uses backward Euler, and the trapezoidal method, to simulate the motion of clothing on a woman model (Hauth and Etzmuß [2001]).

the $f(t_j, y_j)$ term this method is implicit. This is both good and bad. It is good because it helps make the method A-stable (see below). However, it is bad because it can make finding y_j computationally difficult. Unless the problem is simple enough that the difference equation can be solved by hand, it is necessary to use something like Newton's method to solve (1.50), and this must be done for each time step. Some of the issues that arise with this situation are developed in Exercise 1.33.

As for stability (STEP 5), for the radioactive decay equation (1.42) one finds that (1.50) reduces to

$$y_{j+1} = \frac{1}{1 + rk} y_j. \tag{1.52}$$

Assuming $y_0 = \alpha$, then the solution of this finite difference equation is $y_j = \alpha(1 + rk)^{-j}$. This goes to zero as j increases irrespective of the value of k. Consequently, this method is A-stable. Another point to its credit is that the solution decays monotonically to zero, just as does the exact solution. For this reason backward Euler is said to be a monotone method. In contrast, recall that for Euler's method the solution is $y_j = \alpha(1 - rk)^j$. If $-1 < 1 - rk < 0$, which is part of the stability interval for the method, the resulting solution goes to zero, but it oscillates as it does so. In other words, Euler's method is monotone only if $0 < 1 - rk < 1$ (i.e., it is conditionally monotone). The fact that backward Euler preserves the monotonicity of the solution is, for some problems, important, and this is explored in more depth in Exercises 1.11 and 1.12. We will return to this issue of a monotone scheme in Chapter 4, when we investigate how to solve wave propagation problems.

Leapfrog Method

It is natural to expect that a more accurate approximation of the derivative will improve the resulting finite difference approximation of the differential equation. In looking over Table 1.1, the centered difference formula would appear to be a good choice for such an improvement because it has quadratic error (versus linear for the first two formulas listed). Introducing this into (1.18) we obtain

$$y(t_{j+1}) - y(t_{j-1}) + 2k\tau_j = 2kf(t_j, y(t_j)), \tag{1.53}$$

where $\tau_j = O(k^2)$. Dropping the truncation error τ_j, the resulting finite difference approximation is

$$y_{j+1} = y_{j-1} + 2kf(t_j, y_j), \quad \text{for } j = 1, 2, \ldots, M-1. \tag{1.54}$$

This is known as the leapfrog, or explicit midpoint, method. Because this equation uses information from two previous time steps it is an example of a two-step method. In contrast, both Euler methods use information from a single time step back, so they are one-step methods. What this means is that the initial condition (1.19) is not enough information to get leapfrog started, because we also need y_1. This is a relatively minor inconvenience that will be addressed later. It is more interesting right now to concentrate on the truncation error. It would seem that the leapfrog method, with its $O(k^2)$ truncation error, will produce a more accurate numerical solution than either of the two Euler methods. As it turns out, this apparently obvious conclusion could not be farther from the truth. This becomes evident from our stability test. Applying (1.54) to the radioactive decay equation (1.42) yields $y_{j+1} = y_{j-1} - 2rky_j$. This second-order difference equation can be solved by assuming a solution of the form $y_j = s^j$. By doing this it is found that the general solution has the form $y_j = \alpha_0 s_+^j + \alpha_1 s_-^j$, where $s_\pm = -kr \pm \sqrt{1 + k^2 r^2}$ and α_0, α_1 are arbitrary constants(see Exercise 1.31). Because $|s_-| > 1$, it is impossible to find a step size k to satisfy the stability condition. Therefore, the leapfrog method is unstable.

1.3 Methods Obtained from Numerical Quadrature

Another approach to deriving a finite difference approximation of an IVP is to integrate the differential equation and then use a numerical integration

Rule	Integration Formula
Right Box	$\int_{x_i}^{x_{i+1}} f(x)dx = hf(x_{i+1}) + O(h^2)$
Left Box	$\int_{x_i}^{x_{i+1}} f(x)dx = hf(x_i) + O(h^2)$
Midpoint	$\int_{x_{i-1}}^{x_{i+1}} f(x)dx = 2hf(x_i) + \frac{h^3}{3}f''(\eta_i)$
Trapezoidal	$\int_{x_i}^{x_{i+1}} f(x)dx = \frac{h}{2}\left(f(x_i) + f(x_{i+1})\right) - \frac{h^3}{12}f''(\eta_i)$
Simpson	$\int_{x_{i-1}}^{x_{i+1}} f(x)dx = \frac{h}{3}\left(f(x_{i+1}) + 4f(x_i) + f(x_{i-1})\right) - \frac{h^5}{90}f''''(\eta_i)$

Table 1.4. Numerical integration formulas. The points $x_1, x_2, x_3, \ldots$ are equally spaced with step size $h = x_{i+1} - x_i$. The point η_i is located within the interval of integration.

rule. This is a very useful idea that is best explained by working through an example. To get started, a time grid must be introduced, and so STEP 1 is the same as before. However, STEP 2 and STEP 3 differ from what we did earlier.

STEP 2. Integrate the differential equation between two time points. We will take t_j and t_{j+1}, and so from (1.18) we have

$$\int_{t_j}^{t_{j+1}} \frac{dy}{dt} dt = \int_{t_j}^{t_{j+1}} f(t, y(t))dt. \tag{1.55}$$

Using the Fundamental Theorem of Calculus we obtain

$$y(t_{j+1}) - y(t_j) = \int_{t_j}^{t_{j+1}} f(t, y(t))dt. \tag{1.56}$$

STEP 3. Replace the integral in STEP 2 with a finite difference approximation. This is where things get a bit interesting, because there are numerous choices, and they produce different numerical procedures. A few of the most often used possibilities are listed in Table 1.4. We will use the trapezoidal rule, and introducing this into (1.56) yields

$$y(t_{j+1}) - y(t_j) = \frac{k}{2}\left[f(t_{j+1}, y(t_{j+1})) + f(t_j, y(t_j))\right] + O(k^3). \tag{1.57}$$

STEP 4. Drop the big-O term. After dropping the $O(k^3)$ term in (1.57) the resulting equation is

$$y_{j+1} = y_j + \frac{k}{2}(f_{j+1} + f_j), \quad \text{for } j = 0, 1, 2, \ldots, M - 1, \tag{1.58}$$

where $f_j = f(t_j, y_j)$. From the initial condition (1.19) we have that the starting value is

$$y_0 = \alpha. \tag{1.59}$$

The finite difference equation (1.58) is known as the trapezoidal method for solving (1.18). Because of the f_{j+1} term this method is implicit, and it is not hard to show that it is A-stable. To determine the truncation error for the method note that in (1.57) the error at each time step is $O(k^3)$. In taking M time steps to reach $t = T$ the resulting error is therefore $M \times O(k^3) = O(k^2)$. In other words, the truncation error is $\tau_j = O(k^2)$.

One of the attractive features of the quadrature approach is that it involves multiple decision points that can be varied to produce different numerical methods. For example, the integration interval can be changed to, say, $t_{j-1} \leq t \leq t_{j+1}$ and then Simpson's rule used on the resulting integral (see Exercise 1.7). Another option is to not use a quadrature rule but instead replace the function f in the integral in (1.56) with an approximation that can be integrated exactly. The most often used approximations involve interpolating polynomials, and these give rise to what are called Adams methods. As an example, the linear function that interpolates f at t_j and t_{j+1} is

$$f(t, y(t)) = -\frac{t - t_{j+1}}{k} f(t_j, y(t_j)) + \frac{t - t_j}{k} f(t_{j+1}, y(t_{j+1})) + O(k^2). \quad (1.60)$$

Introducing this into (1.56) produces, not unexpectedly, the trapezoidal method (1.58). Because the result is implicit it is an example of what is called an Adams–Moulton method. To obtain something not listed in Table 1.3 one could use a linear function that interpolates f at t_{j-1} and t_j. Working out the details (see Exercise 1.8), one obtains

$$y_{j+1} = y_j + \frac{k}{2}(3f_j - f_{j-1}), \quad \text{for } j = 1, 2, \ldots, M - 1.$$

This is explicit, and for this reason it is an example of an Adams–Bashforth method. It is not the purpose of this book to develop an exhaustive catalogue of methods, so the particulars of each of these variations of the quadrature approach are left for the exercises.

Example
We have derived several methods for solving IVPs, including the Euler, backward Euler, leapfrog, and trapezoidal methods. It is worth taking them out for a test drive to see how they compare, and the logistic equation (1.5) is a good candidate for this. The equation that is solved is

$$\frac{dy}{dt} = 10y(1 - y), \quad \text{for } 0 < t, \tag{1.61}$$

where $y(0) = 0.1$. As before, we take $T = 1$, so the time points are determined from the expression $t_j = jk$, for $j = 0, 1, 2, \ldots, M$ and $k = 1/M$. Because

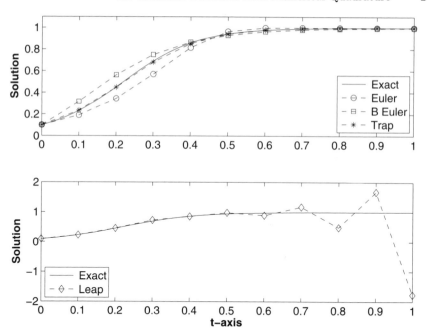

Figure 1.6. Solution of the logistic equation (1.61) using different numerical schemes. The leapfrog method is shown in the lower plot, and the two Euler schemes and the trapezoidal method are in the upper graph.

$f(t, y) = 10y(1 - y)$ our methods reduce to the finite difference equations listed below:

$$\text{Euler: } y_{j+1} = y_j + 10ky_j(1 - y_j),$$
$$\text{Backward Euler: } y_{j+1} = y_j + 10ky_{j+1}(1 - y_{j+1}),$$
$$\text{Leapfrog: } y_{j+1} = y_{j-1} + 20ky_j(1 - y_j),$$
$$\text{Trapezoidal: } y_{j+1} = y_j + 5k\left(y_j(1 - y_j) + y_{j+1}(1 - y_{j+1})\right).$$

The initial condition is $y_0 = 0.1$, and for the leapfrog method it is assumed that $y_1 = y(k)$ (i.e., the exact value at $t = t_1$ is used). Just how well these four expressions do is shown in Figure 1.6 for the case $M = 10$. The first thing one notices is just how badly the leapfrog method does (it had to be given its own graph because it behaves so badly). This is not unexpected, because we know that the method is not A-stable. The other three solution curves also behave as expected. In particular, the two Euler methods are not as accurate as the trapezoidal method and are approximately equal in how far each differs from the exact solution. To quantify just how accurately each method does in solving the problem, in Figure 1.7 the error (1.37) is plotted as a function of the number of grid points used to reach T. Because of its stability problems the leapfrog method is omitted in this figure, and in its place the

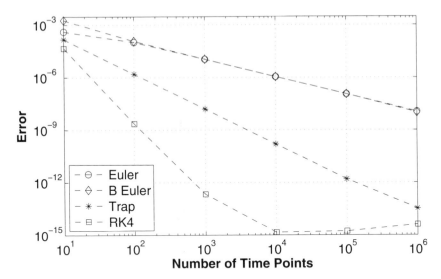

Figure 1.7. Error at $t = 1$ as a function of the number of time steps used to solve the logistic equation (1.61). Each curve decreases as $O(k^n)$, where n is determined from the truncation error for the method.

error obtained using the RK4 method, which is considered in the next section, is included. As predicted, all decrease according to their respective truncation errors. For example, the trapezoidal method decreases as $O(k^2)$ and the two Euler methods as $O(k)$. The only exception to this is RK4, which shows a break from $O(k^4)$ once M gets up above 10^4. The reason for this is that the error has started to reach the level of round-off, and so it is not expected to continue its linear decrease past this point.

1.4 Runge–Kutta Methods

An extraordinarily successful family of numerical approximations for IVPs comes under the general classification of Runge–Kutta (RK) methods. The derivation is based on the question of whether it is possible to determine an explicit method for finding y_{j+1} that only uses the value of the solution at t_j and has a predetermined truncation error. The secret in getting this to work is making a good guess as to what such a formula might look like. To demonstrate, the best single-step explicit method we have so far has a truncation error of $O(k)$. So, suppose we are interested in obtaining one that is $O(k^2)$. We have been able to derive an implicit scheme with this error, and this is the trapezoidal method (1.58). The reason it is implicit is the term $f(t_{j+1}, y_{j+1})$. Suppose we experiment a little and use Euler's method to approximate y_{j+1} in this term with $y_j + kf_j$. The resulting finite difference approximation is

$$y_{j+1} = y_j + \frac{k}{2}[f(t_j, y_j) + f(t_j + k, y_j + kf_j)]. \qquad (1.62)$$

It is not clear whether this explicit method has the desired truncation error. However, it is useful, because it provides insight into what a $O(k^2)$ explicit might look like. Based on this, the Runge–Kutta assumption is that the method has the form

$$y_{j+1} = y_j + k[af(t_j, y_j) + bf(t_j + \alpha k, y_j + \beta k f_j)], \qquad (1.63)$$

where the constants a, b, α, β are chosen to achieve the stated truncation error. To accomplish this the exact solution is substituted into (1.63) and Taylor's theorem is then used to reduce the expression, much as was done in reducing (1.40) to (1.41). Carrying out the calculations, one finds that $a + b = 1$, $\beta = \alpha$, and $2b\alpha = 1$ (see Exercise 1.10). These three equations are called the order conditions, and interestingly, the values for a, b, α, β are not unique. A simple choice is $a = b$, and this yields what is known as Heun's method, which is given in (1.62) and also listed in Table 1.3. To its credit, Heun is explicit and has a truncation error as good as the trapezoidal method. What is lost, however, is unconditional stability.

The one method from the Runge–Kutta family that deserves special attention is RK4, which is listed in Table 1.3. This is used in so many computer codes that it has become the workhorse of IVP solvers. The derivation of RK4 requires a generalization of the assumption in (1.63) and involves considerable more work in reducing the resulting expressions. To motivate how the formula is obtained, suppose the differential equation is $y' = f(t)$. Integrating this as in (1.56) and then using Simpson's rule yields

$$y(t_{j+1}) - y(t_j) = \frac{k}{6}[f(t_j) + 4f(t_j + 0.5k) + f(t_{j+1})] + O(k^5)$$

$$= \frac{k}{6}\left[f(t_j) + 2f\left(t_j + \frac{k}{2}\right) + 2f\left(t_j + \frac{k}{2}\right) + f(t_{j+1})\right] + O(k^5).$$

The $O(k^5)$ term is dropped from this expression and in doing so it should be understood that this is the error that is made in progressing from t_j to t_{j+1}. As explained earlier for the quadrature methods, the total number of steps taken is $M = O(1/k)$ and therefore the resulting truncation error is $\tau_j = M \times O(k^5) = O(k^4)$. With this we obtain the RK4 formula given in Table 1.3, as applied to this particular differential equation.

Example
For the logistic example considered earlier, the RK4 formulas given in Table 1.3 are

$$k_1 = 10ky_j(1 - y_j),$$

$$k_2 = 10k\left(y_j + \frac{1}{2}k_1\right)\left(1 - y_j - \frac{1}{2}k_1\right),$$

$$k_3 = 10k\left(y_j + \frac{1}{2}k_2\right)\left(1 - y_j - \frac{1}{2}k_2\right),$$

$$k_4 = 10k(y_j + k_3)(1 - y_j - k_3).$$

The resulting numerical accuracy of the method is shown in Figure 1.7. RK4 is clearly superior to the others listed, to the point that it achieves an error on the order of round-off far ahead of the others. Given this, you might wonder why the other methods are even discussed, much less used by anyone. Well, there are several reasons for considering other methods, and one is that RK4 is only conditionally A-stable. This is true of all explicit Runge–Kutta methods, and as we will see later, this limits their use for solving partial differential equations such as those considered in Chapter 3. Another reason is that RK4 does not do well in preserving certain properties of the solution, and an important example of this is discussed in Section 1.6.

The ideas developed here can be generalized to produce higher-order RK methods, although the complexity of the derivation can be enormous. For example, in celestial mechanics you occasionally see people use twelfth-order RK methods. Such a scheme is not easy to derive, because it results in 5972 order conditions, and, as occurred earlier, these form an underdetermined nonlinear system. This situation is further complicated by the somewhat unexpected problem that Runge–Kutta methods that are $O(k^p)$ for the scalar equation $y' = f(t, y)$ are not necessarily $O(k^p)$ for the system $\mathbf{y}'(t) = \mathbf{f}(t, \mathbf{y})$ if $p \geq 5$. In other words, to derive a higher order Runge–Kutta method for systems you are not able to simply use a scalar equation and then convert the variables to vectors when you are done. Those interested in deriving higher-order methods, or in a more systematic derivation of RK4, should consult the texts by Butcher [1987] and Lambert [1991].

1.5 Extensions and Ghost Points

The ideas developed in this chapter can be embellished without much difficulty to handle more complex problems, including partial differential equations. To illustrate how this is done, consider the following nonlinear second-order equation

$$\frac{d}{dt}\left[(1 + e^{-3t})\frac{du}{dt}\right] + u^3 = \sin(t), \quad \text{for } 0 < t, \tag{1.64}$$

where the initial conditions are

$$u(0) = 2, \; u'(0) = -5. \tag{1.65}$$

Our objective is to derive a $O(k^2)$ finite difference approximation for this IVP. One option is to rewrite the problem as a first-order system and then use one or more of the methods listed in Table 1.3 (see Exercise 1.19). A variation of this approach is used in the next section, but here we work with the equation directly. To get things started we expand the derivative to obtain

$$(1 + e^{-3t})\frac{d^2u}{dt^2} - 3e^{-3t}\frac{du}{dt} + u^3 = \sin(t).$$

We are now in position to carry out STEP 2, which means we evaluate the differential equation at the time point $t = t_j$ to obtain

$$(1 + e^{-3t_j})\frac{d^2u}{dt^2}(t_j) - 3e^{-3t_j}\frac{du}{dt}(t_j) + u(t_j)^3 = \sin(t_j). \qquad (1.66)$$

To carry out STEP 3, approximations for the derivatives must be selected, and we use the centered $O(k^2)$ approximations listed in Table 1.1. Doing this and then dropping the truncation error term (STEP 4) gives us

$$(1 + e^{-3t_j})\frac{u_{j+1} - 2u_j + u_{j-1}}{k^2} - 3e^{-3t_j}\frac{u_{j+1} - u_{j-1}}{2k} + u_j^3 = \sin(t_j). \quad (1.67)$$

This simplifies to an explicit formula for u_{j+1} that has the form

$$u_{j+1} = a_j u_j + b_j u_j^3 + c_j u_{j-1} + d_j, \quad \text{for } j = 1, 2, 3, \ldots, \qquad (1.68)$$

where a_j, b_j, c_j, d_j depend on t_j but not on the u_j's. To use this expression we need u_0 and u_1. From the initial condition $u(0) = 2$ we have that $u_0 = 2$. To determine u_1 we must replace the initial condition $u'(0) = -5$ with a finite difference approximation. As always, there are options, and one is to use a one-sided difference (see Exercise 1.17). There is, however, another approach, which introduces a useful idea we will have need of occasionally. It starts by introducing the centered difference approximation

$$u'(0) = \frac{u(t_1) - u(t_{-1})}{2k} + O(k^2). \qquad (1.69)$$

At first glance it would seem that we are not able to use this approximation, because it requires a value of the solution outside the original interval for the problem. However, it is possible to extend the solution into this region as long as the solution and its derivatives are continuous at $t = 0$. Assuming this is the case, then $t = t_{-1}$ is called a ghost point, and the resulting finite difference approximation of the initial condition is

$$u_{-1} = 10k + u_1. \qquad (1.70)$$

To use this in our algorithm, the differential equation (1.64) is extended to include $t = 0$. This allows us to let $j = 0$ in (1.68), and from this we obtain

$$u_1 = a_0 u_0 + b_0 u_0^3 + c_0 u_{-1} + d_0.$$

Using (1.70), this reduces to

$$u_{-1} = (1 - c_0)^{-1}(a_0 u_0 + b_0 u_0^3 + d_0 + 10k).$$

Now that we know u_{-1} and u_0, the algorithm is complete.

Aside from introducing the idea of a ghost point, the above example looks to be a routine application of what was developed earlier for first-order equations. However, note that the requirement of consistency had a broader impact here, because it was necessary to introduce approximations into the initial conditions as well as the differential equation. This has repercussions for the truncation error. To produce a $O(k^2)$ method it is necessary to use $O(k^2)$ approximations. If any one of the derivatives in either the differential equation or initial condition were to have been approximated using a $O(k)$ formula, then the best we could guarantee is that the method is $O(k)$. A demonstration of this can be found in Exercise 1.18.

1.6 Conservative Methods

In this section we take up the study of equations obtained from Newton's second law, but without the dependence of the force on time or velocity. In this case (1.8) reduces to

$$m\frac{d^2 y}{dt^2} = F(y). \tag{1.71}$$

Now, we could proceed as in the previous section and introduce a finite difference approximation for the second derivative. Alternatively, we could use the system version given in (1.13) and simply plug into one of the formulas in Table 1.3. However, with the objective of pushing the envelope a bit and maybe learning something in the process, we try a different approach.

Given that this example concerns Newtonian mechanics, and the importance of energy in mechanics, it is worth introducing this into the formulation. For (1.71) the energy is

$$H(t) = \frac{m}{2}(y')^2 + V, \tag{1.72}$$

where the function $V(y)$ satisfies $\frac{dV}{dy} = -F(y)$. The function H is called the Hamiltonian for the system, and it is the sum of the kinetic and potential energies. By multiplying (1.71) by y' and integrating, it is found that H is a constant for this problem, which means that energy is conserved. If the initial conditions are $y(0) = \alpha$ and $y'(0) = \beta$, then $H = \frac{m}{2}\beta^2 + V(\alpha)$. Our goal is to obtain a finite difference approximation that comes very close to keeping the energy in the problem conserved.

1.6.1 Velocity Verlet

To start things off we write (1.71) in system form as

$$y' = v, \tag{1.73}$$

$$v' = \frac{1}{m}F(y), \tag{1.74}$$

where we have introduced the velocity $v = y'$. It is not hard to show that even in the simplest case of when $F = -y$, the only method listed in Table 1.3 that conserves energy is the trapezoidal method (see Exercise 1.24). Using this in (1.73) and (1.74) we obtain

$$y_{j+1} = y_j + \frac{k}{2}(v_{j+1} + v_j), \tag{1.75}$$

$$v_{j+1} = v_j + \frac{k}{2m}(F(y_{j+1}) + F(y_j)). \tag{1.76}$$

As is always the case with the trapezoidal method, the resulting equations are implicit. For a large system such as the one that produced Figure 1.1, implicit methods are computationally intensive and can take excessive cpu time to use. The question therefore arises as to whether it might be possible to tweak the above equations so they are explicit yet still do reasonably well with conserving energy. With this in mind, note that one of the culprits for the implicitness is the v_{j+1} term in (1.75). Can we find an approximation for this term that uses information at earlier time steps? One possibility is to use the Euler method on (1.74), which gives us $v_{j+1} = v_j + \frac{k}{m}F(y_j)$. Introducing this into (1.75) yields

$$y_{j+1} = y_j + kv_j + \frac{1}{2m}k^2 F(y_j), \tag{1.77}$$

$$v_{j+1} = v_j + \frac{k}{2m}\left(F(y_{j+1}) + F(y_j)\right). \tag{1.78}$$

Assuming we first use (1.77) to calculate y_{j+1} and then use (1.78) to find v_{j+1}, the procedure is explicit. It is known as the velocity Verlet method for solving (1.71) and it is used extensively in molecular dynamics and animation applications where real-time computer simulation of objects is required. The reason for its popularity is not just that it is explicit, $O(k^2)$ accurate, and requires only one force evaluation per step. The same is true for other methods. It is the method of choice because it does a better job than most in approximating the energy over long time intervals. The latter is the reason it, or methods similar to Verlet, are used in such areas as computational astrophysics, as illustrated in Figure 1.8.

Example
To demonstrate the effectiveness of velocity Verlet we solve the linear harmonic oscillator problem when $F = -y$ and $H = \frac{1}{2}(y')^2 + \frac{1}{2}y^2$. Assuming that

the initial conditions are $y(0) = 1$ and $y'(0) = 0$, then the exact solution is $y(t) = \cos(t)$ and $v(t) = -\sin(t)$. In this case, $H(t) = \frac{1}{2}$ and the solution, for any value of t, lies on the circle $y^2 + v^2 = 1$. The results of the calculation are shown in Figure 1.9. For comparison, the values obtained using RK4 are also shown. Both methods produce an accurate solution for small values of t, but for larger values the two methods start to differ significantly. For example, the energy decay using RK4 is substantial, whereas Verlet produces a value of H that oscillates but remains very near the exact value over the entire time period. The frequency of this oscillation is such that the Verlet result in Figure 1.9(a) looks to be a solid bar running across the upper portion of the plot. However, over a shorter time interval, as in Figure 1.9(b), the oscillatory nature of the curve is evident. It is also apparent in Figure 1.9(c) that the position and velocity obtained from Verlet, even for large values of t, are very nearly on the circle followed by the exact solution, whereas RK4 provides a very poor approximation. However, all is not perfect with Verlet. Although the computed (y, v) values follow the right path, they move along the path a bit faster than the exact solution. For example, for $0 \le t \le 4000$ the exact solution makes 636 complete circuits around this circle while the velocity Verlet solution makes 643 trips. So, we have a situation in which the computed solution is very close to being on the right path but is just a little ahead of where it is supposed to be. One last point to make is that the computing time for Verlet is significantly less than it is for RK4, and the reason is that Verlet requires fewer function evaluations per time step than RK4.

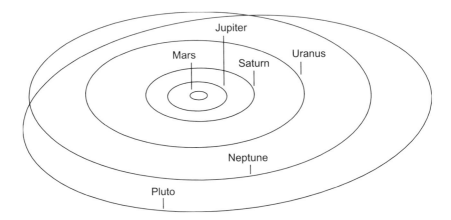

Figure 1.8. The study of the stability of the planetary orbits in the solar system requires accurate energy calculations over very large time intervals (with or without Pluto). One of the more successful approaches uses a symplectic method to study the orbits over 1.1 billion years using a time step of one year (Wisdom and Holman [1991]).

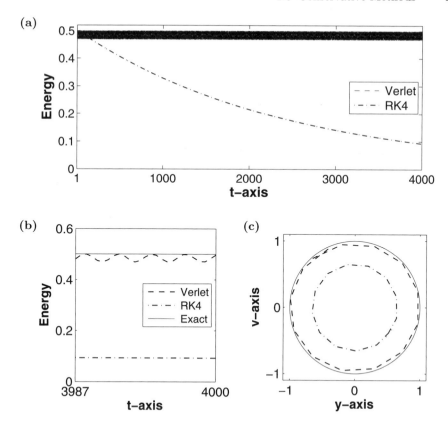

Figure 1.9. In (a) the energy, or Hamiltonian, H computed for the linear harmonic oscillator using the velocity Verlet and RK4 methods are shown. The energy over the smaller time interval $3987 \leq t \leq 4000$ is shown in (b). The corresponding values of (y, v), for $3987 \leq t \leq 4000$, are shown in (c).

1.6.2 Symplectic Methods

It is apparent from the last example that for this problem the velocity Verlet method is better than RK4 even though the latter has a much better truncation error. The question is, did we just get lucky or is it possible to find other methods with properties similar to those of Verlet? To address this, recall that we started out looking for a method that conserves energy. This was the reason for selecting the trapezoidal method, but when Euler's method was used to transform (1.75) into (1.77) we lost energy conservation. The fact that H is not constant using velocity Verlet is evident in Figure 1.9, but it is also clear that the method does a respectable job in determining the energy. The reason is that velocity Verlet possesses a special property connected with preserving area, and orientation, in the phase plane. To explain what this is,

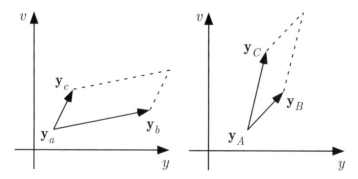

Figure 1.10. Phase plane parallelograms used to introduce a symplectic approximation. Using the method, in one time step, $a \to A$, $b \to B$, and $c \to C$.

suppose that with the initial condition $\mathbf{y}_a = (y_a, v_a)^T$ the method takes one time step and produces the value $\mathbf{y}_A = (y_A, v_A)^T$. For example, using (1.77), $y_A = y_a + kv_a + \frac{1}{2m}k^2 F(y_a)$. In a similar manner, assume using the initial conditions $\mathbf{y}_b, \mathbf{y}_c$ that the method takes one time step and produces the values $\mathbf{y}_B, \mathbf{y}_C$, respectively. The special property we are looking for has to do with the parallelogram formed by $\mathbf{y}_a, \mathbf{y}_b, \mathbf{y}_c$ as compared to the transformed parallelogram formed by $\mathbf{y}_A, \mathbf{y}_B, \mathbf{y}_C$ (see Figure 1.10). It is enough to consider the situation of when $\mathbf{y}_b, \mathbf{y}_c$ are close to $\mathbf{y}_a$, and so it is assumed that both points are within a distance h of $\mathbf{y}_a$, where h is small. In this case, if the two parallelograms in Figure 1.10 have the same area to first order in h^2 and have the same orientation, then the method is said to be symplectic. By carrying out the calculations one can obtain a rather simple test for whether a method has this property.

Theorem 1.1. *Suppose*

$$y_{j+1} = f(y_j, v_i),$$
$$v_{j+1} = g(y_j, v_j),$$

is a finite difference approximation of (1.73), (1.74). The method is symplectic if and only if $f_y g_v - f_v g_y = 1$, $\forall y, v$.

It is a bit easier to remember the equation in this theorem if it is written in matrix form as

$$det \begin{pmatrix} f_y & f_v \\ g_y & g_v \end{pmatrix} = 1, \ \forall y, v. \tag{1.79}$$

Example
As an example, for the velocity Verlet scheme in (1.75), (1.76), we have that

$$f(y,v) = y + kv + \frac{k^2}{2m}F(y),$$

$$g(y,v) = v + \frac{k}{2m}\left[F\left(y + kv + \frac{k^2}{2m}F(y)\right) + F(y)\right].$$

With this, $f_y = 1 + \frac{1}{2m}k^2F'(y)$ and $g_v = 1 + \frac{1}{2m}k^2F'(f)$. After calculating the other two derivatives one finds that $f_yg_v - f_vg_y = 1$, and therefore velocity Verlet is a symplectic method.

There are multiple reasons for introducing symplecticity. The first is that the original problem is symplectic, and so we are simply requiring our numerical method to preserve this property. A second reason is that it is easier to find symplectic methods than it is to find methods that conserve energy. For example, it is possible to tweak even the Euler method so it is symplectic (see Exercise 1.23). The third, and most important, reason for introducing this idea is that even though a symplectic method does not necessarily produce a Hamiltonian that is constant, it does come very close to the exact result. The energy computed by velocity Verlet, which is shown in Figure 1.9, is a typical example. In fact, it is possible to prove that as long as the function $H(y,v)$ is smooth, then a symplectic method provides an accurate approximation of the energy over an exponentially long time interval. Therefore, transferring our objective from energy conservation to symplecticity is worthwhile. The fly in the ointment here is that determining whether a method is symplectic when there are more than two equations is rather involved. There has been considerable research over the last few years for finding methods that preserve geometrical properties of the solution, including first integrals such as energy, and an in-depth discussion of this can be found in Stuart and Humphries [1998] and Hairer et al. [2003].

1.7 Next Steps

The approach that has been taken in this chapter is similar to what often occurs in applied mathematics. Namely, one has a problem to solve and it is difficult enough that analytical methods do not work. For this reason a numerical solution is sought, and to do this an algorithm is derived and then implemented. In the process one pays close attention that the method is consistent and stable, the two cornerstones for convergence. The development has been introductory and leaves open numerous questions and topics for discussion.

One aspect that deserves more attention relates to the theory underlying the methods. Typically, the starting point for this is to generalize the finite difference formulas that have been obtained and consider the general linear p-step formula

$$\mathbf{y}_{j+p} + a_{p-1}\mathbf{y}_{j+p-1} + \ldots + a_0\mathbf{y}_j = k(b_p\mathbf{f}_{j+p} + b_{p-1}\mathbf{f}_{j+p-1} + \ldots + b_0\mathbf{f}_j), \quad (1.80)$$

where $\mathbf{f}_j = \mathbf{f}(t_j, \mathbf{y}_j)$ and the a_j's, b_j's are constants with either a_0 or b_0 nonzero. The formula in (1.80) is linear because it involves only linear combinations of the $\mathbf{y}_j$'s and $\mathbf{f}_j$'s. It results in an explicit method if $b_p = 0$ or if $\mathbf{f}$ does not depend on $\mathbf{y}$, otherwise it is implicit. The question of convergence in this case introduces a form of stability determined using something called the root condition. This is the requirement that the method produces bounded solutions to the problem when $\mathbf{f} = \mathbf{0}$. With this one can prove the following result.

Theorem 1.2. *(Dahlquist Equivalence Theorem) If the approximations for the starting values* $\mathbf{y}_0$, $\mathbf{y}_1$, $\ldots$, $\mathbf{y}_{p-1}$ *are consistent then the p-step formula in (1.80) is convergent if and only if it is consistent and satisfies the root condition.*

The question of the order of the error when using (1.80) can be worked out using Taylor's theorem, as was done for the example involving (1.40). The result is contained in the following theorem.

Theorem 1.3. *The p-step formula in (1.80) has order* $O(k^n)$ *if and only if*

$$\sum_{i=0}^{p} a_i = 0,$$

$$\sum_{i=0}^{p} i^m a_i = m \sum_{i=0}^{p} i^{m-1} b_i, \quad \text{for } m = 1, 2, \ldots, n.$$

It should be remembered that the above theorem applies to (1.80). If one wants the method to have order $O(k^n)$ then the starting values must also be approximated with order $O(k^n)$. Also, both theorems require $\mathbf{f}$ to be sufficiently smooth. A analysis of these two theorems can be found in Isaacson and Keller [1966] and Deuflhard et al. [2002].

In constructing a numerical algorithm it is natural to ask just what is possible in terms of the order of the error using a formula like the one in (1.80). For example, one of the criticisms of the trapezoidal method is that it is implicit and this makes it hard to use. So, one might ask if it is possible to find an explicit method with the same truncation error as the trapezoidal method. This is the same question used when introducing the Runge–Kutta methods, but now we are considering methods that have the form given in (1.80). The answer is contained in the two Dahlquist barrier theorems. The first theorem states that the maximal order of a convergent p-step method is at most p if the method is explicit. If it is implicit then the maximal order is $p + 1$ if p is odd and it is $p + 2$ if p is even. This means there are no one-step explicit methods with a better truncation error than Euler, and there are no one-step implicit methods better than the trapezoidal method. The second barrier theorem states that if a p-step is A-stable then it must be implicit. Moreover, the maximal order for an A-stable method is just two. In other words, the trapezoidal method is as good as it gets for A-stable methods. An extended discussion of these theorems can be found in Hairer et al. [2002].

Another topic that was only partially investigated concerns methods for stiff equations. To explain what these are, one has to look no farther than the logistic equation (1.5). As shown in Figure 1.3, the solution starts near zero, gradually increases, and then levels off at one. If the value of λ is increased the transition from zero to one occurs much faster, and for large values of λ it looks almost vertical. Such rapid changes in the solution are characteristic of stiff equations, and what this means is that the logistic equation becomes stiffer as λ increases. Such problems can be quite challenging to solve numerically. Certainly A-stability and monotone methods play an important role, but other forms of stability are sometimes needed. An example is L-stability, which is introduced in Chapter 3. Another tool for such problems involves using an adaptive procedure, where the time step is modified during the computation to account for the changes in the solution. This requires good ways to estimate how the solution is changing to be able to adjust the time step appropriately, and this goes under the general classification of a posteriori estimates. A discussion of this can be found in Shampine [1994]. A third approach that is often used for such problems is to use an implicit Runge–Kutta method. The derivation of these, as with their explicit counterparts, is most commonly done using a graph-theoretic approach and in some cases collocation is used. A good introduction to this can be found in Lambert [1991] and Hairer and Wanner [2004].

One of the more interesting applications involving IVPs arises with computational nonlinear dynamics. For such problems the conservation of energy and symplecticity play an essential role and these were investigated in this chapter. A topic that was not covered is bifurcation phenomena, where one is interested in the change in the behavior of the solution as one or more parameters are changed. As an example, the flow of water in a pipe will be smooth and laminar if the pressure is not very large. However, if the pressure is increased there is a value p_{cr} at which the flow will become irregular and turbulent. In this case p_{cr} is a bifurcation point and the structure of the solution changes dramatically as the pressure passes through p_{cr}. Developing numerical methods for such situations is challenging and one of the more common approaches is to use a continuation method. This is discussed in Govaerts [2000].

Exercises

1.1. Determine whether the following produce consistent approximations (i.e., the truncation term $\tau_j \to 0$ as $k \to 0$). For those that are consistent give the order of the truncation term.
(a) $y(t_j) = \frac{1}{2}(y(t_{j+1}) + y(t_{j-1})) + \tau_j$.
(b) $y'(t_j) = \frac{1}{3k}[y(t_{j+1}) + y(t_j) - 2y(t_{j-1})] + \tau_j$.

(c) $y'(t_j) = y(t_{j+1}) - 2y(t_j) + \tau_j$.

(d) $y'(t_j) = \frac{1}{2k}[(\alpha - 3)y(t_j) + 2(2 - \alpha)y(t_{j+1}) - (1 - \alpha)y(t_{j+2})] + \tau_j$.

1.2. Assuming that the following are used to solve $y' = f(t, y)$, determine the order of the truncation error.

(a) $y_{j+1} = y_j + \frac{k}{2}(3f_j - f_{j-1})$.

(b) $y_{j+1} = -y_j + 2y_{j-1} + 3kf_j$.

(c) $y_{j+1} = \frac{4}{3}y_j - \frac{1}{3}y_{j-1} + \frac{2k}{3}f_j$.

(d) $y_{j+1} = y_j + kf(t_j + (1 - \alpha)k, \alpha y_j + (1 - \alpha)y_{j+1})$, where $0 \le \alpha \le 1$.

1.3. Determine whether the methods in Exercise 1.2 are A-stable or conditionally A-stable.

1.4. Find a finite difference approximation of the following problems using a method that is, at least conditionally, A-stable and has a truncation error of $O(k^2)$. If a method other than one listed in Table 1.3 is used then the stability and truncation error requirements must be justified.

(a) $y' = 1 + 2e^{-t}y - y^3$, where $y(0) = 3$.

(b) $\frac{d}{dt}\left(\frac{1}{t+y}\right) = \sin(y)$, where $y(0) = 5$.

(c) $2y' = -1 - \sqrt{y^2 + (y')^2}$, where $y(0) = 1$.

(d) $y' = e^{-y'} + t$, where $y(0) = 1$.

(e) $y' = y - e^{-2t}v$ and $v' = y^2 + 3v$, where $y(0) = 1, v(0) = -2$.

(f) $y'' + (1 - y^2)y' + y = \cos(t)$, where $y(0) = 1, y'(0) = 0$.

(g) $\frac{d}{dt}\left(\frac{y'}{\sqrt{1+(y')^2}}\right) + y' + y = 0$, where $y(0) = 0, y'(0) = 1$.

(h) $\theta_1'' = -\mu\theta_1 + \lambda\theta_2$ and $\theta_2'' = \lambda\theta_1 - \mu\theta_2$, where $\theta_1(0) = 1, \theta_1'(0) = \theta_2(0) = \theta_2'(0) = 0$.

(i) $ye^y = t$ for $t \ge 0$.

(j) $y(t) = (1 + t^2) \int_0^t (y(s) - \sin(s^2))ds$ for $t \ge 0$.

(k) $y' = a(t) - y^3$, where $a(t) = \int_0^t e^{2(t-s)}y(s)ds$ and $y(0) = -1$.

(l) Calculating the value of $y(t) = \sum_{n=0}^{\infty} \frac{1}{2^n n!}t^{2n}$, for $t \ge 0$.

1.5. The θ-method for solving (1.18) involves picking a value of θ satisfying $0 \le \theta \le 1$ and then using the difference equation

$$y_{j+1} = y_j + k\theta f_{j+1} + k(1 - \theta)f_j.$$

(a) What method is obtained if $\theta = 0$, or $\theta = \frac{1}{2}$, or $\theta = 1$?

(b) Explain how this method can be obtained using the Euler and backward Euler methods.

(c) Show how this method can be derived using numerical quadrature.

(d) For what values of θ is the method A-stable?

(e) For what values of θ is the method implicit?

(f) Determine whether the method is monotone or conditionally monotone (see Exercise 1.11).

1.6. This is a problem about the A-stability of some of the methods discussed in this chapter.
(a) Show that the leapfrog method is not A-stable.
(b) Show that the trapezoidal method is A-stable.
(c) Show that RK2 is conditionally A-stable. Is the k interval for stability larger or smaller than it is for the Euler method?
(d) Show that RK4 is conditionally A-stable and the k interval for stability is somewhat larger than it is for the Euler method.

1.7. This problem uses the integration method to derive various finite difference approximations.
(a) Using a one-point integration rule for the integral in (1.56) derive the Euler method and the order of the truncation error.
(b) Using a one-point integration rule for the integral in (1.56) derive the backward Euler method and the order of the truncation error.
(c) In STEP 2 of the integration method, suppose one integrates the equation from t_{j-1} to t_{j+1}. It is possible in this case to use Simpson's rule to approximate the integral. What finite difference equation results from this and what is the order of the truncation error?

1.8. As an approximation suppose the function $f(t, y(t))$ in the integral in (1.56) is replaced with the linear function of t that interpolates the function at the two points $(t_{j-1}, f(t_{j-1}, y(t_{j-1})))$, $(t_j, f(t_j, y(t_j)))$.
(a) Integrate the approximation for $f(t, y)$ to obtain the finite difference equation

$$y_{j+1} = y_j + \frac{k}{2}(3f_j - f_{j-1}), \quad \text{for } j = 1, 2, \ldots, M - 1.$$

Because this is explicit it is an example of an Adams–Bashforth method.
(b) Is the method A-stable?
(c) What is the order of the error of the linear approximation for f for $t_j \leq t \leq t_{j+1}$? What is the order of the truncation error for the finite difference method in (a)?

1.9. As an approximation suppose the function $f(t, y(t))$ in the integral in (1.56) is replaced with the quadratic polynomial that interpolates the function at the three points $(t_{j-1}, f(t_{j-1}, y(t_{j-1})))$, $(t_j, f(t_j, y(t_j)))$, and $(t_{j+1}, f(t_{j+1}, y(t_{j+1})))$.
(a) Integrate the approximation for $f(t, y)$ to obtain the finite difference equation

$$y_{j+1} = y_j + \frac{k}{12}(5f_{j+1} + 8f_j - f_{j-1}), \quad \text{for } j = 1, 2, \ldots, M - 1.$$

Because this is implicit it is an example of an Adams–Moulton method.
(b) Is the method A-stable?
(c) What is the order of the error of the quadratic approximation for f for $t_j \leq t \leq t_{j+1}$? What is the order of the truncation error for the finite difference method in (a)?

1.10. This problem develops a systematic approach for deriving Runge–Kutta methods, with error $O(k^2)$, for solving $y' = f(t, y)$. The start-off assumption is that the method has the general form given in (1.63), and the goal is to determine the constants a, b, α, β that produce the best truncation error.
(a) Assuming small k show that

$$f(t + \alpha k, y + \beta k f) = f + \alpha k f_t + \beta k f f_y$$
$$+ \frac{1}{2}(\alpha k)^2 f_{tt} + \alpha \beta k^2 f f_{ty} + \frac{1}{2}(\beta k)^2 f^2 f_{yy} + O(k^3),$$

where $f = f(t, y)$.
(b) Substitute the exact solution $y(t)$ into (1.63) and then expand the result for small k using part (a). Use this to show that the best truncation error is $O(k^2)$ and this is obtained by taking $a + b = 1$, $\beta = \alpha$, $2b\alpha = 1$. These are the order conditions for the method.
(c) Show how Heun's method is obtained from the result from part (b).
(d) There is one undetermined constant in the solution in part (b). Can this constant be determined by finding which value produces the largest stability region? What about using a requirement of computational simplicity to determine the constant?
(e) Can monotonicity (see Exercise 1.11) be used to determine the undetermined constant in (b)?

1.11. It is desirable that if the exact solution is monotonically increasing (or decreasing) then the numerical solution does the same thing. The test case for this property is the usual radioactive decay IVP given in (1.2) and (1.3), where $\alpha > 0$. A numerical method is said to be monotone if it produces a monotonically decreasing solution to this problem. An example demonstrating the importance of monotonicity is given in Exercise 1.15.
(a) Show that backward Euler is monotone.
(b) Show that Euler, trapezoidal, and RK4 are conditionally monotone (i.e., they are monotone only if the step size is small enough) and that RK4 has the largest monotone interval of the three.

1.12. Backward difference methods (BDFs) use one-sided backward differences to derive finite difference approximations. An example is the backward Euler method, which is a one-step BDF. The reason for considering such an approximation is that BDFs usually have better monotone properties.
(a) Derive a two-step BDF for (1.18). What is the truncation error for the method?
(b) Determine whether the method is A-stable.
(c) Determine whether the method is monotone or conditionally monotone (see Exercise 1.11).

1.13. This problem concerns the IVP involving the Bernoulli equation

$$y' + y^3 = \frac{y}{a + t}, \quad \text{for } t > 0,$$

where $y(0) = 1$. You are to solve this problem using the backward Euler, trapezoidal, and RK4 methods.

(a) Verify that the exact solution is

$$y = \frac{a+t}{\sqrt{\beta + \frac{2}{3}(a+t)^3}}.$$

(b) Assuming $a = 0.01$, on the same axes plot the exact and the three numerical solutions for $0 \le t \le 3$ in the case $M = 80$.

(c) Redo (b) for $M = 20$, $M = 40$, and $M = 160$. If one or more of the methods is unstable you can exclude it from the plot (for that value of M) but make sure to state this in your write-up.

(d) Plot the max error e_∞ as a function of M for each method, using $M = 40, 80, 160, 320, 640$. The four curves should be in the same log-log plot.

(e) Compare the four methods based on your results from parts (a)–(d). This includes ease of use, speed of calculation, accuracy of results, and apparent stability. Also comment on the sensitivity of the two implicit methods to the stopping condition used in the nonlinear equation solver (e.g., could you use a relatively large tolerance without affecting accuracy and that enables the calculation go faster?).

1.14. This problem concerns the equation

$$y' = \alpha(h(t) - y) + h'(t), \quad \text{for } t > 0,$$

where $y(0) = \beta$. You are to solve this problem using the backward Euler, trapezoidal, and RK4 methods.

(a) Assuming $h(0) = 0$, show that $y = h(t) + \beta e^{-\alpha t}$ is the solution of the problem.

(b) Assume $\alpha = 20$, $\beta = 1$, and $h(t) = \sin(t)$. On the same axes plot the exact and the three numerical solutions for $0 \le t \le 3$ in the case $M = 40$.

(c) Redo (b) for $M = 80$ and for $M = 20$. If one or more of the methods is unstable for either case you can exclude it from the plot (for that value of M) but make sure to state this in your write-up.

(d) Based on your results from (a)–(c) explain why a better measure of the error in this problem is e_∞ rather than e_M. Plot e_∞ as a function of M for each method, using $M = 80, 160, 320, 640, 1280, 2560$. The three curves should be in the same log-log plot.

(e) Compare the three methods based on your results from (a)–(d). This includes ease of use, speed of calculation, accuracy of results, and apparent stability.

1.15. This problem concerns the deceptively simple equation

$$\epsilon y' = (1-t)y - y^2, \quad \text{for } t > 0,$$

where $y(0) = 1$. You are to solve this problem using the backward Euler, trapezoidal, and RK4 methods in the case $\epsilon = 0.01$ and $T = 2$.

(a) On the same axes plot the exact and four numerical solutions for $0 \le t \le T$ in the case $M = 60$.
(b) Redo (a) for $M = 30$ and for $M = 120$. If one or more of the methods is unstable you can exclude it from the plot for that M value but make sure to state this in your write-up.
(c) Compare the three methods based on your results from parts (a) and (b). This includes ease of use, speed of calculation, accuracy of results, and apparent stability.

1.16. This problem concerns solving the Lane–Emden equation

$$y'' + \frac{2}{t}y' + y^n = 0, \quad \text{for } t > 0,$$

where n is a nonnegative integer. Assume that the initial conditions are $y(0) = 1$ and $y'(0) = 0$. This equation arises in studying the temperature distribution within a star.
(a) Verify that (i) $y = \frac{\sin(t)}{t}$ is the solution when $n = 1$, and (ii) $y = (1 + \frac{1}{3}t^2)^{-1/2}$ is the solution when $n = 5$.
(b) Derive a second-order method for this problem.
(c) Assuming $n = 1$, on the same axes plot the exact and numerical solutions for $0 \le t \le 10$ in the case $M = 40$.
(d) Assuming $n = 5$, on the same axes plot the exact and numerical solutions for $0 \le t \le 10$ in the case $M = 40$.

1.17. This problem investigates using a one-sided approximation of an initial condition rather than a ghost point.
(a) Derive a one-sided difference approximation of the initial condition in (1.65) that has truncation error $O(k^2)$.
(b) With the approximation from part (a) and the finite difference equation (1.68), state what problem must be solved to find u_1 and u_2. Also, explain why the method is no longer explicit.

1.18. This problem investigates how the approximation of the initial conditions can affect the accuracy of the entire calculation. The equation is

$$y'' + 2\alpha y' + \alpha^2 y = 0, \quad \text{for } 0 < t,$$

where the initial conditions are $y(0) = 0, y'(0) = 1$. The exact solution in this case is $y = te^{-\alpha t}$.
(a) Using centered differences, derive a second-order accurate approximation for the differential equation.
(b) Use the fourth formula in Table 1.1 to derive a second-order accurate approximation for the initial conditions.
(c) Use the first formula in Table 1.1 to derive a first-order accurate approximation for the initial conditions.

(d) Let $T = 0.5$ and $\alpha = 2$. Plot the error $|y(T) - \bar{y}_M|$ as a function of M, for $10 \le M \le 10^4$, using the methods from (a) and (b).

(e) Redo (d) using the methods from (a) and (c).

(f) What is the consequence of using a lower-order approximation at just one point?

1.19. This problem explores using the system formulation of the IVP given in (1.64), (1.65).

(a) Write the IVP as a first-order system.

(b) Derive an explicit finite difference approximation of the system in part (a) that has truncation error $O(k^2)$.

(c) Comment on the differences between the system and direct approaches for solving the problem (e.g., ease of use, computational complexity, satisfying the initial conditions, etc.).

1.20. Write the following IVPs as first-order systems.

(a) The van der Pol equation $y'' + a(1 - y^2)y' + y = 0$, assuming $y(0) = 1, y'(0) = 0$.

(b) $\frac{d}{dt}(e^y \frac{dy}{dt}) - ty = 0$, where $y(0) = 3, y'(0) = -1$.

(c) $y'''' - y'' + \mu(t)y = t + 1$, where $y(0) = 0, y'(0) = -1, y''(0) = 4, y'''(0) = 1$.

(d) A reduced form of the equations for planetary motion is

$$r'' - r\theta' = -\frac{K}{r},$$
$$r\theta'' + 2r'\theta' = 0.$$

Assume $r(0) = a, r'(0) = b, \theta(0) = c, \theta'(0) = d$.

1.21. There are various ways to rewrite equations in system form, and this exercise demonstrates this for the problem

$$y'' + g(y)y' + y = 0, \text{where } y(0) = 1, y'(0) = 0.$$

(a) Rewrite this in system form using the change of variables in (1.10), (1.11).

(b) Suppose it is possible to find a function $G(y)$ such that $\frac{dG}{dy} = g(y)$. In this case the Lienard transform is $u = y, v = y' + G(y)$. What is the resulting first-order system for $\mathbf{u} = (u, v)^T$?

(c) Explain why there is no significant difference between the systems in (a) and (b) if the problem is linear as in the mass–spring–dashpot equation (1.16). Can the same be said if the equation is nonlinear as it is for the van der Pol equation $y'' + a(1 - y^2)y' + y = 0$?

1.22. This problem examines different ways to derive, and express, the velocity Verlet method.

(a) Derive (1.77) directly from (1.73), (1.74) by using Taylor's theorem on $y(t_j + k)$.

(b) The Verlet method is obtained by using a centered $O(k^2)$ approximation in (1.71). What is the resulting finite difference approximation?

(c) Show that velocity Verlet can be reduced to the Verlet method derived in (b).

1.23. A symplectic Euler method for solving (1.71) is

$$y_{j+1} = y_j + kv_j,$$

$$v_{j+1} = v_j + \frac{k}{m}F(y_{j+1}).$$

(a) Derive this method and show that it is first order. Also show that the method is symplectic.

(b) Use this method to solve the problem used for Figure 1.9. Also solve the problem using the standard Euler method. With this, on the same axes in the yv-plane, plot the computed points (y, v) for $0 \le t \le 18$ using $M = 180$. Also plot the exact solution on the same axes.

(c) Plot the computed energy H obtained from the two methods in (b), for $0 \le t \le 18$ using $M = 180$.

(d) Based on your results from (b) and (c) comment on the effectiveness of the two methods.

1.24. The equation for a linear harmonic oscillator can be written in system form as $y' = v, v' = -y$, and the Hamiltonian is $H(t) = \frac{1}{2}y^2 + \frac{1}{2}v^2$.

(a) If the θ-method is used (see Exercise 1.5) what are the resulting finite difference equations for y and v? The same θ should be used in each equation.

(b) Letting $H_j = \frac{1}{2}y_j^2 + \frac{1}{2}v_j^2$, show that the θ-method results in $H_{j+1} = \gamma H_j$, where $\gamma = 1 + k^2(1 - 2\theta)/(k^2\theta + 1)$. From this explain why the only value of θ that conserves energy corresponds to the trapezoidal method. Also, compare this with what happens to the energy as j increases using Euler or backward Euler.

(c) What if RK2 is used on both equations, is energy conserved?

1.25. This problem considers the equation for a pendulum $\ell \frac{d^2\theta}{dt^2} = -g \sin(\theta)$, where $\theta(0) = \frac{\pi}{4}, \theta'(0) = 0$, and ℓ, g are positive constants

(a) By introducing the angular velocity θ' write the equation as a first-order system.

(b) Write down the velocity Verlet method for the resulting system.

(c) The Hamiltonian for the pendulum is $H = \frac{1}{2}\ell(\theta')^2 - g\cos(\theta)$. Show that this is constant and determine its value.

(d) Use RK4 on the system in (a) and the Verlet method from (b) to solve the problem for $0 \le t \le 100$. Use these methods to plot, on the same axes, H as a function of t. Use $M = 200$ and take $\ell = g = 1$.

(e) In comparison to the exact result, how do the two methods do in approximating the Hamiltonian over this time interval? What if one uses $M = 400$ or $M = 100$?

1.26. Suppose to solve (1.71) one uses an implicit scheme of the form

$$y_{j+1} = p(y_j, v_j, y_{j+1}, v_{j+1}),$$
$$v_{j+1} = q(y_j, v_j, y_{j+1}, v_{j+1}).$$

Assume that these can be solved at least implicitly with the result that $y_{j+1} = f(y_j, v_j)$, $v_{j+1} = g(y_j, v_j)$.

(a) Writing $p = p(y, v, r, s)$, $q = q(y, v, r, s)$ show that the method is symplectic if and only if $p_y q_v - p_v q_y + p_s q_r - p_r q_s + p_r + q_s = 1$, $\forall y, v, r, s$.

(b) Suppose backward Euler is used to solve both (1.73) and (1.74). Show that this produces a method that is not symplectic.

(c) Show that the trapezoidal method (1.75) and (1.76) is not symplectic.

(d) The implicit midpoint method applied to (1.73), (1.74) is $y_{j+1} = y_j + \frac{1}{2}k(v_j + v_{j+1})$, $v_{j+1} = v_j + \frac{1}{m}kF((y_j + y_{j+1})/2)$. Show that this method is symplectic.

(e) Show that the method in part (d) has truncation error $O(k^2)$.

1.27. Show that $y_{j+1} = -y_j$, $v_{j+1} = v_j$ conserves area. Explain in geometric terms why it is not symplectic.

1.28. The equations of motion (1.73), (1.74) are time reversible. To define this for a numerical method, suppose that with the initial condition $\mathbf{y} = (y_a, v_a)^T$ the method takes one time step and produces the value $\mathbf{y} = (y_A, v_A)^T$. The method is time reversible if it will then take the initial condition $\mathbf{y} = (y_A, -v_A)^T$ and in one time step produce the value $\mathbf{y} = (y_a, -v_a)^T$. The idea here is that you can get back to your original position y_a by simply reversing your velocity.

(a) Show that Verlet is time reversible.

(b) Show that Euler is not time reversible.

(c) Is the symplectic Euler method in Exercise 1.23 time reversible?

1.29. On several occasions in this book we end up having to solve a first-order system of the form

$$\mathbf{A}\mathbf{y}'(t) = \mathbf{B}\mathbf{y}(t),$$

where the $n \times n$ matrix $\mathbf{A}$ is invertible. Neither $\mathbf{A}$ nor $\mathbf{B}$ depends on t.

(a) Show that the Euler, trapezoidal, and RK4 methods reduce to solving one or more matrix equations at each time step.

(b) Discuss the differences in the computational effort per time step in the three methods.

1.30. Suppose in Table 1.1 that the grid points are not necessarily equally spaced and that $h_i = x_i - x_{i-1}$ and $h_{i+1} = x_{i+1} - x_i$.

(a) Show that the centered difference formula for the first derivative becomes

$$f'(x_i) = \frac{f(x_{i+1}) - f(x_{i-1})}{h_{i+1} + h_i} + O\left(\frac{h_{i+1}^2}{h_{i+1} + h_i}\right) + O\left(\frac{h_i^2}{h_{i+1} + h_i}\right).$$

(b) Show that the centered difference formula for the second derivative becomes

$$f''(x_i) = 2\frac{h_i f(x_{i+1}) - (h_{i+1} + h_i)f(x_i) + h_{i+1}f(x_{i-1})}{h_{i+1}h_i(h_{i+1} + h_i)}$$
$$+ O\left(\frac{h_{i+1}^2}{h_{i+1} + h_i}\right) + O\left(\frac{h_i^2}{h_{i+1} + h_i}\right).$$

1.31. Consider the second-order difference equation $y_{j+1} + 2\alpha y_j + \beta y_{j-1} = 0$, where α, β are known constants with $\beta \neq \alpha^2$.
(a) By assuming a solution of the form $y_j = r^j$, show that the general solution of the equation has the form $y_j = Ar_+^j + Br_-^j$, where A, B are arbitrary constants and $r_\pm = -\alpha \pm \sqrt{\alpha^2 - \beta}$.
(b) Show that if $\alpha^2 < \beta$ then the solution in (a) can be written as $y_j = As^j \cos(\omega j + a)$, where A, a are arbitrary constants and s, ω depend on α, β.
(c) Show that if $\alpha^2 = \beta$ then the solution has the form $y_j = (A + Bj)r^j$.
(d) What is the solution if the equation is $y_j + 2\alpha y_{j-1} + \beta y_{j-2} = 0$?

1.32. This problem considers how to check to see whether a numerical algorithm is correct. The equation to be solved is

$$y' + f(t)y = g(t), \quad \text{for } 0 < t,$$

where the initial condition is $y(0) = a$.
(a) Derive a second-order accurate approximation for the IVP.
(b) Usually one is given f and g and then is expected to solve for y. However, suppose one were to identify, or pick, the solution first. Explain how to determine f or g in this case along with a.
(c) Explain how to use the idea in (b) to test that the computer code for your algorithm in (a) is correct. Your numerical test(s) should include check(s) on the rate of convergence, on correctly coding the functions f and g, as well as handling the initial condition correctly.
(d) Are your tests in (c) extendable to more complicated equations? For example, are they extendable to a nonlinear equation such as $y' + f(t)y + h(t)y^3 = g(t)$?

1.33. This problem considers how to solve the nonlinear equation obtained in going from t_j to t_{j+1} using an implicit method. The trapezoidal method is used as a prototype for this situation.
(a) For the trapezoidal method show that finding y_{j+1}, given y_j, is the same as finding $z = y_{j+1}$ that satisfies $g(z) = 0$, where $g(z) \equiv kf(t_{j+1}, z) - 2z + 2y_j + kf_j$.
(b) Newton's method for solving $g(z) = 0$ involves picking z_0 and then letting

$$z_{i+1} = z_i - \frac{g(z_i)}{g'(z_i)}, \quad \text{for } i = 0, 1, 2, 3, \dots.$$

Assume that the stopping condition for this iteration is $|z_{i+1} - z_i| \le \epsilon$, where ϵ is given. What is $g'(z_i)$?
(c) For convergence, z_0 should be close to y_{j+1}. Explain how to pick z_0. Also, explain how ϵ should be chosen in relation to k and why one should not pick ϵ too small.
(d) What are $g(z)$ and $g'(z_i)$ when using the backward Euler method?

1.34. The idea of A-stability was introduced by Dahlquist [1963]. Write a short essay explaining why he used the letter A in the name of this property.

2

Two-Point Boundary Value Problems

2.1 Introduction

In this chapter we investigate how to find the numerical solution of what are called two-point boundary value problems (BVPs). The most apparent difference between these problems and the IVPs studied in the previous chapter is that BVPs involve only spatial derivatives. What this means is that we consider how to solve a differential equation in an interval $0 < x < \ell$, where the solution is required to satisfy conditions at the two endpoints $x = 0, \ell$. Examples of such problems are below.

2.1.1 Birds on a Wire

A string, or cord, strung between two points is what effectively occurs with telephone wires running between two telephone poles. The cord deflects due to its weight and other forces it is subjected to (e.g., birds sitting on the wire as in Figure 2.1). Assuming that the poles are located at $x = 0, \ell$ and letting $y(x)$ designate the vertical deflection of the cord, then the mathematical problem has the form

$$\frac{d^2y}{dx^2} = f(x), \quad \text{for } 0 < x < \ell, \tag{2.1}$$

where

$$y(0) = y(\ell) = 0. \tag{2.2}$$

In the differential equation (2.1), the function $f(x)$ accounts for the various forces on the cord. The two boundary conditions in (2.2) represent the fact that the cord is held fixed at the left and right ends.

2.1.2 Chemical Kinetics

In the study of reactions of certain chemical mixtures one comes across the following problem

Figure 2.1. Animated example demonstrating the vertical deflection of a string due to loading (Pixar Animation Studios [2000]). Note the asymmetry in the loading function in this example.

$$\frac{d^2y}{dx^2} = -\gamma e^y, \quad \text{for } 0 < x < \ell, \tag{2.3}$$

where $y(0) = y(\ell) = 0$. This is known as Bratu's equation, and one of its distinguishing features is that it is nonlinear. A consequence of this is that the problem has no solution if $\gamma > \gamma_c$, where $\gamma_c \approx 3.5$. On the other hand, if $\gamma = \gamma_c$ then it has one solution, and if $\gamma < \gamma_c$ then it has exactly two solutions. This situation is interesting from a numerical point of view. For example, one question that arises is whether it is possible to construct an algorithm that is capable of finding both solutions if $\gamma < \gamma_c$. This is not easy to answer and we return to this issue later in the chapter.

2.2 Derivative Approximation Methods

To introduce the ideas we start with the problem of finding the function $y(x)$ that satisfies

$$\frac{d^2y}{dx^2} + p(x)\frac{dy}{dx} + q(x)y = f(x), \quad \text{for } 0 < x < \ell, \tag{2.4}$$

where

$$y(0) = \alpha \text{ and } y(\ell) = \beta. \tag{2.5}$$

In this BVP, the functions $p(x), q(x), f(x)$ are given, as are the constants ℓ, α, β. One can prove that if the functions $p(x), q(x), f(x)$ are continuous, with $q(x) \leq 0$, for $0 \leq x \leq \ell$, then this problem has a unique solution (Keller [1992]). It is assumed throughout this chapter that this holds.

We will drive a numerical approximation for the BVP in (2.4), (2.5) using a sequence of steps much like what we used for IVPs. There are, however, significant differences that will be apparent once the finite difference approximation has been completed.

STEP 1. We first introduce the spatial points where we will compute the solution. These points are labeled sequentially as $x_0, x_1, x_2, \ldots, x_{N+1}$, and a schematic drawing indicating their location along the x-axis is shown in Figure 2.2. We confine our attention to a uniform grid with step size h, so we have the following formula for the grid points

$$x_i = ih, \quad \text{for } i = 0, 1, 2, \ldots, N+1. \tag{2.6}$$

The labeling system used here has $x_0 = 0$ and $x_{N+1} = \ell$, with step size $h = \ell/(N+1)$.

STEP 2. Evaluate the differential equation at the grid point $x = x_i$ to obtain

$$y''(x_i) + p(x_i)y'(x_i) + q(x_i)y(x_i) = f(x_i). \tag{2.7}$$

STEP 3. Replace the derivative terms in STEP 2 with finite difference approximations. As with IVPs, there are numerous choices and a few are listed in Table 1.1. We choose the centered difference formulas, which means we use the following expressions

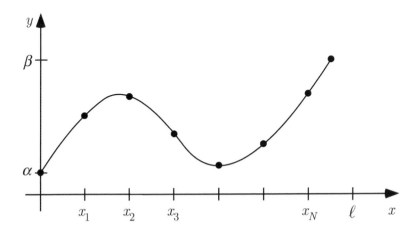

Figure 2.2. Grid system used to derive a finite difference approximation of the BVP.

$$y'(x_i) = \frac{y(x_{i+1}) - y(x_{i-1})}{2h} - \frac{1}{6}h^2 y'''(\eta_i), \qquad (2.8)$$

$$y''(x_i) = \frac{y(x_{i+1}) - 2y(x_i) + y(x_{i-1})}{h^2} - \frac{1}{12}h^2 y''''(\bar{\eta}_i), \qquad (2.9)$$

where η_i and $\bar{\eta}_i$ are between x_{i-1} and x_{i+1}. Introducing these into (2.7), we obtain

$$\frac{y(x_{i+1}) - 2y(x_i) + y(x_{i-1})}{h^2} + p(x_i)\frac{y(x_{i+1}) - y(x_{i-1})}{2h} + q(x_i)y(x_i) + \tau_i = f(x_i). \qquad (2.10)$$

The truncation error appearing in the above equation is

$$\tau_i = -\frac{1}{12}h^2 y''''(\bar{\eta}_i) - \frac{1}{6}h^2 p(x_i)y'''(\eta_i)$$
$$= O(h^2). \qquad (2.11)$$

Simplifying (2.10) yields

$$c_i y(x_{i+1}) + a_i y(x_i) + b_i y(x_{i-1}) + h^2 \tau_i = h^2 f(x_i), \qquad (2.12)$$

where

$$a_i = -2 + h^2 q(x_i), \qquad (2.13)$$

$$b_i = 1 - \frac{h}{2}p(x_i), \qquad (2.14)$$

$$c_i = 1 + \frac{h}{2}p(x_i). \qquad (2.15)$$

STEP 4. Drop the truncation error. This is the step where we go from an exact problem to one that is, hopefully, an accurate approximation of the original. After dropping τ_i in (2.12), the resulting finite difference equation is

$$c_i y_{i+1} + a_i y_i + b_i y_{i-1} = h^2 f(x_i), \quad \text{for } i = 1, 2, \ldots, N. \qquad (2.16)$$

From the boundary conditions (2.5) we also have that

$$y_0 = \alpha \text{ and } y_{N+1} = \beta. \qquad (2.17)$$

We have completed the derivation of the finite difference approximation for the BVP. One important question left unanswered is, how accurate is this approximation? For example, to derive (2.16) from (2.12) we dropped a term that is $O(h^4)$. Does this mean that we can expect y_i to be within $O(h^4)$ of $y(x_i)$ or is it the truncation error that determines this difference? This question will be answered shortly.

2.2.1 Matrix Problem

With the finite difference equation in (2.16) we have ended up with N equations in N unknowns. This system can be written in matrix form by collecting the unknowns into a vector as follows

$$\mathbf{y} = \begin{pmatrix} y_1 \\ y_2 \\ \vdots \\ y_N \end{pmatrix}. \tag{2.18}$$

In this case, (2.16) and (2.17) combine to give the matrix equation

$$\mathbf{A}\mathbf{y} = \mathbf{z}, \tag{2.19}$$

where

$$\mathbf{z} = \begin{pmatrix} h^2 f_1 - \alpha b_1 \\ h^2 f_2 \\ \vdots \\ h^2 f_{N-1} \\ h^2 f_N - \beta c_N \end{pmatrix} \tag{2.20}$$

and

$$\mathbf{A} = \begin{pmatrix} a_1 & c_1 & & & & \\ b_2 & a_2 & c_2 & & \mathbf{0} & \\ & b_3 & a_3 & c_3 & & \\ & & \ddots & \ddots & \ddots & \\ & \mathbf{0} & & & & c_{N-1} \\ & & & & b_N & a_N \end{pmatrix}. \tag{2.21}$$

The matrix $\mathbf{A}$ is tridiagonal, with the superdiagonal formed from the c_i's and the subdiagonal formed from the b_i's.

Our task is now to solve the matrix equation (2.19). The tridiagonal structure of $\mathbf{A}$ can be used to advantage to reduce the computational effort needed to solve the problem. If one carries out an LU factorization of $\mathbf{A}$ and keeps track of the zeros in the matrix, a particularly simple algorithm is produced (Isaacson and Keller [1966]). The procedure, given in Table 2.1, takes $8N - 7$ flops (floating-point operations), which is significantly less than the $\frac{2}{3}N^3 + O(N^2)$ flop count one gets with a full LU factorization. Another advantage is that it requires substantially less storage, needing approximately six N-vectors. The question left open, however, is whether the matrix equation even has a solution. For this we need a bit of theory.

Set: $w = a_1$, $y_1 = \dfrac{z_1}{w}$

For $i = 2, 3, \ldots, N$

$$v_i = \frac{c_{i-1}}{w}$$

$$w = a_i - b_i v_i$$

$$y_i = \frac{z_i - b_i y_{i-1}}{w}$$

End

For $j = N - 1, N - 2, \ldots, 1$

$$y_j = y_j - v_{j+1} y_{j+1}$$

End

Table 2.1. Algorithm for solving $\mathbf{A}\mathbf{y} = \mathbf{z}$ when $\mathbf{A}$ is the tridiagonal matrix given in (2.21).

2.2.2 Tridiagonal Matrices

Two issues that need to be addressed before running off and coding the method are whether the matrix equation (2.19) has a unique solution and even if it does whether the algorithm in Table 2.1 can be used to find the solution of the BVP. With this in mind, we introduce the row sums r_i for $\mathbf{A}$, defined as

$$r_i = \begin{cases} |b_i| + |c_i| & \text{if } i \neq 1, N, \\ |c_1| & \text{if } i = 1, \\ |b_N| & \text{if } i = N. \end{cases} \tag{2.22}$$

The reason for considering r_i is that one of the more popular, and easier, tests for invertibility of a matrix is to compare r_i with the value of the corresponding diagonal entry a_i. This idea gives rise to the following definition.

Definition 2.1. *The tridiagonal matrix* $\mathbf{A}$ *in (2.21) is diagonally dominant if* $|a_i| \geq r_i$, $\forall i$. *It is strict diagonally dominant if strict inequality holds* $\forall i$.

As an example consider the following tridiagonal matrices

$$\mathbf{A}_1 = \begin{pmatrix} 2 & 1 & 0 \\ 1 & 1 & 1 \\ 0 & 1 & 4 \end{pmatrix}, \quad \mathbf{A}_2 = \begin{pmatrix} 2 & 1 & 0 \\ 1 & -2 & 1 \\ 0 & 1 & 4 \end{pmatrix}, \quad \mathbf{A}_3 = \begin{pmatrix} 2 & 1 & 0 \\ 1 & 3 & 1 \\ 0 & 1 & 4 \end{pmatrix}.$$

In this case, $\mathbf{A}_1$ is not diagonally dominant, $\mathbf{A}_2$ is diagonally dominant but it is not strict diagonally dominant, and $\mathbf{A}_3$ is strict diagonally dominant.

It is shown in most books on numerical linear algebra that a strict diagonally dominant matrix is invertible. Unfortunately, our matrix does not qualify (e.g., let $p = q = 0$), and so we need a more refined test. What is needed is a condition that guarantees that the variable w in Table 2.1 is always nonzero. The theorem below identifies when this happens.

Theorem 2.1. *The tridiagonal matrix* $\mathbf{A}$ *in (2.21) is invertible, and the algorithm in Table 2.1 can be used to solve* $\mathbf{Ay} = \mathbf{z}$ *if either one of the following holds:*

1. $\mathbf{A}$ *is strict diagonally dominant,*
2. $\mathbf{A}$ *is diagonally dominant,* $c_i \neq 0$ $\forall i$, *and* $|b_N| < |a_N|$.

The second condition is the one we will use and it is therefore worth understanding why the stated conditions are needed. As mentioned earlier, the only operation of concern in the algorithm is the division by w, and so the majority of the proof consists of showing this cannot happen. The first thing to notice is that for $i = 2$, $|v_2| = |c_1/a_1| \leq 1$, where the inequality holds because the matrix is diagonally dominant. With this $|w| = |a_2 - b_2 v_2| \geq |a_2| - |b_2 v_2| \geq |a_2| - |b_2| > 0$, where the last inequality holds because the matrix is diagonally dominant and $c_2 \neq 0$. Continuing this argument, using induction, it is not hard to show that $|v_i| \leq 1$ and $|w| \geq |a_i| - |b_i| > 0$. As before, the last inequality holds, except for the last row of the matrix, because we are assuming $|a_i| \geq |b_i| + |c_i|$ and $c_i \neq 0$. The fact that it holds for $i = N$ is because we have explicitly assumed that $|b_N| < |a_N|$. Therefore, w is never zero and the algorithm can be used to solve the matrix equation. The proof for the first condition is similar and can be found in Süli and Mayers [2003].

The above theorem will be used to establish conditions that ensure that (2.19) has a unique solution. Before doing this, however, there is another useful bit of information from linear algebra related to the eigenvalues of a tridiagonal matrix. For this it is assumed that the entries of $\mathbf{A}$ are constant, that is, it has the form

$$\mathbf{A} = \begin{pmatrix} a & c & & & & \\ b & a & c & & \mathbf{0} & \\ & b & a & c & & \\ & & \ddots & \ddots & \ddots & \\ \mathbf{0} & & & b & a & c \\ & & & & b & a \end{pmatrix}. \tag{2.23}$$

Given the formulas in (2.14) and (2.15), for smaller values of h the coefficients satisfy $bc > 0$. In this case one finds that the eigenvalues λ_i of (2.23) are given as (see Exercise 2.28)

$$\lambda_i = a + 2\sqrt{bc} \cos\left(\frac{i\pi}{N+1}\right), \quad \text{for } i = 1, 2, \ldots, N. \tag{2.24}$$

The usefulness of this result comes from the fact that the condition number $\text{cond}_2(\mathbf{A})$ of the matrix, using the Euclidean norm, is the ratio of the largest and smallest eigenvalues (in absolute value). To determine these values note that for small values of h it follows from Taylor's theorem, and the formulas in (2.13)–(2.15), that $a + 2\sqrt{bc} = (q - \frac{1}{4}p^2)h^2 + O(h^4)$. Given our assumption that $q \leq 0$ then $a + 2\sqrt{bc} \leq 0$ for small values h. Assuming this holds then, using (2.24), we have that $\max_{i=1,\dots,N} |\lambda_i| = |\lambda_N|$ and $\min_{i=1,\dots,N} |\lambda_i| = |\lambda_1|$. Consequently,

$$\text{cond}_2(\mathbf{A}) = \frac{\left| a + 2\sqrt{bc} \, \cos\left(\frac{N\pi}{N+1}\right) \right|}{\left| a + 2\sqrt{bc} \, \cos\left(\frac{\pi}{N+1}\right) \right|}. \tag{2.25}$$

2.2.3 Matrix Problem Revisited

We now return to the question of whether the finite difference approximation for the BVP has a unique solution. To answer this we use the second option in Theorem 2.1. Because we are assuming $q(x) \leq 0$, to use this theorem we need $2 - h^2 q_i \geq |1 + hp_i/2| + |1 - hp_i/2|$ and $1 + hp_i/2 \neq 0$. As it turns out, both of these conditions will hold if $|hp(x_i)| < 2$, $\forall i$. With this we obtain the following theorem.

Theorem 2.2. *If the functions $p(x), q(x), f(x)$ are continuous for $0 \leq x \leq \ell$, with $q(x) \leq 0$, then the finite difference approximation as expressed in (2.19) has a unique solution if the step size satisfies $hp_\infty < 2$, where $p_\infty = \max_{0 \leq x \leq \ell} |p(x)|$.*

With the question of existence of a solution out of the way, the next issue is whether round-off error is a problem. For this we limit the conversation to the special case of when p and q are constants. To make use of the formula in (2.25), from Taylor's theorem, if N is large,

$$\cos\left(\frac{N\pi}{N+1}\right) = -1 + \frac{\pi^2}{2N^2} + O\left(\frac{1}{N^3}\right),$$

$$\cos\left(\frac{\pi}{N+1}\right) = 1 - \frac{\pi^2}{2N^2} + O\left(\frac{1}{N^3}\right).$$

With these and the formulas for the coefficients in (2.13)–(2.15), we have that as the number of grid points increases,

$$\text{cond}_2(\mathbf{A}) = \frac{16N^2}{4\pi^2 + (p^2 - 4q)\ell^2} + O(N). \tag{2.26}$$

For many of the problems we consider, N is usually no larger than about 10^2, and so the condition number of the matrix is not particularly large. In other words, the problems are reasonably well conditioned. Even so, (2.26) shows that the condition number of the problem increases quickly with N, and it is necessary to keep in mind if a large number of grid points are needed to solve the problem.

A result such as Theorem 2.2 increases a person's comfort level when calculating the numerical solution in the sense that one knows what conditions are needed to guarantee that a solution exists. However, there are certainly step sizes and functions $p(x), q(x), f(x)$ that don't satisfy these conditions, yet the code, and BVP, are just fine. More important, what the theorem does not address is just how accurately the computed solution approximates the exact solution of the BVP. One approach to answering this question involves using the method to actually solve a problem, and that brings us to the following examples.

Example 1
Consider the following BVP

$$y'' - y = -\sin(2\pi x), \quad \text{for } 0 < x < 1, \tag{2.27}$$

where $y(0) = y(1) = 0$. The exact solution in this case is

$$y = \frac{\sin(2\pi x)}{1 + 4\pi^2}. \tag{2.28}$$

The finite difference equation in (2.16) takes the form

$$y_{i+1} - (2 + h^2)y_i + y_{i-1} = -h^2 \sin(2\pi x_i), \quad \text{for } i = 1, 2, \dots, N, \tag{2.29}$$

where $y_0 = y_{N+1} = 0$. The exact solution is shown in Figure 2.3 along with two solutions computed using (2.29). The computed solution is shown for a coarse grid, $N = 2$, and a somewhat finer grid, $N = 6$. Given the improvement obtained by increasing N, it would appear that the method does indeed

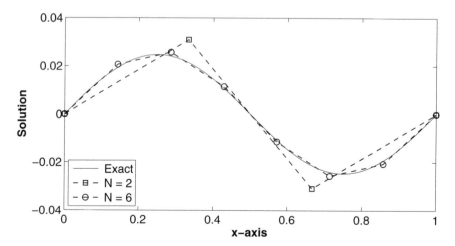

Figure 2.3. Exact solution of (2.27) along with the numerical solution obtained using (2.29) when $N = 2, 6$. The symbols are the computed values, and the dashed lines are drawn by the plotting program simply to connect the points.

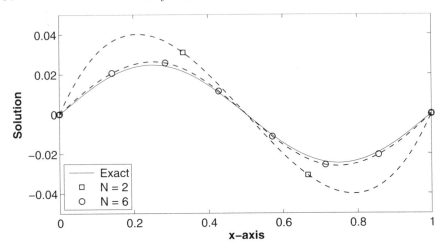

Figure 2.4. The data from Figure 2.3 are replotted using spline interpolation to connect the computed values.

converge to the exact solution. Whether this actually happens is considered shortly. The reason for the postponement is that there is something unsatisfactory about the computed solutions in Figure 2.3. It is not unreasonable to want to know the solution throughout the interval, and not just at the few points used in the calculation. For example, even for the birds on a wire in Figure 2.1, the entire curve is needed to draw the picture. One option is to simply use piecewise linear interpolation, as was done by the plotting program used to draw Figure 2.3. This is marginally OK, but this generally means that a relatively large number of grid points will be needed so that the resulting curve appears smooth. Another option is to use a method such as spline interpolation to connect the computed values. This is fairly painless, because most plotting programs have this capability, and to illustrate what results, the data are replotted using spline interpolation in Figure 2.4. The $N = 2$ curve is not so great but the $N = 6$ result is a respectable reproduction of the exact solution. A third option for connecting the points is to incorporate splines at the beginning and use them when deriving the approximation of the BVP. A variation of this idea is explored later in the chapter when we discuss residual methods.

Example 2
As a second example we consider the problem of solving

$$\epsilon y'' - x^2 y' - y = 0, \quad \text{for } 0 < x < 1, \tag{2.30}$$

where $y(0) = y(1) = 1$ and $\epsilon = 10^{-2}$. It is possible to solve this equation using hypergeometric functions, but a simple closed-form expression is not available (Holmes [1995]). So, turning to a numerical solution, the finite difference approximation is given in (2.16), where $a_i = -2 - h^2/\epsilon$, $b_i = 1 + hx_i^2/(2\epsilon)$, $c_i = 1 - hx_i^2/(2\epsilon)$, and $f(x_i) = 0$. According to Theorem 2.2 we should pick a step size that satisfies $h < 2\epsilon$. However, ignoring this useful piece of information, various attempts at finding a numerical solution are shown in Figure 2.5. The plots for $N = 10$ and $N = 20$ show that the method has difficulty computing the solution in the vicinity of $x = 1$. The reason is the rapid change in the solution in this region, which is an example of what is called a boundary layer. It is necessary to have at least a couple of grid points within this layer before the numerical solution is capable of giving a reasonable answer, and that is why $N = 120$ is successful. It is possible to derive finite difference approximations for layer problems that do not require such a small step size and an example can be found in Exercise 2.3. There are also much more sophisticated methods, and the book by Miller et al. [1996] should be consulted for more information about this.

2.2.4 Error Analysis

The above examples demonstrate the effectiveness of the method, and so it is now time to determine how the accuracy depends on the spatial step size h. To examine this issue we introduce the usual three solutions, which are

$$y(x_i) \equiv \text{exact solution of the BVP at } x = x_i; \tag{2.31}$$

$$y_i \equiv \text{exact solution of finite difference equation at } x = x_i; \tag{2.32}$$

$$\overline{y}_i \equiv \text{solution of difference equation at } x = x_i \text{ calculated}$$
$$\text{by the computer.} \tag{2.33}$$

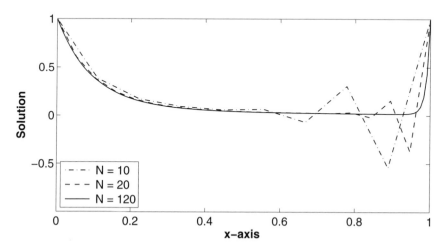

Figure 2.5. Computed solution of (2.30) when $N = 10, 20, 120$. The steep rise in the solution near $x = 1$ is known as a boundary layer.

We are interested in the difference between the exact solution of the BVP and the values we end up computing using our algorithm. Therefore, we are interested in the error $e_i = |y(x_i) - \bar{y}_i|$. The question we are going to ask is, if we increase N in (2.19) will e_i converge to zero or at least decrease down to the level of the round-off? We want the answer to this question to be yes and that it is true no matter what choice we make for i. If this holds then the method is convergent. To help make it more apparent what is contributing to the error we rewrite it as follows

$$e_i = |y(x_i) - y_i + y_i - \bar{y}_i|. \tag{2.34}$$

From this the error can be considered as coming from the following two sources:

$y(x_i) - y_i$: This is the error, at $x = x_i$, arising from approximating the BVP with a finite difference equation. As with IVPs, we should be able to estimate its size using the truncation error, and this is demonstrated in the next example.

$y_i - \bar{y}_i$: This is the error, at $x = x_i$, due to round-off when using floating-point calculations to compute the solution of the matrix equation. As shown in numerical linear algebra, a matrix with a large condition number $\text{cond}(\mathbf{A})$ can be so sensitive to round-off that it is impossible to calculate the solution using floating-point arithmetic (Demmel [1997]). To quantify what this means, when using double-precision, if $\text{cond}(\mathbf{A}) \approx 10^n$, then a rule of thumb is that one should not expect $\bar{y}_i$ to be correct to any more than about $15 - n$ digits. From our earlier estimate for the condition number given in (2.26) we have that round-off is not of particular concern certainly if N is no more than 10^2 or even 10^3.

From the above discussion, we see that the remaining mystery surrounding the accuracy of the computation concerns how the difference $y(x_i) - y_i$ depends on the truncation error. This is the topic of the next example.

Example 3
Suppose the BVP is

$$\frac{d^2y}{dx^2} - \gamma y = f(x), \quad \text{for } 0 < x < \ell, \tag{2.35}$$

where $y(0) = \alpha$ and $y(\ell) = \beta$. It is also assumed that $\gamma > 0$. To investigate how the error depends on the accuracy of the approximation let $E_i = y(x_i) - y_i$. From (2.12) and (2.16) one finds that

$$E_{i+1} - (2 + \gamma h^2)E_i + E_{i-1} = -h^2 \tau_i, \quad \text{for } i = 1, 2, \ldots, N, \tag{2.36}$$

where $E_0 = 0$ and $E_{N+1} = 0$. From this we get $(2 + \gamma h^2)E_i = E_{i+1} + E_{i-1} + h^2 \tau_i$. Letting $E_\infty = \max |E_i|$ and $\tau_\infty = \max |\tau_i|$, we have that

$$|(2 + \gamma h^2)E_i| = |E_{i+1} + E_{i-1} + h^2 \tau_i|$$
$$\leq |E_{i+1}| + |E_{i-1}| + h^2 |\tau_i|$$
$$\leq E_\infty + E_\infty + h^2 \tau_\infty$$
$$= 2E_\infty + h^2 \tau_\infty.$$

Because this holds $\forall i$, then $(2 + \gamma h^2)E_\infty \leq 2E_\infty + h^2 \tau_\infty$, and from this we conclude that $\gamma E_\infty \leq \tau_\infty$. Recalling that $\tau_\infty = O(h^2)$, it follows that the maximum error satisfies $E_\infty = O(h^2)$. Therefore, the pointwise error satisfies $y(x_i) - y_i = O(h^2)$.

It is possible to extend the argument used in the above example and prove the following theorem.

Theorem 2.3. *In addition to the conditions assumed in Theorem 2.2, if* $q(x) \leq -\gamma < 0$ *for* $0 \leq x \leq \ell$ *then*

$$|y(x_i) - y_i| \leq C \left[\tau_\infty + \max \left\{ |E_0|, |E_{N+1}| \right\} \right], \tag{2.37}$$

where $C = \max \{1, 1/\gamma\}$.

With the approximation in (2.16), we have $\tau_\infty = O(h^2)$ and $E_0 = E_{N+1} = 0$. Therefore, from the above theorem we can conclude that the approximate solution y_i converges to the exact solution $y(x_i)$ as $h \to 0$ and the rate of convergence is $O(h^2)$. Another useful conclusion we can derive from this theorem concerns situations in which we do not satisfy the boundary conditions exactly but end up with either $E_0 = O(h)$ or $E_{N+1} = O(h)$. In this case, even though we have $O(h^2)$ approximations everywhere else, we end up with $|y(x_i) - y_i| = O(h)$. This same observation was made for IVPs, as demonstrated in Exercise 1.18.

Armed with this new information we return to the two examples considered earlier.

Example 1 (cont'd.)
To demonstrate that the convergence is quadratic for this example, in Figure 2.6 the error is plotted as a function of the number of grid points used in the calculation. Actually, two different measures of the error are plotted. One is the pointwise error $e_i = |y(x_i) - \bar{y}_i|$ at $x_i = \frac{1}{3}$ and the other is the maximum error $e_\infty = \max |y(x_i) - \bar{y}_i|$. Both graphs show that by increasing N from 10 to 10^2, the error decreases from about 10^{-3} to about 10^{-5}, and this rate of decrease continues as N increases. In other words, the method does indeed converge quadratically.

Example 2 (cont'd.)
It is of interest to note that Theorem 2.2 states that the finite difference equation has a well-defined solution if $h < 2\epsilon$. Because $h = 1/(N+1)$ and $\epsilon = 10^{-2}$,

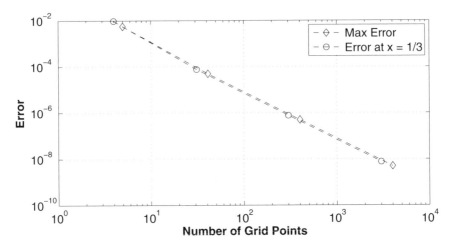

Figure 2.6. Difference between the exact and computed solutions of (2.27) as a function of the number of grid points used. Shown are the values of the maximum error $e_\infty = \max|y(x_i) - \bar{y}_i|$ as well as the pointwise error $e_i = |y(x_i) - \bar{y}_i|$ at $x_i = \frac{1}{3}$.

the requirement for this problem is $N > 49$. Perhaps more telling is Theorem 2.3, which for this problem gives us that $|y(x_i) - y_i| \le \tau_\infty/\epsilon$. Consequently, the fact that ϵ is small has an impact on what we would use as an estimate for guaranteeing that y_i is close to $y(x_i)$.

To recap what we have learned about the error, as shown in (2.34), the difference between the exact and computed solutions depends on $y(x_i) - y_i$ as well as on the error arising from round-off. The latter, however, is relatively small if $\mathbf{A}$ is well conditioned. In this case the computed solution will converge to the exact solution as h is decreased at a rate equal to the truncation error. This holds until the error gets down to the level of the error coming from round-off, with the exact level depending on the condition number of $\mathbf{A}$.

2.2.5 Extensions

The ideas used to construct the finite difference approximation to the linear BVP are easily extended to more complex problems.

Boundary Conditions

It is not uncommon to find real-world situations in which the boundary conditions involve specifying a flux or stress condition. A consequence of this for the mathematical problem is that one or more of the boundary conditions involve derivatives. To demonstrate how to handle such situations suppose the condition at $x = 0$ is replaced with

$$ay(0) + by'(0) = \alpha. \tag{2.38}$$

Proceeding as before, we still obtain (2.16). The difference is that we must now introduce a finite difference approximation into the boundary condition. Our objective is an approximation of the BVP that has an error that is $O(h^2)$, so this limits our choices for what to use to approximate $y'(0)$. One possibility is to introduce a ghost point x_{-1}, and this is explored in Exercise 2.4. Another approach is to use a one-sided approximation that is $O(h^2)$. One is listed in Table 1.1, and substituting it into (2.38) leads to the equation

$$ay_0 + \frac{b}{2h}(-y_2 + 4y_1 - 3y_0) = \alpha, \tag{2.39}$$

or equivalently

$$y_0 = \frac{b}{2ah - 3b}(y_2 - 4y_1) + \frac{2\alpha h}{2ah - 3b}. \tag{2.40}$$

The resulting finite difference approximation for the BVP is given by the tridiagonal matrix equation in (2.19), except that the first row is modified. In particular, a_1 is replaced with $a_1 - 4b_1 b/(2ah - 3b)$, c_1 is replaced with $c_1 + b_1 b/(2ah - 3b)$, and z_1 is replaced with $h^2 f_1 - 2\alpha h b_1/(2ah - 3b)$. This is assuming, of course, that $2ah \neq 3b$.

Nonlinear BVPs

The extension of the finite difference method to nonlinear BVPs is relatively straightforward. The difficulty is solving the resulting system of nonlinear algebraic equations. To demonstrate what is involved, suppose the differential equation is

$$y'' = f(x, y, y'), \quad \text{for } 0 < x < \ell, \tag{2.41}$$

where

$$y(0) = \alpha \text{ and } y(\ell) = \beta. \tag{2.42}$$

This problem includes the linear equation in (2.4), with $f(x, y, y') = -p(x)y' - q(x)y + f(x)$. It also includes nonlinear equations such as $f(x, y, y') = 1 + y^3$ and $f(x, y, y') = yy' + x$, as well as the chemical kinetics problem in (2.3). In formulating nonlinear BVPs, the question of the existence and uniqueness of the solution becomes challenging. The usual theorem quoted to guarantee that this occurs requires the function $f(x, y, z)$ to be smooth with $\frac{\partial f}{\partial z}$ bounded and $\frac{\partial f}{\partial y}$ bounded and positive. As it turns out, it is not uncommon for applications to produce smooth f's that do not satisfy one or more of these conditions, and an example is the chemical kinetics equation given in (2.3). It is beyond the scope of this text to pursue the theory underlying nonlinear BVPs, and the reader is referred to Keller [1992] for a detailed discussion of this topic. The point of view taken here is that there are one or more smooth solutions of the problem and we are going to derive a method for calculating them.

Using the centered $O(h^2)$ approximations in (2.8), (2.9) we obtain the following nonlinear finite difference equation

$$y_{i+1} - 2y_i + y_{i-1} = h^2 f\left(x_i, y_i, \frac{y_{i+1} - y_{i-1}}{2h}\right), \quad \text{for } i = 1, 2, \dots, N, \quad (2.43)$$

where

$$y_0 = \alpha \text{ and } y_{N+1} = \beta. \quad (2.44)$$

By setting

$$\mathbf{y} = \begin{pmatrix} y_1 \\ y_2 \\ \vdots \\ y_N \end{pmatrix}, \quad (2.45)$$

the equations in (2.43) can be written in vector form as $\mathbf{F}(\mathbf{y}) = \mathbf{0}$, where

$$F_i \equiv y_{i+1} - 2y_i + y_{i-1} - h^2 f\left(x_i, y_i, \frac{y_{i+1} - y_{i-1}}{2h}\right). \quad (2.46)$$

We are now faced with solving a nonlinear system of equations, and the conventional approach is to use Newton's method. To explain what this entails, recall that Newton's method starts with a user-specified approximation $\mathbf{z}_0$ of the solution and then constructs a sequence $\mathbf{z}_1, \mathbf{z}_2, \mathbf{z}_3, \dots$ that hopefully converges to a solution of the equation. The word "hopefully" is used here because one is usually not sure whether the proximity requirement that $\mathbf{z}_0$ be close to the solution is actually satisfied. We won't worry about this right now and instead concentrate on what form the method takes for this example. Using Newton's method, the $\mathbf{z}_k$'s are determined using the equation

$$\mathbf{z}_{k+1} = \mathbf{z}_k - \mathbf{J}_k^{-1}\mathbf{F}_k, \quad \text{for } k = 0, 1, 2, \dots, \quad (2.47)$$

where $\mathbf{J}_k$ is the Jacobian of $\mathbf{F}$ evaluated at $\mathbf{z}_k$ and $\mathbf{F}_k = \mathbf{F}(\mathbf{z}_k)$. For the problem at hand, as seen in (2.46), F_i depends only on three components of $\mathbf{y}$, so $\mathbf{J}_k$ contains mostly zeros. Carrying out the needed calculations, one finds, in fact, that the Jacobian is the tridiagonal matrix

$$\mathbf{J} = \begin{pmatrix} \bar{a}_1 & \bar{c}_1 & & & \\ \bar{b}_2 & \bar{a}_2 & \bar{c}_2 & & \mathbf{0} \\ & \bar{b}_3 & \bar{a}_3 & \bar{c}_3 & \\ & & \ddots & \ddots & \ddots \\ \mathbf{0} & & & & \bar{c}_{N-1} \\ & & & \bar{b}_N & \bar{a}_N \end{pmatrix}, \quad (2.48)$$

where

$$\bar{a}_i = -2 - h^2 \frac{\partial f}{\partial y}(x_i, y_i, z_i), \tag{2.49}$$

$$\bar{b}_i = 1 - \frac{h}{2} \frac{\partial f}{\partial z}(x_i, y_i, z_i), \tag{2.50}$$

$$\bar{c}_i = 1 + \frac{h}{2} \frac{\partial f}{\partial z}(x_i, y_i, z_i), \tag{2.51}$$

$$z_i = \frac{y_{i+1} - y_{i-1}}{2h}. \tag{2.52}$$

In the above equation, the z derivative indicates differentiation with respect to the third component of f, i.e., $\frac{\partial f}{\partial z} = \frac{\partial f}{\partial z}(x, y, z)$. With this (2.47) can be written as the matrix equation

$$\mathbf{J}_k \mathbf{z}_{k+1} = \mathbf{b}_k, \quad \text{for } k = 0, 1, 2, \ldots, \tag{2.53}$$

where $\mathbf{b}_k = \mathbf{J}_k \mathbf{z}_k - \mathbf{F}_k$. To use this it is necessary to select a start-off vector $\mathbf{z}_0$. Although it is not uncommon to see people simply take $\mathbf{z}_0 = \mathbf{0}$, it is generally better to try to pick something that contains a bit more information about the solution. With this in mind, remember that $\mathbf{z}_0$ serves as the initial approximation of the solution vector $\mathbf{y}$. For this reason, it is worth having this approximation satisfy at least the boundary conditions, and one possibility is to use a linear function. Other approximations are certainly worth considering, and this is demonstrated in the example below.

In looking at (2.53) it is seen that Newton's method for this problem reduces to solving a tridiagonal matrix equation that must be solved multiple times to find the solution of the BVP. There is no doubt that this is more computational work than is needed for the linear problem, but it is not overwhelming because of the efficiency of the tridiagonal solver.

Example 4
As an example consider Bratu's equation (2.3). Taking $\gamma = 1$ the problem becomes

$$y'' = -e^y, \quad \text{for } 0 < x < 1, \tag{2.54}$$

where $y(0) = y(1) = 0$. For this BVP, $f(x, y, z) = -e^y$. From this one finds that the entries in the Jacobian (2.48) are $\bar{a}_i = -2 + h^2 e^{y_i}$ and $\bar{b}_i = \bar{c}_i = 1$. As stated earlier, there are two solutions of this problem. The general form for each solution is

$$y = -2 \ln \left[\frac{\cosh(c(1 - 2x))}{\cosh(c)} \right], \tag{2.55}$$

where c satisfies $\cosh(c) = 2\sqrt{2}c$. There are two solutions of this equation, and they are $c_1 = 0.379 \cdots$ and $c_2 = 2.73 \cdots$. These give rise to the two solutions of the BVP, and our algorithm is capable of computing either of them. This is done by picking $\mathbf{z}_0$ in an appropriate way. In deciding what choice to make, note that the linear function connecting the two boundary conditions leads only to $\mathbf{z}_0 = \mathbf{0}$. To find a better $\mathbf{z}_0$ it is seen from (2.54) that the solution must

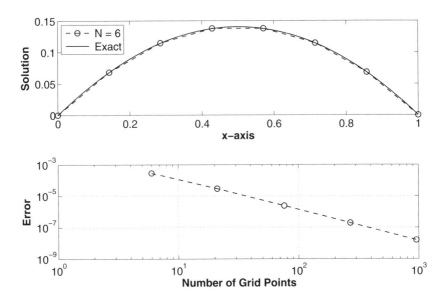

Figure 2.7. Numerical solution of (2.54) obtained using the finite difference approximation in (2.43), with $N = 6$, and the exact solution (2.55), when $c = c_1$. Also shown is the maximum error $e_\infty = \max |y(x_i) - \overline{y}_i|$, which shows the decrease expected from a $O(h^2)$ approximation.

be concave down. The simplest such function that also satisfies the boundary conditions is the quadratic $\mu x(1 - x)$, where μ is a positive constant. Taking $\mu = 1$ leads to a choice for $\mathbf{z}_0$ that converges to the solution corresponding to $c = c_1$ in (2.55). Similarly, $\mu = 16$ leads to the solution with $c = c_2$. To demonstrate the effectiveness of the method the computed and exact solutions are shown in Figure 2.7 in the case $c = c_1$. Also shown is the maximum error as a function of the number of grid points used, and the $O(h^2)$ convergence is clearly seen in this figure.

2.3 Residual Methods

There are several interesting, and effective, alternatives to the derivative approximation method for finding approximate solutions of BVPs. It is easier to explain what the possibilities are by using an example problem, so we return to the linear BVP in (2.4). Our approach using finite differences was first to find an accurate approximation of the problem. After that, the resulting finite difference equation was solved, and from this we obtained the approximate solution. We now reverse this approach and start off by approximating the solution using a function, or functions, of our choosing. With this we then attempt to have the approximation solve the BVP as accurately as possible.

The first step is to select a function $Y(x)$ to serve as the approximation. To keep our options open for the moment we write this approximating function as

$$Y(x) = \sum_{k=0}^{N+1} a_k \phi_k(x), \qquad (2.56)$$

where the ϕ_k's are basis functions that we will specify shortly. The a_k's are "fitting coefficients" that we will use so Y comes as close to solving the BVP as we can make it. In the same way, N can be thought of as the "quality" parameter, the idea being that as N increases, the error in the approximation is reduced.

There is enough flexibility in this idea to allow for a wide variety of methods. For example, there are various ways the solution can be approximated and there are different ways we can require them to satisfy the BVP. Before getting into the details of how this might be done, it is worth introducing a few examples of basis functions.

2.3.1 Basis Functions

Polynomial Approximation. One of the simplest choices for an approximating function is a polynomial, which can be obtained by taking $\phi_k(x) = x^k$. The result is $Y(x) = a_0 + a_1 x + \ldots + a_{N+1} x^{N+1}$. This certainly produces a smooth function, but it suffers the same limitations as Lagrange interpolation. A particularly significant flaw is that this choice need not converge to $y(x)$ as N increases.

B-Splines Approximation. One way to avoid the difficulties of Lagrange interpolation is to use piecewise polynomial functions. The simplest example in this category is piecewise linear interpolation. The difficulty with this is that we will eventually substitute $Y(x)$ into the BVP, and this will require that Y'' exist and be continuous. Unfortunately, piecewise linear interpolation produces corners, which means it is not differentiable. To avoid this one can use an approximation function $Y(x)$ that is piecewise cubic. There are a couple of ways to construct such functions and the most straightforward is to start by subdividing the interval using grid points just as we did in Step 1 for the finite difference approximation. Between adjacent grid points x_i and x_{i+1} the solution is approximated with a cubic, each subinterval in this way getting its own cubic polynomial. The coefficients of the cubics are determined from the requirement that at the grid points Y is smooth, which means Y and its first two derivatives are continuous. This construction is what the plotting program did when it produced Figure 2.4. To be able to use this type of function to solve a BVP it is convenient to express it in terms of what are called cubic B-splines $B_k(x)$. As defined, B_k is a piecewise cubic that is centered at the grid point x_k, is zero everywhere except if $x_{k-2} < x < x_{k+2}$, and has a continuous second derivative. The graph of $B_k(x)$ is shown in Figure

2.8. These conditions mean that $B_k(x)$ consists of a cubic between x_{k-2} and x_{k-1}, another cubic between x_{k-1} and x_k, another between x_k and x_{k+1}, and then another between x_{k+1} and x_{k+2}. The transition between cubics must be smooth and this limits what form they can have. Working out the details it is found that

$$
B_k(x) = \begin{cases}
0 & \text{if } x \leq x_{k-2}, \\[2mm]
\dfrac{1}{6h^3}(x - x_{k-2})^3 & \text{if } x_{k-2} \leq x \leq x_{k-1}, \\[2mm]
\dfrac{1}{6} + \dfrac{1}{2h}(x - x_{k-1}) & \\
\quad + \dfrac{1}{2h^2}(x - x_{k-1})^2 - \dfrac{1}{2h^3}(x - x_{k-1})^3 & \text{if } x_{k-1} \leq x \leq x_k, \\[2mm]
\dfrac{1}{6} - \dfrac{1}{2h}(x - x_{k+1}) & \\
\quad + \dfrac{1}{2h^2}(x - x_{k+1})^2 + \dfrac{1}{2h^3}(x - x_{k+1})^3 & \text{if } x_k \leq x \leq x_{k+1}, \\[2mm]
-\dfrac{1}{6h^3}(x - x_{k+2})^3 & \text{if } x_{k+1} \leq x \leq x_{k+2}, \\[2mm]
0 & \text{if } x_{k+2} \leq x.
\end{cases}
$$

$$(2.57)$$

At first glance the above expression looks complicated, but it should be remembered that B_k is just a piecewise cubic that produces the simple-looking curve in Figure 2.8.

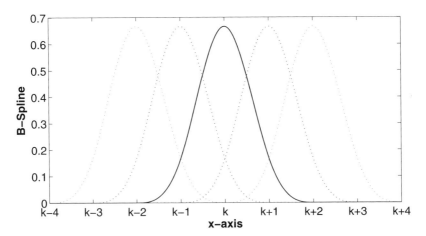

Figure 2.8. Plot of cubic B-splines $B_{k-2}(x), B_{k-1}(x), B_k(x), B_{k+1}(x), B_{k+2}(x)$. The solid curve is $B_k(x)$. On the x-axis, the point x_k is labeled as k (and the same holds for the other x_i's).

	x_{k-1}	x_k	x_{k+1}	$x_{k \pm m}$ for $m > 1$
B_k	$\frac{1}{6}$	$\frac{2}{3}$	$\frac{1}{6}$	0
B_k'	$\frac{1}{2h}$	0	$-\frac{1}{2h}$	0
B_k''	$\frac{1}{h^2}$	$-\frac{2}{h^2}$	$\frac{1}{h^2}$	0

Table 2.2. Values of the B-spline $B_k(x)$, as defined in (2.57), at the grid points used in its construction.

The approximating function in this case is

$$Y(x) = \sum_{k=-1}^{N+2} a_k B_k(x), \quad \text{for } 0 \le x \le \ell, \tag{2.58}$$

where the basis function $B_k(x)$ is given in (2.57). In terms of solving BVPs, it is significant that the nonzero part of B_k is localized to a small neighborhood of x_k, namely the interval $x_{k-2} < x < x_{k+2}$. Because of this, out of all the terms in (2.58), only B_{i-1}, B_i, and B_{i+1} contribute to the value of Y at $x = x_i$. In particular, using the values given in Table 2.2 one finds that

$$Y(x_i) = \frac{1}{6}(a_{i-1} + 4a_i + a_{i+1}), \tag{2.59}$$

$$Y'(x_i) = \frac{1}{2h}(a_{i-1} - a_{i+1}), \tag{2.60}$$

$$Y''(x_i) = \frac{1}{h^2}(a_{i-1} - 2a_i + a_{i+1}). \tag{2.61}$$

This is also the reason for including $k = -1$ and $k = N + 2$ in (2.58). For example, B_{-1} contributes to the solution at $x = 0$ and it therefore needs to be included in the summation.

Fourier Approximation. Another often used set of approximation methods is based on Fourier series. An example is a sine series, given as

$$Y(x) = \sum_{k=1}^{N+1} a_k S_k(x), \tag{2.62}$$

where $S_k(x) = \sin(\pi k x/\ell)$. Because $S_k(0) = S_k(\ell) = 0$, this expansion requires the boundary conditions $y(0) = y(\ell) = 0$. This is not much of a restriction, because one can always make the change of variables $y(x) = u(x) + \alpha + (\beta - \alpha)x/\ell$. The transformed BVP for $u(x)$ has homogeneous boundary conditions, as required when using (2.62).

2.3.2 Residual

Now that we have identified various choices for an approximating function, the next question is how we can use it to satisfy the BVP. We want Y to come as close to satisfying the BVP as possible, and with this goal in mind we introduce the residual $r(x)$, defined as

$$r(x) \equiv Y'' + p(x)Y' + q(x)Y - f(x). \tag{2.63}$$

It is assumed that Y satisfies the two boundary conditions, so how well we do in solving the BVP translates into how close to zero we can get $r(x)$. Now, up to two of the a_k's in (2.56) are needed to satisfy the boundary conditions. Therefore, the remaining a_k's are used to get $r(x)$ close to zero. Two of the more commonly used methods for producing a small residual are presented below.

Collocation

The idea in collocation is simply to select points in the interval and then require the residual to be zero at those locations. The number of collocation points selected corresponds to the number of free parameters in the approximating function. For example, to use the B-spline expansion in (2.58) it is necessary to select $N + 2$ points from the interval. The easiest choice is to pick the same points used to construct the B-splines, and this means we set $r(x_i) = 0$. Using the formulas in (2.59)–(2.61), this results in the equations (see Exercise 2.18)

$$(6 + 3hp_i + h^2 q_i)a_{i+1} + 4(-3 + h^2 q_i)a_i + (6 - 3hp_i + h^2 q_i)a_{i-1}$$
$$= 6h^2 f_i, \quad \text{for } i = 0, 1, \ldots, N + 1. \tag{2.64}$$

In addition, from the boundary conditions it is required that $a_{-1} + 4a_0 + a_1 = 6\alpha$ and $a_N + 4a_{N+1} + a_{N+2} = 6\beta$. Therefore, the method reduces to solving a matrix equation $\mathbf{Ba} = \mathbf{w}$, where $\mathbf{B}$ is an $(N+2) \times (N+2)$ tridiagonal matrix of the form

$$\mathbf{B} = \begin{pmatrix} g_0 & h_0 & & & \\ d_1 & g_1 & h_1 & & \mathbf{0} \\ & d_2 & g_2 & h_2 & \\ & & \ddots & \ddots & \ddots \\ \mathbf{0} & & & & h_N \\ & & & d_{N+1} & g_{N+1} \end{pmatrix}. \tag{2.65}$$

In this expression $g_0 = 2(-3 + hp_0), h_0 = hp_0, g_{N+1} = -2(3 + hp_{N+1}), d_{N+1} = -hp_{N+1}$. For the other entries, $d_i = 6 - 3hp_i + h^2 q_i, g_i = 4(-3 + h^2 q_i), h_i = (6 + 3hp_i + h^2 q_i)$. The vectors in this equation are

$$\mathbf{a} = \begin{pmatrix} a_0 \\ a_1 \\ \vdots \\ a_N \\ a_{N+1} \end{pmatrix} \quad \text{and} \quad \mathbf{w} = \begin{pmatrix} \dfrac{h^2 f_0 - \alpha(6 - 3hp_0 + h^2 q_0)}{6h^2 f_1} \\ \vdots \\ 6h^2 f_N \\ h^2 f_{N+1} - \beta(6 + 3hp_{N+1} + h^2 q_{N+1}) \end{pmatrix}. \quad (2.66)$$

In terms of error one can show that $Y(x) - y(x) = O(h^2)$, and so the method is comparable in terms of accuracy and computational effort to the finite difference approximation in (2.16).

Least Squares

Another approach to reducing the residual is to introduce the least squares error function

$$E = \int_0^\ell r^2(x)dx. \quad (2.67)$$

The value of E depends on the a_i's, and the requirement is that the coefficients minimize E. This gives us the equations

$$\frac{\partial E}{\partial a_i} = 0. \quad (2.68)$$

Assuming that $Y(x)$ is given in (2.58), then (2.68) reduces to

$$\sum_{k=0}^{N+1} \alpha_{ik} a_k = \beta_i, \quad (2.69)$$

where $\alpha_{ik} = \int_0^\ell \chi_i \chi_k dx$, $\beta_i = \int_0^\ell \chi_i f(x)dx$, and $\chi_i(x) = \phi_i'' + p(x)\phi_i' + q(x)\phi_i$. This is where the computational cost of this method starts to become evident. Unless the basis functions ϕ_i have very special properties, the matrix equation that comes from (2.69) for the a_i's will involve a full matrix. Even for the B-splines, which are zero over most of the interval, the matrix has a bandwidth of three, compared to a bandwidth of one that is obtained using collocation. Nevertheless, there are reasons why this method is used, as shown in the next example.

Example 5
To compare the various versions of the residual method that have been discussed suppose the approximating function is $Y(x) = a_0 + a_1 x + a_2 x^2$. Also, suppose $p(x) = q(x) = 0$, so the BVP is $y'' = f(x)$ with $y(0) = \alpha$ and $y(\ell) = \beta$. The residual in this case is $r(x) = 2a_2 - f(x)$. To determine the coefficients of the approximating function note that the boundary conditions give us $a_0 = \alpha$ and $a_1 \ell + a_2 \ell^2 = \beta - \alpha$. To use the collocation method to find a_2 we need to select a point from the interval, and the choice is $x_1 = \ell/2$. Setting $r(x_1) = 0$

gives us $a_2 = \frac{1}{2} f(x_1)$. On the other hand, using the least squares approach we find that

$$E = 4\ell a_2^2 - 4a_2 \int_0^\ell f dx + \int_0^\ell f^2 dx. \qquad (2.70)$$

Setting $\frac{\partial E}{\partial a_2} = 0$ it follows that $a_2 = \frac{1}{2\ell} \int_0^\ell f dx$. The resulting approximations are shown in Figure 2.9 in the case $f(x) = 4(x-1)e^{-2x}$. The exact solution of the BVP is $y(x) = xe^{-2x}$ and this is also shown in the figure. In terms of the pros and cons of these two methods, the least squares approach produces a more stable answer. For example, the collocation method depends entirely on the value of f at one point, whereas the least squares value involves the average of the function over the interval. On the other hand, the possibility of picking different points gives the collocation method a useful flexibility (see Exercise 2.21). As a final comment it might seem that both methods are flawed given the rather large differences in the curves in Figure 2.9. However, only three points were used here, and as demonstrated in the next example, when more points are used the accuracy improves significantly.

Example 6
As a second example, suppose the problem is $y'' + y' - 2y = -3e^{-2x}$, for $0 < x < 1$, with $y(0) = 0$ and $y(1) = e^{-2}$. The solution in this case is $y(x) = xe^{-2x}$. We use the B-spline expansion in (2.58) as the approximating function and collocation is used to determine the coefficients. The problem in this case reduces to solving (2.64). The resulting solution is shown in Figure 2.10 for various values of N. It is apparent from this plot that the method converges as N increases. This observation is supported by the lower graph in Figure 2.10, which gives the maximum error as a function of N. For comparison the error

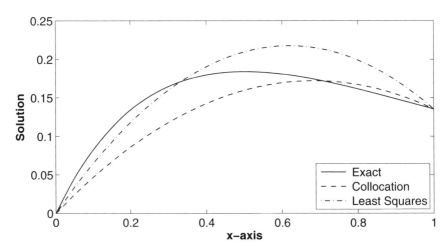

Figure 2.9. Comparison of the collocation and least squares methods when a quadratic approximation function is used.

curve using the finite difference approximation (2.16) is also shown. They have the same rate of convergence, $O(h^2)$, and produce comparable results for this problem.

2.4 Shooting Methods

Even though the steps used to derive finite difference approximations for IVPs and BVPs are similar, they produce different mathematical problems to solve. With IVPs we obtained an algorithm in which the solution is calculated one time step at a time. For BVPs, on the other hand, we ended up with an algorithm in which all the unknowns are calculated at once. As it turns out, there are numerical methods for BVPs that resemble those obtained for IVPs. To explain how this is done we start with the nonlinear BVP in (2.41), (2.42). By introducing the vector $\mathbf{y}(x) = (y_1, y_2)^T$, where $y_1 = y$ and $y_2 = y'$, the differential equation can be written as a first-order system

$$\mathbf{y}' = \mathbf{f}(x, \mathbf{y}), \tag{2.71}$$

where

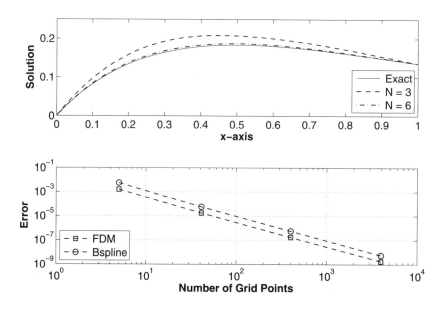

Figure 2.10. In the upper graph the solution obtained using cubic B-splines, with collocation, is shown along with the exact solution. In the lower graph the maximum error using B-splines is plotted as a function of the number of points used. For comparison the values obtained using a centered finite difference approximation is also shown.

$$\mathbf{f}(x, \mathbf{y}) = \begin{pmatrix} y_2 \\ f(x, y_1, y_2) \end{pmatrix}.$$ (2.72)

To use IVP methods to solve (2.71) we need to have initial conditions that are equivalent to the original boundary conditions. From the information we currently have,

$$\mathbf{y}(0) = \begin{pmatrix} \alpha \\ s \end{pmatrix},$$ (2.73)

where $s = y'(0)$ is unknown. So the question is, can we find the value of s for which the solution of the resulting IVP is identical to the solution of the BVP? Put another way, what value of s results in the solution of the IVP satisfying $y(\ell) = \beta$? The idea underlying shooting is to pick a value of s, then use an IVP method to march over to $x = \ell$ and see whether $y(\ell) = \beta$. If not, then adjust the value of s and use the IVP method again and see how much closer $y(\ell)$ is to β. This is continued until $|y(\ell) - \beta|$ is sufficiently small. The convergence of this procedure is similar to what was obtained for the matrix method as given in Theorem 2.3. Namely, if the IVP solver has truncation error $O(h^p)$, then $|y(x_i) - y_i| = O(h^p) + O(|y(\ell) - \beta|)$. It is assumed in this case that the step size satisfies the solver's stability condition and $f_y \geq 0$. The proof of this result can be found in Keller [1992]. It is important to point out that this says nothing about how sensitive the method is to the condition of the problem. As shown in one of the examples to follow, shooting is more prone to having problems with this than the matrix method.

Perhaps the most pressing question in this procedure is how to adjust s so $y(\ell)$ ends up close to β. To address this set $g(s) = y(\ell) - \beta$. This function enables us to express the question of getting $y(\ell)$ close to β in terms of finding the value of s such that $g = 0$. Put this way, then, we can use something such as the secant or Newton's method to improve the value of s. For example, to use the secant method we need to specify two values for s, say s_1 and s_2. In this case the subsequent values for s are determined using the secant formula

$$s_{j+1} = s_j - \frac{g(s_j)}{g(s_j) - g(s_{j-1})}(s_j - s_{j-1}), \quad \text{for } j = 2, 3, 4, \ldots.$$ (2.74)

As usual for iterative methods, for this to work the start-off values s_1 and s_2 need to be close to the exact solution. One last point to make is that the method developed here for finding s works whether (2.71) is linear or nonlinear. It is possible to simplify the procedure a bit for linear problems, and this is explored in Exercise 2.22.

By opening the door to IVP solvers we have the opportunity to use a higher-order method such as RK4. To see how well this works we solve the BVPs used earlier when we were investigating the matrix approach.

Example 1 (cont'd.)
The exact solution of the BVP from Example 1, given in (2.27), is shown in Figure 2.11 along with two solutions computed using shooting with RK4.

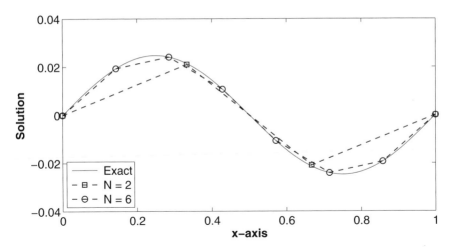

Figure 2.11. Exact solution of (2.27) along with the numerical solution obtained using shooting with RK4 when $N = 2, 6$.

Comparing this result with Figure 2.3, it is clear that shooting with RK4 produces a more accurate result, at least for the number of grid points used in this example.

Example 2 (cont'd.)
The BVP from Example 2, given in (2.30), is a more challenging test. Our shooting algorithm requires start values s_1 and s_2, and so suppose we take $s_1 = 0$. In this case the shooting method using RK4, with $N = 120$, calculates the value at $x = 1$ to be $y = 1.86 \times 10^{15}$, which misses the exact value $y(1) = 1$ by a rather large margin. Similarly, if $s_2 = 1$ then one obtains $y = 2.06 \times 10^{15}$. The magnitudes of these two values are a concern, and a hint that there might be a problem with ill-conditioning. Nevertheless, going ahead with the procedure one finds that the method concludes that $y(1) = 0.7509$. If one tries to improve the result and takes $N = 1200$, the method finds $y(1) = 2.5145$. Making N larger makes things even worse; for example, $N = 2400$ yields $y(1) = 3.5509$. Due to this ill-conditioning, shooting is not competitive with the matrix approach on this problem.

The last example demonstrates the limitation of using a simple shooting method when rapid changes are present in the solution. Namely, the ill-conditioned nature of the procedure makes it unusable on such problems. It is possible in some cases to makes modifications so it can be used, and one such approach involves multiple shooting over subintervals. This is beyond the scope of this text, and the interested reader is referred to Stoer and Bulirsch [2002] and Ascher et al. [1995] for a discussion of this topic.

2.5 Next Steps

With the chapter coming to an end it is worth taking a few moments to identify some of the topics not covered. As usual, the theoretical underpinnings of the methods were considered in the development but not to the extent found in a typical numerical analysis text. To expand on this comment, consider the problem of solving

$$\frac{d^2y}{dx^2} + p(x)\frac{dy}{dx} + q(x)y = -\lambda y, \quad \text{for } 0 < x < \ell, \tag{2.75}$$

with the boundary conditions in (2.2). Certainly $y = 0$ is a solution, and so the question is whether there are values of λ for which there are nonzero solutions. In other words, this is an eigenvalue problem and λ is the eigenvalue. In conjunction with this there is the associated inhomogeneous problem, which consists of the same boundary conditions but the differential equation is

$$\frac{d^2y}{dx^2} + p(x)\frac{dy}{dx} + q(x)y = -\lambda y + f(x), \quad \text{for } 0 < x < \ell. \tag{2.76}$$

The above equation reduces to the one we studied by taking $\lambda = 0$. In fact, the question is whether $\lambda = 0$ yields a well-posed mathematical problem. The theorem below contains the answer.

Theorem 2.4. *If p, q, and f are continuous, for $0 \le x \le \ell$, then the following hold:*
1. *There are an infinite number of eigenvalues $\lambda_1 < \lambda_2 < \lambda_3 < \ldots$, each corresponding to an eigenfunction $y_1(x), y_2(x), y_3(x), \ldots$.*
2. *If λ is not equal to one of the eigenvalues then the solution of (2.76) is unique.*
3. *If λ is equal to one of the eigenvalues, say $\lambda = \lambda_j$, then (2.76) has a solution only if*

$$\int_0^\ell f(x)y_j(x)e^{\int_0^x p(s)ds}dx = 0.$$

For example, if p and q are constant then the eigenvalues are

$$\lambda_j = -q + \frac{1}{4}\left[p^2 + \left(\frac{j\pi}{\ell}\right)^2\right].$$

It follows from this that if $q \le 0$ then the eigenvalues are positive. In other words, $\lambda = 0$ is not an eigenvalue and from Conclusion 2 of the theorem this means it produces a problem with a unique solution. This helps explain why it was assumed in the formulation of the BVP in (2.4), (2.5) that $q(x) \le 0$, for $0 \le x \le \ell$. The study of eigenvalues and their role for solving linear BVPs is part of Sturm–Liouville theory, and an introduction of this material can be found in Boyce and DiPrima [2004] and Haberman [2003].

From a theoretical viewpoint, knowing the eigenvalues determines when and if the BVP has a solution. With this comes the natural question of whether it is possible to compute these values. There has been considerable work on this problem and a survey of this topic can be found in Pryce [1994]. This question is also addressed in Exercise 2.27. In applications an interesting twist to this problem often arises. The eigenvalues typically correspond to the natural frequencies of a physical object, such as the vibrational modes of a guitar string. As such they are easily obtained experimentally and are therefore known. The question that then arises is, what is the object made of that causes it to generate these particular eigenvalues? In other words, if you know the λ_j's in (2.75) can you use this information to determine the coefficients p and q? This is an example of an inverse problem and it raises several numerical challenges on how to approximate and then compute these functions. This is an active area of research and an introduction can be found in Vogel [2002]. As an aside, the question of whether you can use the eigenvalues in this way was the reason Marc Kac [1966] wrote the paper, "Can you hear the shape of a drum?" The history of this interesting question and its solution can be found in Cipra [1993].

As with IVPs, one of the topics only briefly considered concerns stiff equations. The solution in Figure 2.5 is a typical example, where the solution changes very rapidly over a small spatial interval. Because of the prominence of such problems in applications, numerical methods for stiff equations have been studied extensively. The most widely used approach involves nonuniform grids, with the mesh spacing adjusted according to how rapidly the solution changes. For example, in Figure 2.5 when using $N = 10$ the jump in the solution between successive grid points is rather large at the right end of the interval. An algorithm that uses an automated refinement scheme would notice this and respond by placing a few more grid points in this region and then proceed to recalculate the solution. This refinement process would be continued until the jumps were sufficiently small and the rapid changes in the solution were well resolved. This is a fairly simple idea but it requires some care in deciding when and where to refine, or to unrefine, the grid. A survey of the basic ideas for finite difference approximations, and collocation, can be found in Ascher et al. [1995].

A topic that did not receive any consideration is finite element methods. These are based on rewriting the differential equation in integral form and then using an approximation much like the one in (2.56) developed for residual methods. There are various ways to transform the problem into integral form and one is used in Exercise 2.15, where the BVP is transformed into finding the function that minimizes an energy integral. Having a minimization formulation such as this opens up a number of interesting possibilities on constructing a numerical solution. This idea is used, for example, in Chapter 6 to solve the matrix equations that are obtained when approximating multidimensional BVPs. An introduction to how this is used for finite elements is nicely described in Hughes [2000].

Exercises

2.1. Consider the BVP

$$y'' - \alpha(2x - 1)y' - 2\alpha y = 0, \quad \text{for } 0 < x < 1,$$

where $y(0) = y(1) = 1$.
(a) Verify that $y = e^{-\alpha x(1-x)}$ is the exact solution.
(b) Write down the finite difference equation (2.16) and the corresponding matrix equation (2.19) for this problem. What is the limitation on the step size according to Theorem 2.2?
(c) Plot, on the same axes, the numerical and exact solutions for $N = 10$ and $\alpha = 10$.
(d) Graph, using a log-log plot, the maximum error as a function of N for $N = 10, 20, 40, 80$ and $\alpha = 10$. Explain the result in terms of the truncation error.

2.2. Consider the BVP

$$x^2 y'' + xy' - 4y = 20x^3, \quad \text{for } 1 < x < 2,$$

where $y(1) = 0$ and $y(2) = 31$.
(a) Verify that $y = 4(x^3 - 1/x^2)$ is the exact solution.
(b) Write down the finite difference equation (2.16) and the corresponding matrix equation (2.19) for this problem. What is the limitation on the step size according to Theorem 2.2?
(c) Plot, on the same axes, the numerical and exact solutions for $N = 3$.
(d) Graph, using a log-log plot, the maximum error as a function of N for $N = 3, 6, 12, 24, 48$. Explain the result in terms of the truncation error.

2.3. This problem concerns the problem

$$\epsilon y'' - y' = -1, \quad \text{for } 0 < x < 1,$$

where $y(0) = 1, y(1) = 3$.
(a) Verify that $y = 1 + x + \beta(e^{x/\epsilon} - 1)$, where $\beta = (e^{1/\epsilon} - 1)^{-1}$ is the solution.
(b) Suppose (2.9) is used to approximate the second derivative but the backward difference $y'(x_i) = \frac{1}{h}(y(x_i) - y(x_{i-1})) + O(h)$ is used for the first derivative. What is the resulting finite difference equation? What is the matrix form of the equation? What is the truncation error for the method?
(c) Show that the matrix in part (b) is invertible irrespective of the value of h. According to Theorem 2.2, what is the limitation on h if (2.16) is used?
(d) Suppose $\epsilon = 0.1$. For $N = 10$ plot, on the same axes, the solution obtained using the method from (b), the exact solution, and the solution using (2.16). Do the same thing for $N = 20$ and for $N = 40$.
(e) Redo (d) for $\epsilon = 0.01$.

(f) Discuss the merits of the two methods for this problem. Which do you recommend if ϵ becomes even smaller?

2.4. Instead of using one-sided approximations to handle derivatives in the boundary conditions one can introduce ghost points that are located outside the interval of the original problem. To investigate this idea suppose that the BVP is given in (2.4), (2.5) except that the condition at $x = 0$ is replaced with $ay(0) + by'(0) = \alpha$, where $b \neq 0$.

(a) Use (2.8) to derive a finite difference approximation of the boundary condition at $x = 0$. This will require you to introduce a ghost point $x = x_{-1}$ and the corresponding solution at this location.

(b) A consequence of introducing a ghost point is that for the finite difference equation (2.16) the index range is $i = 0, 1, 2, \ldots, N$. Using the result from part (a) what is the $i = 0$ equation? How is the matrix equation (2.19) changed? In the remainder of the problem assume $p = 4, q = -1, a = 1, b = 1, \ell = 1$, and the exact solution is $y(x) = e^{-2x} \cos\left(\frac{3\pi}{2}x\right)$.

(c) Graph, using a log-log plot, the maximum error as a function of N for $N = 5, 25, 625, 3125$. On the same axes plot the error when the problem is solved using (2.40).

(d) Are there any advantages, or disadvantages, of using a ghost point rather than the one-sided approximation in (2.40)?

2.5. This problem considers how to handle different boundary conditions when solving (2.4). You are to find a finite difference approximation of the problem that has truncation error $O(h^2)$, and you should express the difference equation in matrix form.

(a) Suppose the boundary conditions are $y(0) = \alpha$ and $y(\ell) + \gamma y'(\ell) = \beta$, where γ is nonzero.

(b) Suppose one uses periodic boundary conditions of the form $y(0) = y(\ell)$ and $y'(0) = y'(\ell)$.

2.6. Consider the problem of solving $y'' = \lambda^2 y$, for $-1 < x < 1$, where $y(-1) = y(1) = 1$ and $\lambda > 0$.

(a) Show that the exact solution is

$$y(x) = \frac{\cosh(\lambda x)}{\cosh(\lambda)}.$$

(b) Assuming that the difference formula in (2.9) is used, what is the resulting finite difference equation for the BVP? Show that the exact solution of this equation is

$$y_i = \frac{\cosh(\mu x_i)}{\cosh(\mu)},$$

where $\mu = \frac{1}{h} \cosh^{-1}(1 + \frac{1}{2}h^2\lambda^2)$.

(c) Using (a) and (b) prove that $y(x_i) - y_i = O(h^2)$.

2.7. This problem shows how to obtain a $O(h^4)$ accurate approximation of the solution of particular BVPs.

(a) Use Taylor's theorem to show that $y(x_{i+1}) - 2y(x_i) + y(x_{i-1}) = h^2 y''(x_i) + \frac{1}{12} h^4 y''''(x_i) + O(h^6)$ and from this show that $y(x_{i+1}) - 2y(x_i) + y(x_{i-1}) = \frac{1}{12} h^2 [y''(x_{i+1}) + 10y''(x_i) + y''(x_{i-1})] + O(h^6)$.

(b) Assuming that $y(x)$ satisfies the differential equation $y'' = F(x, y)$, use the result from (a) to derive the finite difference approximation

$$y_{i+1} - 2y_i + y_{i+1} = \frac{h^2}{12}(F_{i+1} + 10F_i + F_{i-1}).$$

This is known as Numerov's method.

(c) What does the approximation in (b) reduce to when $F(x,t) = f(x) - q(x)y$? Express the result in matrix form.

(d) Show that the change of variables $y = u(x) \exp(-\frac{1}{2} \int^x p \, dx)$ transforms (2.4) into an equation of the form considered in (b). Are there any conditions that would limit using the method in (b) to solve (2.4)?

2.8. Consider the following problem

$$\frac{d}{dx}\left(D(x)\frac{dy}{dx}\right) - y = f(x), \quad \text{for } 0 < x < \ell,$$

where

$$y(0) = \alpha \text{ and } y(\ell) = \beta.$$

The functions $D(x), f(x)$ are assumed to be smooth and $D(x)$ is positive. This is a *symmetric*, or self-adjoint, BVP, and the question is, should we make an effort to make sure the finite difference approximation also is symmetric?

(a) By expanding the derivative term reduce the equation to the form in (2.4). From this write down the resulting matrix equation that comes from the finite difference approximation of this problem. Is the matrix symmetric?

(b) As another approach integrate the equation from $x = x_{i-1/2}$ to $x = x_{i+1/2}$ and show that

$$D(x_{i+1/2})y'(x_{i+1/2}) - D(x_{i-1/2})y'(x_{i-1/2}) - \int_{x_{i-1/2}}^{x_{i+1/2}} y(x)dx$$

$$= \int_{x_{i-1/2}}^{x_{i+1/2}} f(x)dx.$$

In this expression $x_{i\pm1/2} = x_i \pm \frac{h}{2} = (i \pm \frac{1}{2})h$. Use centered differences to approximate the two derivates in the above expression and use the midpoint rule to approximate the two integrals. This will produce a finite difference equation involving y_{i-1}, y_i, y_{i+1} (although $D(x)$ will be evaluated at $x_{i\pm1/2}$). What is the resulting matrix equation? Is it symmetric? What is the order of the truncation error?

(c) Suppose $D(x) = \epsilon(1 + x)e^{x/\epsilon}$ and $f(x) = 1 - e^{-x/\epsilon}$, with $\alpha = \beta = 1$ and $\epsilon = 1/20$. Plot the computed solution using the methods from (a) and (b) when 20 grid points are used along the x-axis. Also plot the two solutions when 40 points are used. Which method appears to produce the most accurate results? Comment: before doing part (c) it is recommended you test your code on a known solution to make sure it is working correctly (e.g., if $D = 1 + x$ and $f = x(6 + 9x - x^2)$, with $\ell = 1$, $\alpha = 0$, and $\beta = 1$, then $y = x^3$).

2.9. This problem derives what is known as the box scheme for the BVP in (2.4).

(a) Using the system form of the differential equation given in (2.71), evaluate the system at $x = x_{i+1/2}$ and use a centered difference for $\mathbf{y}'(x_{i+1/2})$. Note that $x_{i\pm1/2} = (i \pm \frac{1}{2})h$.

(b) Show that $\mathbf{y}(x_{i+1/2}) = \frac{1}{2}(\mathbf{y}(x_{i+1}) + \mathbf{y}(x_i)) + O(h^2)$.

(c) Use part (b) to rewrite the result from (a) in terms of $\mathbf{y}(x_{i+1})$ and $\mathbf{y}(x_i)$.

(d) Express the result in (c) as a finite difference equation for y_i. What is the truncation error for this approximation?

2.10. This problem concerns the nonlinear problem $2y'' + (y')^2 + a^2y^2 = 2a^2$, for $0 < x < 1$, where $y(0) = 1, y(1) = 0$. Assume $a = \frac{5\pi}{2}$.

(a) Verify that $y = 1 - \sin(ax)$ is the solution.

(b) In (2.47) and (2.53) what are $\mathbf{J}$ and $\mathbf{F}$ for this problem?

(c) Plot the exact and numerical solutions on the same axes in the case $N = 4$. Do the same thing for $N = 9$. Make sure to explain how you pick $\mathbf{z}_0$.

(d) Graph, using a log-log plot, the maximum error as a function of N, taking $N = 10, 20, 40, 80, 160$. Does the error behave as expected?

2.11. The vertical displacement of a cable satisfies

$$\frac{d^2y}{dx^2} = \mu\sqrt{1 + \left(\frac{dy}{dx}\right)^2}, \quad \text{for } 0 < x < \ell,$$

where μ is a positive constant and the boundary conditions are $y(0) = \alpha, y(1) = \beta$.

(a) In (2.47) and (2.53) what are $\mathbf{J}$ and $\mathbf{F}$ for this problem?

(b) The method requires a guess for the solution. Write down a function that you believe is a reasonable choice and explain your reasoning why this should satisfy Newton's proximity requirement.

(c) Plot the numerical solution in the case $N = 9$. In this case take $\mu = 9, \alpha = 8, \beta = 6, \ell = 1$.

(d) The exact solution has the form $y(x) = B + \frac{1}{\mu}\cosh(\mu x + A)$, where A, B are determined from the boundary conditions. With this, plot the exact solution and the numerical solution from part (c) on the same axes.

(e) Graph, using a log-log plot, the maximum error as a function of N, taking $N = 10, 20, 40, 80, 160$. Does the error behave as expected?

2.12. In fluid dynamics one comes across having to solve the Falkner–Skan equation

$$y''' + yy' + \beta[1 - (y')^2] = 0, \quad \text{for } 0 < x < \ell,$$

where $y(0) = y'(0) = 0$ and $y'(\ell) = 1$. Derive a finite difference approximation of this BVP that has $O(h^2)$ truncation error.

2.13. In the example shown in Figure 2.9 the collocation point was selected to be the midpoint of the interval. Assuming $y(x) = xe^{-2x}$, is there a better choice for x_1?

2.14. Consider the BVP given by $y'' = f(x)$, with $y(0) = \alpha$ and $y(\ell) = \beta$, and assume that the approximating function is $Y(x) = a_0 + a_1x + a_2x^2 + a_3x^3$.
(a) What are the coefficients of Y if collocation is used?
(b) What are the coefficients of Y if least squares is used?
(c) Compare the two approximations from (a), (b) in the case where the exact solution is $y(x) = xe^{-2x}$. Make sure to explain how you select the two collocation points.

2.15. Another approach for finding the approximation function involves minimizing the energy. For the BVP given by $y'' = f(x)$, with $y(0) = y(1) = 0$, it can be shown that the solution is the one and only function that both minimizes the energy integral $F(y) = \int_0^1 (\frac{1}{2}y_x^2 + yf)dx$ and also satisfies the given boundary conditions. Assume the approximating function is $Y(x) = a_0 + a_1x + a_2x^2$.
(a) Find the coefficients so that Y satisfies the boundary conditions and also minimizes the energy integral.
(b) What are the coefficients of Y if collocation is used?
(c) What are the coefficients of Y if least squares is used?
(d) Suppose the exact solution is $y(x) = x(1 - x)e^{2x}$. On the same axes plot $y(x)$ and the approximations from (a)–(c). Does any method appear to do better than the others?

2.16. Consider the BVP given by $y'' = f(x)$, with $y(0) = \alpha$ and $y(1) = \beta$, and assume that the approximating function is $Y(x) = \sum_{k=-1}^{2} a_k B_k(x)$. In this case $x_{-1} = -1, x_0 = 0, x_1 = 1$, and $x_2 = 2$.
(a) On the same axes sketch the four B-splines in this sum.
(b) Assuming that $0 \le x \le 1$, show the residual is $r(x) = a_{-1}(1 - x) + a_0(3x - 2) + a_1(-3x + 1) + a_2x - f(x)$.
(c) Assuming that the two collocation points are $x = 0$ and $x = \ell$, write down the equations that uniquely determine the coefficients of Y.
(d) Redo (c) for the equally spaced collocation points $x = \frac{1}{3}$ and $x = \frac{2}{3}$.
(e) Redo (c) for the Gaussian collocation points $x = \frac{1}{2}(1 - \frac{1}{\sqrt{3}})$ and $x = \frac{1}{2}(1 + \frac{1}{\sqrt{3}})$.
(f) Suppose the exact solution is $y(x) = xe^{-2x}$. On the same axes plot $y(x)$ and the approximations from (c)–(e). Which method is best?

2.17. Explain how to use B-splines and collocation to solve (2.4) when the boundary conditions are $ay(0) + by'(0) = \alpha, y(\ell) = \beta$.

2.18. This problem develops some of the properties of cubic B-splines.
(a) Show that (2.58) can be written as

$$Y(x) = \sum_{k=-1}^{N+2} a_k B\left(\frac{x - x_k}{h}\right), \quad \text{for } 0 \le x \le \ell,$$

where

$$B(x) = \begin{cases} 0 & \text{if } 2 \le |x|, \\ \frac{1}{6}(2 - |x|)^3 & \text{if } 1 < |x| < 2, \\ \frac{1}{6}(4 - 6x^2 + 3|x|^3) & \text{if } 0 \le |x| \le 1. \end{cases}$$

(b) Show that $B_k(x_k) = 2/3$, $B_k(x_{k\pm1}) = 1/6$, and $B_k(x_j) = 0$ if $j \ne k, k\pm1$.
(c) Show that $B'_k(x_k) = 0$, $B'_k(x_{k\pm1}) = \mp1/2h$, and $B'_k(x_j) = 0$ if $j \ne k, k\pm1$.
(d) Show that $B''_k(x_k) = -2/h^2$, $B''_k(x_{k\pm1}) = 1/h^2$, and $B''_k(x_j) = 0$ if $j \ne k, k \pm 1$.
(e) Use the results from (b)–(d) to derive (2.64).

2.19. A natural question to ask is why a simpler piecewise cubic is not used to define the B-spline basis function (2.57). For example, a potential candidate is the piecewise cubic

$$\overline{B}_k(x) = \begin{cases} 0 & \text{if } x \le x_{k-1}, \\ a_1 + b_1 x + c_1 x^2 + d_1 x^3 & \text{if } x_{k-1} \le x \le x_k, \\ a_2 + b_2 x + c_2 x^2 + d_2 x^3 & \text{if } x_k \le x \le x_{k+1}, \\ 0 & \text{if } x_{k+1} \le x. \end{cases}$$

Explain why the required continuity of $\overline{B}_k$, $\overline{B}'_k$, and $\overline{B}''_k$ at the points x_{k-1}, x_k, x_{k+1} makes this choice untenable.

2.20. Although the B-spline method, with collocation, has the same $O(h^2)$ convergence rate as the centered finite difference approximation, some people feel it is a better method. In this problem you are to see whether there is any merit to this opinion using the BVP in (2.4), (2.5) when p, q are constant, $f = 0$, $\ell = 1$, $\alpha = 0$, and $\beta = 1$. Assuming $p^2 > 4q$, then the exact solution is

$$y(x) = \frac{e^{ax} - e^{bx}}{e^a - e^b},$$

where $a = (-p + \sqrt{p^2 - 4q})/2$ and $b = (-p - \sqrt{p^2 - 4q})/2$.
(a) Compare the two numerical methods when $p = 1$ and $q = -2$. Do this by plotting the error and, if possible, the computing time for each method as a function of N.

(b) Compare the two numerical methods for the BVP for two or three other choices of p, q. Pick something interesting (i.e., something that challenges the codes), such as $p = 100, q = -1000$ or $p = -2, q = -8$.

(c) What is your conclusion? Does the B-spline method appear to do better than the finite difference method?

2.21. It is not necessary to select the collocation points to be the same points used to construct the B-splines, as was done to obtain (2.64). Instead, suppose the points selected are $\bar{x}_i$ for $i = 0, 1, 2, \ldots, N + 1$, where $\bar{x}_i < \bar{x}_{i+1}$, $\bar{x}_0 = 0$, and $\bar{x}_{N+1} = \ell$. What does $r(\bar{x}_i) = 0$ reduce to in this case? Is the resulting matrix still tridiagonal?

2.22. This problem develops a version of the shooting method that works on the linear BVPs given in (2.4).

(a) Assume $f = 0$. Suppose $y_a(x)$ is the solution of (2.4) with $y_a(0) = 1$ and $y_a'(0) = 0$. Similarly, let $y_b(x)$ be the solution of (2.4) with $y_b(0) = 0$ and $y_b'(0) = 1$. Show how the solution of (2.4), (2.5) can be written in terms of $y_a(x)$ and $y_b(x)$.

(b) Still assuming $f = 0$, explain how to use the result in part (a), and shooting, to solve (2.4), (2.5). Your method should not require numerically solving $g(s) = 0$ as in (2.74).

(c) Show how to modify the ideas in parts (a) and (b) to solve the BVP when $f \neq 0$.

(d) There is not much advantage in using the procedure in part (b) as compared to using the method based on (2.74). Why not?

2.23. This problem examines shooting for particular linear BVPs. Assume that ω is positive.

(a) Solve the IVP $y'' - \omega^2 y = 0$, where $y(0) = \alpha$ and $y'(0) = s$.

(b) Show if ω is large that a small change in the value of s results in a relatively large change in the value of the solution at $x = \ell$. This means that the procedure is ill-conditioned for such problems.

(c) What does s have to be so that the solution satisfies the BVP $y'' - \omega^2 y = 0$, for $0 < x < \ell$, where $y(0) = \alpha$ and $y(\ell) = \beta$?

(d) Redo (a) and (c) for the BVP $y'' + \omega^2 y = 0$, for $0 < x < \ell$, where $y(0) = \alpha$ and $y(\ell) = \beta$. Also explain why the procedure is not ill-conditioned in this case.

2.24. This problem examines extensions of the shooting method.

(a) Explain how to use shooting to solve $y'' = f(x, y, y')$ when the boundary conditions are $y'(0) = \alpha$ and $y(\ell) = \beta$.

(b) Explain how to use shooting to solve $y'' = f(x, y, y')$ when the boundary conditions are $y(0) = \alpha$ and $y'(\ell) = \beta$.

(c) Explain how to use shooting to solve the Falkner–Skan problem given in Exercise 2.12.

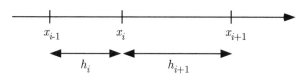

Figure 2.12. Nonuniform grid used in Exercise 2.25.

2.25. For some situations a nonuniform step size is useful and this problem examines how such an approximation is derived.

(a) Suppose a nonuniform step size is used (see Figure 2.12). Find a finite difference approximation for the first and second derivatives at x_i and give the truncation error for each. Your approximations must be consistent, which means that if h_i and h_{i+1} both go to zero, the truncation error also approaches zero.

(b) Using your results in part (a) find the matrix equation that comes from the difference equation corresponding to the BVP in (2.4), (2.5).

2.26. Consider the problem

$$\frac{d^2y}{dx^2} + q(x)y = f(x), \quad \text{for } 0 < x < 1,$$

where, instead of the usual boundary conditions, it is required that

$$\int_0^1 y\,dx = \alpha \quad \text{and} \quad \int_0^1 xy\,dx = \beta.$$

(a) Find a finite difference approximation for this problem that has a truncation error of $O(h^2)$. Write the difference equation in matrix form.

(b) If $q(x) = -1$, $f(x) = -100x$, $\alpha = 0$, and $\beta = 1$, then the exact solution has the form $y = 100x + ae^x + be^{-x}$. Find the coefficients a, b. These values are to be used in parts (c) and (d).

(c) Plot the numerical and exact solutions using a step size of $h = \frac{1}{4}$, $h = \frac{1}{8}$, and $h = \frac{1}{16}$.

(d) Plot the error $E_0 = |y(0) - \bar{y}_0|$, which is the difference between the exact and computed solutions at $x = 0$, as a function of N. Does the curve show the expected $O(h^2)$ convergence?

2.27. An often occurring problem is to find the eigenvalues λ in the BVP

$$\frac{d^2y}{dx^2} + p(x)\frac{dy}{dx} + q(x)y = -\lambda r(x)y, \quad \text{for } 0 < x < \ell,$$

where $y(0) = y(\ell) = 0$. The functions p, q, r are given and the question is what values of λ, if any, result in a nonzero solution of this BVP.

(a) Use finite differences to transform this into a matrix eigenvalue problem of the form $\mathbf{Ay} = \lambda\mathbf{Dy}$, where $\mathbf{D}$ is a diagonal matrix. The truncation error should be $O(h^2)$.

In the remainder of the problem assume $p = q = 0$, $r = 1$, and $\ell = 1$.

(b) What does the matrix equation in (a) reduce to in this case?

(c) Verify that the eigenvalues of the matrix equation are $2\left(1 - \cos\left(\frac{p\pi}{N+1}\right)\right)$
    for $p = 1, 2, 3, \ldots, N$. What are the corresponding eigenvectors?

(d) Find the eigenvalues, and corresponding eigenfunctions, of the BVP.

(e) Show that the difference between the smallest eigenvalue from (c) and the smallest eigenvalue from (d) is $O(h^2)$.

(f) Show that the truncation error for the pth eigenfunction is $O(p^4 h^2)$. From this, comment on how many grid points should be used to calculate, say, the first ten eigenvalues using the approximation in (b). Could you anticipate the number of grid points simply from the graph of the first ten eigenfunctions?

2.28. For the matrix in (2.23), where $bc > 0$, write the eigenvalue equation $\mathbf{Ax} = \lambda\mathbf{x}$ in component form. Solve the resulting difference equation by assuming $x_i = r^i$ and from this show that the eigenvalues are given in (2.24).

2.29. This problem considers whether there are any computational benefits to being able to solve a BVP exactly. The equation is

$$y'' + cy' = f(x), \quad \text{for } 0 < x < 1,$$

where $y(0) = y(1) = 0$.

(a) Show that

$$u(x) = \frac{g(x)}{g(1)}\int_0^1 f(s)g(1-s)ds - \int_0^x f(s)g(x-s)ds,$$

where $g(x) = (e^{-cx} - 1)/c$, is the solution of the above BVP.

(b) Assuming that $f(x)$ is not an elementary function, it will be necessary to use numerical integration to evaluate the formula in (a). Derive an algorithm that evaluates the solution at the grid points in (2.6). Your method should calculate each $y(x_i)$ with an error that is at least $O(h^2)$.

(c) Write down a finite difference approximation for the BVP that has truncation error $O(h^2)$.

(d) Compare the two methods in (b) and (c), and this includes differences in flops, accuracy, and ease of use.

3

Diffusion Problems

3.1 Introduction

To begin the study of finding numerical solutions of partial differential equations we begin with diffusion problems. In physical terms these are problems that involve motion or transport of particles (ions, molecules, etc.) from areas of higher concentration to areas of lower concentration. Simple examples are the spread of a drop of ink dropped into water and the melting of an ice cube. Diffusion is also a key component in the formation of dendrites when liquid metal cools, as well as in the chemical signals responsible for pattern formation (Figure 3.1). Other interesting applications of diffusion arise in the study of financial assets as expressed by the Black–Scholes theory for options pricing and in the spread of infectious diseases (Ross [2002], Brauer and Castillo-Chavez [2001]).

Given a diffusion problem as in the above examples, how should one proceed to calculate the solution? To address this question we consider linear diffusion equations that have the form

$$a(x,t)\frac{\partial^2 u}{\partial x^2} + b(x,t)\frac{\partial u}{\partial x} + c(x,t)u = \frac{\partial u}{\partial t} + f(x,t), \quad \text{for} \quad \begin{cases} 0 < x < \ell, \\ 0 < t. \end{cases} \quad (3.1)$$

It is assumed that the coefficient functions a, b, c, f are smooth with $a > 0$ and $c \leq 0$. To complete the formulation of the problem, boundary and initial conditions need to be specified. These are given when we study particular examples.

3.1.1 Heat Equation

The study of heat conduction in a rod produces the prototypical diffusion equation. To state the mathematical problem that arises, consider the situation of a straight rod oriented so that $x = 0$ and $x = \ell$ designate its two ends. Assuming that the lateral surface is insulated, then the temperature $u(x,t)$ in

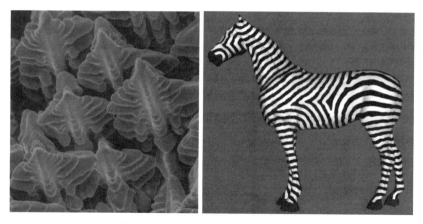

Figure 3.1. Examples where diffusion plays a significant role. On the left are dendrites that appear during the solidification of liquid medal (David et al. [2003]). The picture on the right shows the stripe patterns formed on a zebra obtained from numerically solving nonlinear diffusion equations (Turk [1991]).

the rod depends on the spatial location x and time t. The resulting equation that determines the temperature in the rod is the heat equation, given as

$$D\frac{\partial^2 u}{\partial x^2} = \frac{\partial u}{\partial t}, \quad \text{for} \quad \begin{cases} 0 < x < \ell, \\ 0 < t. \end{cases} \tag{3.2}$$

In this equation, the positive constant D is the thermal diffusivity, and its value depends on what material the rod is composed of. To complete the problem we assume that the left end is at a prescribed temperature $u_L(t)$ and the right end is at a prescribed temperature $u_R(t)$. Together these produce the boundary conditions

$$u(0,t) = u_L(t) \quad \text{and} \quad u(\ell,t) = u_R(t), \quad \text{for} \ \ 0 < t. \tag{3.3}$$

We also need information about the starting temperature, and the initial condition we use is

$$u(x,0) = g(x), \quad \text{for} \ \ 0 \le x \le \ell. \tag{3.4}$$

With (3.2)–(3.4) we have a complete heat conduction problem. The objective of this chapter is to develop and analyze numerical methods that can be used to solve this problem as well as others like it. In preparation for this it is worth examining the properties of the solution and to do this by working through a couple of examples.

Example
Suppose the temperature at the two ends is held constant, and in fact, suppose $u_L(t) = u_R(t) = 0$. To simplify things we also take $D = 1$ and $\ell = 1$. Using the method of separation of variables one finds that the solution is

$$u(x,t) = \sum_{n=1}^{\infty} A_n e^{-\lambda_n^2 t} \sin(\lambda_n x), \qquad (3.5)$$

where $\lambda_n = n\pi$ and

$$A_n = 2 \int_0^1 g(x) \sin(\lambda_n x) dx. \qquad (3.6)$$

There are two special cases that will be used in deciding how well the numerical solutions fare in solving this problem.

Case 1: If $g(x) = \sin(2\pi x)$ then the series in (3.5) reduces to

$$u(x,t) = e^{-4\pi^2 t} \sin(2\pi x). \qquad (3.7)$$

This solution is shown in Figure 3.2, both as time slices and as the solution surface for $0 \le t \le 0.1$.

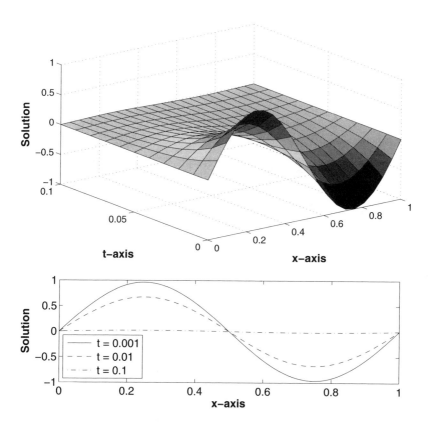

Figure 3.2. (M) Solution of the heat equation when $g(x) = \sin(2\pi x)$. Shown is the solution surface as well as the solution profiles at specific time values.

Case 2: The second example of interest is when $g(x)$ contains jumps, such as the function

$$g(x) = \begin{cases} 1 & \text{if } a \leq x \leq b, \\ 0 & \text{otherwise}, \end{cases} \tag{3.8}$$

where $0 \leq a < b \leq 1$. In this case the solution is given in (3.5) with

$$A_n = \frac{2}{\pi n} (\cos(a\pi n) - \cos(b\pi n)). \tag{3.9}$$

The solution (3.5) is shown in Figure 3.3 in the case of when $a = \frac{1}{4}$ and $b = \frac{3}{4}$ for $0 \leq t \leq 0.1$.

There are properties of the solution evident in Figures 3.2–3.3 that play important roles in analyzing the numerical methods we will use to solve the problem. Of particular importance are the following:

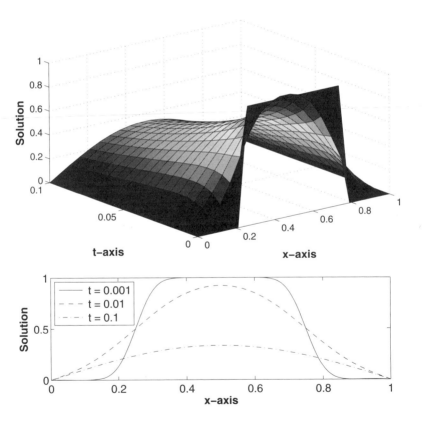

Figure 3.3. (M) Solution of the heat equation when $g(x)$ is given in (3.8), with $a = \frac{1}{4}$ and $b = \frac{3}{4}$. Shown is the solution surface as well as the solution profiles at specific time values.

Smoothness: As seen in Figure 3.3, even with an initial condition that contains jumps, the solution on the interior of the domain (where $0 < x < 1$ and $0 < t$) is smooth. Certainly one can question whether this apparent smoothness is due to the inability of the plotting program to resolve rough spots in the solution. However, one can prove that even with a finite number of jumps in the initial or boundary conditions, the solution of the heat equation, or its generalization in (3.1), is smooth on the interior of the solution domain.

Maximum and Minimum Principles: In looking at the solutions in Figures 3.2 and 3.3 it is evident that the maximum and minimum values of the solution occur on the boundary. Again, this is not an artifact of the plots but a property of the solution of the heat equation. The precise statement is that the maximum and minimum values of the solution $u(x,t)$ over the region $0 \le x \le 1, 0 \le t \le T$ occur on the outer boundary of the region, where either $t = 0$, $x = 0$, or $x = \ell$. One consequence of this is that the heat equation problem is stable in the sense that if small changes are made in the boundary or initial conditions then the resulting changes in the solution are small over the entire domain. This follows because if u_1 and u_2 are solutions of the heat equation then their difference $u_1 - u_2$ is also a solution. Therefore, the maximum and minimum values of $u_1 - u_2$ are determined by the differences between u_1 and u_2 around the outer boundary of the region. If these two functions differ by only a small amount around the outer boundary then their difference $u_1 - u_2$ will be small throughout the entire domain.

Instant Messaging: In Figure 3.3 the solution is zero at $t = 0$ in the intervals $0 < x < \frac{1}{4}$ and $\frac{3}{4} < x < 1$. One might expect that at a point where the solution is initially zero (e.g., $x = \frac{7}{8}$), it will remain zero for a short time. The expectation here is that it should take a few moments for the nonzero portion of the solution to move through the interval, and the farther away one is from the nonzero region the longer it takes. However, this is not true. For this problem the solution is nonzero for $0 < x < 1$ once $t > 0$. What this means is that the heat equation transmits information instantly. This property might cause one to question how such an equation could ever describe any real-world situation. After all, the temperature in Denver would not seem to be instantly affected by the temperature in, say, London. However, as seen in Figure 3.3, the degree to which the nonzero part of $g(x)$ affects the initially zero intervals decreases rapidly with distance. So, even though information propagates infinitely fast, the degree, or strength, of the signal drops very quickly with distance from the source. Just how fast is investigated in Exercise 3.26.

The value of the above three observations is more qualitative than quantitative when computing the solution of a diffusion problem. For example, if the computed solution does not obey the maximum principle then we should seriously question whether the answer is correct. Similarly, if during the derivation of a finite difference approximation of the heat equation it is seen that the instant messaging property does not hold, then one should be concerned about how well the method will work. Such observational tests are easy and can prove invaluable when developing and testing a numerical method.

3.2 Derivative Approximation Methods

To introduce the ideas we start with the problem of finding the function $u(x,t)$ that satisfies the inhomogeneous heat equation, given as

$$\frac{\partial^2 u}{\partial x^2} = \frac{\partial u}{\partial t} + f(x,t), \quad \text{for} \quad \begin{cases} 0 < x < \ell, \\ 0 < t, \end{cases} \tag{3.10}$$

where the boundary conditions are

$$u(0,t) = 0 \ \text{ and } \ u(\ell,t) = 0, \quad \text{for } 0 < t, \tag{3.11}$$

and the initial condition is

$$u(x,0) = g(x), \quad \text{for } 0 < x < \ell. \tag{3.12}$$

In this problem, the functions $f(x,t), g(x)$ are assumed to be given, as is the constant ℓ. Our approach to solving this problem will not appear to be very innovative, since it will closely follow the steps used for IVPs and BVPs. More specifically, we will replace the derivatives with approximations chosen from Table 1.1 and then solve the resulting finite difference equation. Complications do arise, and we will address them when they do. One point to make before beginning is that the computer cannot run forever, so we must decide just how large a time interval to take over which the solution will be computed. It is assumed in what follows that the interval is $0 \leq t \leq T$.

STEP 1. As usual, we first introduce the points where we will compute the solution. Unlike the situation with either IVPs or BVPs, the region over which we solve the problem is two-dimensional. So our points have the form (x_i, t_j), and together they form a lattice as shown in Figure 3.4. We use the formulas introduced earlier, namely,

$$t_j = jk, \quad \text{for } j = 0, 1, 2, \ldots, M, \tag{3.13}$$
$$x_i = ih, \quad \text{for } i = 0, 1, 2, \ldots, N+1. \tag{3.14}$$

As before, the step sizes are $h = \ell/(N+1)$ and $k = T/M$. For the moment we assume that h and k can be chosen independently. As shown later, there are a few situations in which they must be related to each other for the method to work, but these are dealt with as they arise.

STEP 2. Evaluate the differential equation at the grid point $(x,t) = (x_i, t_j)$ to obtain

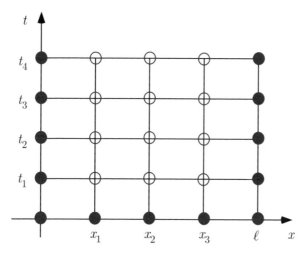

Figure 3.4. Example of the grid system used to find the numerical solution of the heat equation. As shown there are 12 points, the hollow dots, where the solution is to be computed. The values of the solution at the grid points marked with solid dots are known from the given initial and boundary conditions. As shown, $N = 3$ and $M = 4$.

$$u_{xx}(x_i, t_j) = u_t(x_i, t_j) + f(x_i, t_j). \tag{3.15}$$

We also need to evaluate the boundary and initial conditions at their respective grid points, but this will be done later once we have taken care of the differential equation.

STEP 3. Replace the derivative terms in STEP 2 with finite difference approximations. A few possibilities are listed in Table 1.1, and different choices result in different numerical procedures. Also, as before, not all choices will work. It goes without saying that whatever approximations are used, they are consistent, which means that the truncation error goes to zero as k and h approach zero. To start we use a forward difference for the time derivative and a centered difference for the spatial derivative. This gives us

$$u_t(x_i, t_j) = \frac{u(x_i, t_{j+1}) - u(x_i, t_j)}{k} - \frac{1}{2}k u_{tt}(x_i, \eta_j) \tag{3.16}$$

and

$$u_{xx}(x_i, t_j) = \frac{u(x_{i+1}, t_j) - 2u(x_i, t_j) + u(x_{i-1}, t_j)}{h^2} - \frac{1}{12}h^2 u_{xxxx}(\bar{\eta}_i, t_j), \tag{3.17}$$

where η_j is between t_j and t_{j+1} and $\bar{\eta}_i$ is between x_{i-1} and x_{i+1}. Substituting these into (3.15) yields

$$\frac{u(x_{i+1}, t_j) - 2u(x_i, t_j) + u(x_{i-1}, t_j)}{h^2} + \tau_{ij}$$
$$= \frac{u(x_i, t_{j+1}) - u(x_i, t_j)}{k} + f(x_i, t_j), \tag{3.18}$$

where the truncation error is

$$\tau_{ij} = \frac{1}{2} k u_{tt}(x_i, \eta_j) - \frac{1}{12} h^2 u_{xxxx}(\bar\eta_i, t_j)$$
$$= O(k) + O(h^2). \tag{3.19}$$

The above equation can be rewritten as

$$u(x_i, t_{j+1}) = \lambda u(x_{i+1}, t_j) + (1 - 2\lambda) u(x_i, t_j)$$
$$+ \lambda u(x_{i-1}, t_j) - k f(x_i, t_j) + k \tau_{ij}, \tag{3.20}$$

where

$$\lambda = \frac{k}{h^2}. \tag{3.21}$$

The grid points used in (3.20) are shown in Figure 3.5. A diagram such as this is called the stencil for the method.

STEP 4. Drop the truncation error. This is the step where we go from an exact problem to one that is, hopefully, an accurate approximation of the original. After dropping τ_{ij} in (3.20) the resulting equation is

$$u_{i,j+1} = \lambda u_{i+1,j} + (1 - 2\lambda) u_{ij} + \lambda u_{i-1,j} - k f_{ij}, \quad \text{for} \quad \begin{cases} i = 1, 2, \ldots, N, \\ j = 0, 1, 2, \ldots, M - 1. \end{cases} \tag{3.22}$$

The understanding here is that the solution at the previous time step $(t = t_j)$ has been determined and the formula in (3.22) is used to calculate the solution at the most advanced time level $(t = t_{j+1})$. To complete the derivation we need to consider the boundary and initial conditions. From the boundary condition $u(0, t) = 0$ we get that

$$u_{0,j} = 0, \quad \text{for} \quad j = 0, 1, 2, \ldots, M. \tag{3.23}$$

Similarly, the boundary condition $u(\ell, t) = 0$ gives us

$$u_{N+1,j} = 0, \quad \text{for} \quad j = 0, 1, 2, \ldots, M. \tag{3.24}$$

Finally, from the initial condition (3.4) we have

$$u_{i,0} = g_i, \quad \text{for} \quad i = 0, 1, 2, \ldots, N + 1. \tag{3.25}$$

The derivation of the finite difference approximation of the heat equation problem is now complete. As written in (3.22) the method is explicit in the sense that $u_{i,j+1}$ is given explicitly in terms of known quantities, or said

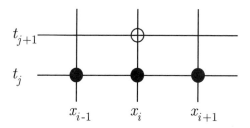

Figure 3.5. Stencil for the explicit method (3.22) indicating the grid points contributing in the formula. The convention used here is that a solid dot indicates a point where the solution has already been determined and a hollow dot indicates a point where the solution is going to be calculated using the given method.

another way, $u_{i,j+1}$ does not depend on the other unknowns at time level t_{j+1}. Although there are certainly other explicit methods, (3.22) is referred to as *the* explicit method for the heat equation.

It is convenient to rewrite the difference equation in (3.22) in matrix form. This can be done by collecting all the unknowns at time level $t = t_j$ into a vector as follows

$$\mathbf{u}_j = \begin{pmatrix} u_{1,j} \\ u_{2,j} \\ \vdots \\ u_{N,j} \end{pmatrix}. \tag{3.26}$$

In this case, (3.22) can be written as

$$\mathbf{u}_{j+1} = \mathbf{A}\mathbf{u}_j - k\mathbf{f}_j, \quad \text{for } j = 0, 1, 2, \ldots, M-1, \tag{3.27}$$

where $\mathbf{u}_0 = \mathbf{g}$. In this equation $\mathbf{A}$ is the $N \times N$ symmetric tridiagonal matrix

$$\mathbf{A} = \begin{pmatrix} 1-2\lambda & \lambda & & & \\ \lambda & 1-2\lambda & \lambda & & 0 \\ & \lambda & 1-2\lambda & \lambda & \\ & & \ddots & \ddots & \ddots \\ 0 & & & & \lambda \\ & & & \lambda & 1-2\lambda \end{pmatrix}, \tag{3.28}$$

and the two vectors are

$$\mathbf{f}_j = \begin{pmatrix} f_{1,j} \\ f_{2,j} \\ \vdots \\ f_{N,j} \end{pmatrix} \quad \text{and} \quad \mathbf{g} = \begin{pmatrix} g_1 \\ g_2 \\ \vdots \\ g_N \end{pmatrix}. \tag{3.29}$$

With (3.22), or (3.27), we have produced an explicit method with truncation error $\tau_{ij} = O(h^2) + O(k)$. One can think of it as Euler's method extended to the heat equation in the sense that a forward difference approximation is used for the time derivative. How well does it do in solving the problem? Well, we will look at a few examples and do a little analysis to address this question. First, though, a question for you to think about. Earlier we saw that the heat equation has an instant messaging property. From the stencil in Figure 3.5, however, it can be seen that the numerical solution at (x_i, t_{j+1}) is determined using only three points from time level $t = t_j$. This means, for example, the value of the solution at $x = x_{i+2}$ $(t = t_j)$ contributes to the exact solution at (x_i, t_{j+1}), but it is incapable of doing so in our numerical method. Is this bad? Is this going to mean that this particular method won't work? As it turns out, this observation does have repercussions for the effectiveness of the method, and this will be evident in the example below. Determining mathematically what the repercussions are will come later.

Example
Our first example is the case $f(x,t) = 0$ and $g(x) = \sin(2\pi x)$. The exact solution is $u(x,t) = e^{-4\pi^2 t}\sin(2\pi x)$, and this function is plotted in Figure 3.2. This surface is drawn using 22 points along the x-axis and 11 on the t-axis. This is brought up because to employ the explicit method we need to decide on how many grid points to use. A reasonable place to start is with the boundary and initial conditions. For example, the initial condition $u(x,0) = \sin(2\pi x)$ can be described reasonably well using 22 points along the x-axis, and so, we take $N = 20$. It is not clear how many points to take along the t-axis, and so we try different values. In particular, taking $T = 0.1$, we try $M = 5, 10, 20$. The resulting numerical solutions at $t = 0.02$, $t = 0.04$, and $t = 0.1$ are shown in Figure 3.6. At $t = 0.02$ and $t = 0.04$ the situation is what we might expect. For example, the computed solution using $M = 20$ is more accurate than when $M = 10$, and the latter is better than when $M = 5$. However, something is horribly wrong at $t = 0.1$. The large-amplitude oscillation obtained at this time level is the sort of behavior seen with unstable methods, which means we should look into this before going any further. One last observation is that it is interesting that the value of M that produces the most accurate solution at $t = 0.02$ and $t = 0.04$ is the one that causes problems at $t = 0.1$. Any idea why?

Convergence

The last example demonstrates that we need to spend some time investigating what is needed to guarantee that the method works. As with IVPs, the issue is stability and how the error depends on the step sizes.

Stability
We start with stability. For IVPs stability was determined by examining how the method fared on a test problem, which happened to correspond to the

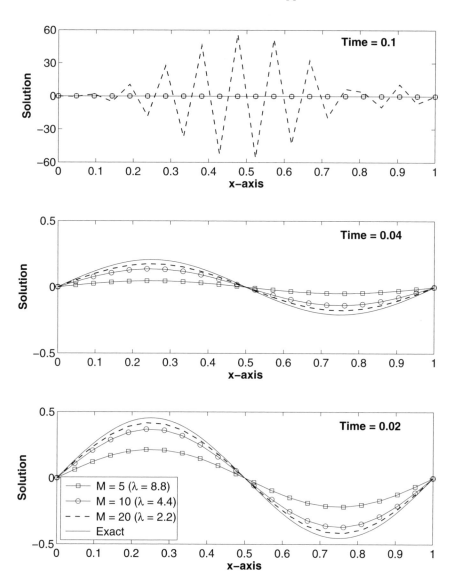

Figure 3.6. (M) Solution of the heat equation obtained using the explicit method (3.27) in the case where the initial condition is $u(x,0) = \sin(2\pi x)$. The method is conditionally stable and problems with stability are evident at $t = 0.1$. For comparison the exact solution is given at each time level.

Euler's Formula	$e^{I\theta} = \cos(\theta) + I\sin(\theta)$	
Trig Form	$\cos(\theta) = \frac{1}{2}(e^{I\theta} + e^{-I\theta})$	$\sin(\theta) = \frac{1}{2I}(e^{I\theta} - e^{-I\theta})$
Half-Angle	$1 + \cos(\theta) = 2\cos^2\left(\frac{1}{2}\theta\right)$	$1 - \cos(\theta) = 2\sin^2\left(\frac{1}{2}\theta\right)$

Table 3.1. Useful identities for stability analysis. In these expressions, $I = \sqrt{-1}$.

equation that arises for radioactive decay. The approach used for partial differential equations is different, and instead we investigate how the method does with a test solution. The problem considered is the homogeneous version of the differential equation. So, in (3.10) we set $f(x, t) = 0$. To explain how the test solution is selected, one sees in (3.5) that the separation of variables solution consists of the superposition of functions that are oscillatory in x. In deciding whether a numerical method is stable we determine how well the method does with such solutions. In particular, the start-off assumption to decide on stability is that the solution of the finite difference equation has the form

$$u_{i,j} = w_j e^{r x_i I}, \tag{3.30}$$

where $I = \sqrt{-1}$. The constant r is used here in place of the λ_n appearing in (3.5). It is understood that r, like λ_n, can take on values over the entire positive real axis. The function w_j is determined from the difference equation by substituting (3.30) into (3.22). Keeping in mind the identities in Table 3.1, we obtain

$$\begin{aligned}
w_{j+1} e^{r x_i I} &= \lambda w_j e^{r x_{i+1} I} + (1 - 2\lambda) w_j e^{r x_i I} + \lambda w_j e^{r x_{i-1} I} \\
&= \lambda w_j e^{r(x_i + h)I} + (1 - 2\lambda) w_j e^{r x_i I} + \lambda w_j e^{r(x_i - h)I} \\
&= (\lambda e^{rhI} + 1 - 2\lambda + \lambda e^{-rhI}) w_j e^{r x_i I} \\
&= (1 - 2\lambda + 2\lambda \cos(rh)) w_j e^{r x_i I} \\
&= \left[1 - 4\lambda \sin^2\left(\frac{rh}{2}\right)\right] w_j e^{r x_i I}.
\end{aligned} \tag{3.31}$$

Canceling the exponential term, we have that $w_{j+1} = \kappa w_j$, where

$$\kappa = 1 - 4\lambda \sin^2\left(\frac{rh}{2}\right) \tag{3.32}$$

is known as the amplification factor for the method. The solution is therefore $w_j = \kappa^j w_0$. What is important is how w_j depends on the time index j. If the method is to be stable, w_j must remain bounded as j increases no matter what the value of r. In other words, the stability requirement is $|\kappa| \le 1$. With (3.32) this breaks down into the inequalities $-1 \le 1 - 4\lambda \sin^2(rh/2) \le 1$. The upper inequality holds because λ is positive, so stability is determined by whether or not $2\lambda \sin^2(rh/2) \le 1$. This holds, irrespective of the value of r, if

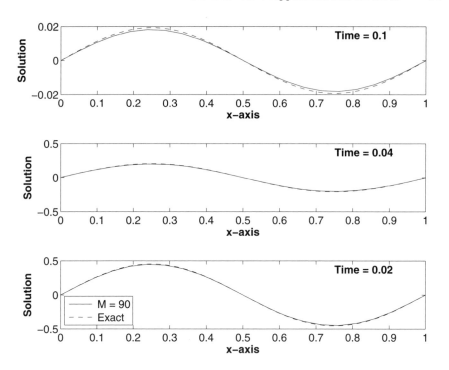

Figure 3.7. (M) Solution of the heat equation using the explicit method (3.27) when the stability condition is satisfied. For this calculation, $\lambda = 0.49$ and $g(x) = \sin(2\pi x)$.

$$\lambda \leq \frac{1}{2}, \tag{3.33}$$

or equivalently,

$$2k \leq h^2. \tag{3.34}$$

Therefore, this explicit method is conditionally stable and the stability condition is given in (3.33).

Example
Armed with this new information we return to the last example, the results of which are shown in Figure 3.6. The corresponding values of λ are given in this figure. Clearly, we did not make good choices for M when it comes to satisfying the stability condition. As it turns out, with $T = 0.1$ and $N = 20$, to satisfy (3.33) we need $M > 88.2$. To verify that all is well, the computed solution is shown in Figure 3.7 when $M = 90$, and it is seen that the oscillatory response obtained earlier is no longer present.

Example
As a second demonstration of the importance of the stability condition sup-

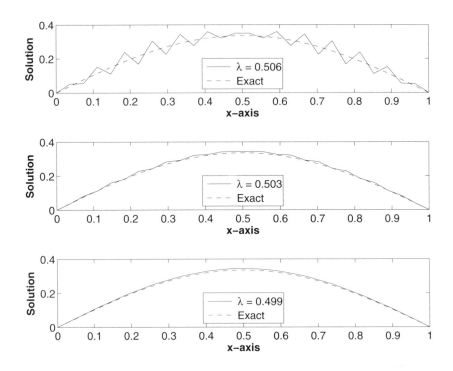

Figure 3.8. Solution of the heat equation at $t = 0.1$ using the explicit method (3.27). The differences in the $\lambda = 0.499$ and $\lambda = 0.506$ plots demonstrate the importance of the stability condition $\lambda \le \frac{1}{2}$. The initial condition is the same one used to produce the solution in Figure 3.3.

pose the initial condition contains jumps as in (3.8). Let's see how well the explicit method does in reproducing the solution curve shown in Figure 3.3 when $t = 0.1$. In the calculations, $N = 26$, so the stability condition in (3.33) gives us the requirement that $M > 145.8$. The computed solution is shown in Figure 3.8 when $M = 144$ ($\lambda = 0.506$), $M = 145$ ($\lambda = 0.503$), and $M = 146$ ($\lambda = 0.499$). Without the condition in (3.33) it would be very difficult to explain why in the world the method fails when one uses 145 time steps to go from $t = 0$ to $t = 0.1$ but it works just fine when one uses 147 steps.

It's nice that we have resolved the stability problem, but in the above examples it has required us to take more time steps than should really be necessary for this problem. This situation becomes even worse if we need to resolve the spatial direction any more finely. For example, if the initial condition is $g(x) = \sin(20\pi x)$ then something on the order of about 200 points would be needed along the x-axis. The stability condition in this case would require at least 7,900 points in the time direction. Given such a formidable computing effort it is worth our time to find another way to solve this problem. Before doing so, however, we have to take care of one detail. Namely, we need to determine how the error depends on the truncation error.

Error Analysis
Based on the above examples we might postulate that as long as the stability condition is satisfied, the numerical solution will converge to the exact solution as the step sizes are reduced. Of course, it is understood that when using floating-point arithmetic, we cannot expect to get closer than round-off. To investigate the situation we introduce the usual three solutions at $(x,t) = (x_i, t_j)$, which are

$$u(x_i, t_j) \equiv \text{exact solution of the problem at } (x_i, t_j); \tag{3.35}$$
$$u_{i,j} \equiv \text{exact solution of finite difference equation at } (x_i, t_j); \tag{3.36}$$
$$\overline{u}_{i,j} \equiv \text{solution of difference equation at } (x_i, t_j) \text{ calculated}$$
$$\text{by the computer.} \tag{3.37}$$

We are interested in the difference between the exact solution of the problem and the values we end up computing using our algorithm. Therefore, we are interested in the error $e_{i,j} = |u(x_i, t_j) - \overline{u}_{i,j}|$. To help make it more apparent what is contributing to the error we rewrite it as follows

$$e_{i,j} = |u(x_i, t_j) - u_{i,j} + u_{i,j} - \overline{u}_{i,j}|. \tag{3.38}$$

From this the error can be considered as coming from the following two sources:

$u(x_i, t_j) - u_{i,j}$: This is the error at (x_i, t_j) between the exact solution of the problem and the exact solution of the finite difference equation. This should be the major contributor to the error until k and h are small enough that this difference gets down to approximately that of the round-off.

$u_{i,j} - \overline{u}_{i,j}$: This is the error at (x_i, t_j) that originates from round-off when floating-point calculations are used to compute the solution of the difference equation. Getting values of 10^{-15} or 10^{-16} is about as good as can be expected using double precision.

Given our experience with IVPs and BVPs, it is expected that $u(x_i, t_j) - u_{i,j}$ is determined by the truncation error. To see if this happens with the heat equation let $E_{i,j} = u(x_i, t_j) - u_{i,j}$. From (3.20) and (3.22) one finds that

$$E_{i,j+1} = \lambda E_{i+1,j} + (1 - 2\lambda)E_{i,j} + \lambda E_{i-1,j} + k\tau_{i,j}, \tag{3.39}$$

where $E_{0,j} = 0$, $E_{N+1,j} = 0$, and $E_{i,0} = 0$. Assuming $2\lambda \leq 1$, setting $E_j = \max_{i=0,1,2,\dots,N+1} |E_{i,j}|$ and $\tau_\infty = \max_{i,j} |\tau_{i,j}|$, we have the following

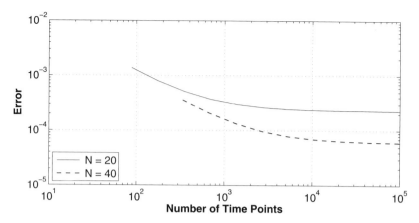

Figure 3.9. Error function $\tau_\infty = 0.1h^2 + k$ as a function of the number of time points used in the calculation.

$$
\begin{aligned}
|E_{i,j+1}| &= |\lambda E_{i+1,j} + (1 - 2\lambda)E_{i,j} + \lambda E_{i-1,j} + k\tau_{i,j}| \\
&\leq \lambda|E_{i+1,j}| + (1 - 2\lambda)|E_{i,j}| + \lambda|E_{i-1,j}| + k|\tau_{i,j}| \\
&\leq \lambda E_j + (1 - 2\lambda)E_j + \lambda E_j + k\tau_\infty \\
&= E_j + k\tau_\infty.
\end{aligned}
$$

This holds $\forall i$, and so it must be that $E_{j+1} \leq E_j + k\tau_\infty$. Iterating back to $j = 0$, we conclude $E_{j+1} \leq E_0 + (j + 1)k\tau_\infty$. Because $E_{i,0} = 0$ and $(j + 1)k = t_{j+1} \leq T$ it follows that $E_{j+1} \leq \tau_\infty T$. Therefore, since $\tau_\infty = O(h^2) + O(k)$ it follows that $u(x_i, t_j) - u_{i,j} = O(h^2) + O(k)$. In other words, if the stability condition holds then the explicit method converges and the difference $u(x_i, t_j) - u_{i,j}$ is bounded by the truncation error. This conclusion is why consistency and stability are both required to guarantee convergence of the method.

To translate the above discussion into what is expected from a computation using the explicit method suppose it is found that $\tau_\infty = 0.1h^2 + k$. This expression is plotted in Figure 3.9 as a function of the number of time points, for two different values of N. The smallest value of M used for each curve is determined by the stability condition, that is, the value of M where $k = \frac{1}{2}h^2$. Both curves decrease linearly with M and then level off for large values of M at a value determined by the spatial error term $0.1h^2$. This can be checked by noting that the limiting (large M) values for the two curves differ by a factor of 4, which they should given the quadratic dependence on h. Is this what is seen in an actual computation? As the next example demonstrates, the real world is a bit more interesting.

Example

To demonstrate the connection between the truncation error and convergence rate we return to our first example, as shown in Figure 3.6. We will determine

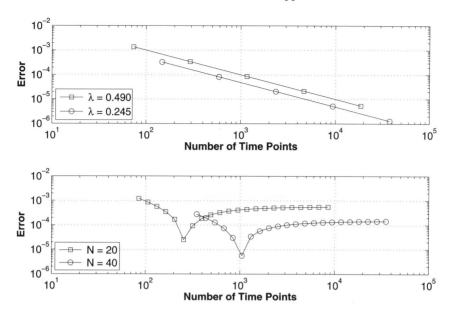

Figure 3.10. Error in the numerical solution at $t = T$ when the explicit method (3.27) is used to solve the heat equation. In the lower graph the error is given for two values of N, and in upper graph the error is given for two values of λ. The initial condition is $u(x, 0) = \sin(2\pi x)$, and in the calculation, $T = 0.1$ and $\ell = 1$.

E_M, which is the error at time level $t = T$. This variable depends on the number of points used in the space and time directions, and we are interested in how quickly it decreases as these numbers are increased. In the lower graph in Figure 3.10, E_M is plotted as a function of the number of time points, for two different values of N. You don't have to look at these curves too long to decide that they are not exactly what was seen earlier in Figure 3.9. The error term is $O(h^2) + O(k)$, so it would be expected that the error decreases like $O(k)$ until E_M gets down to the level of $O(h^2)$. Because h is fixed, the curves in Figure 3.10 should level off for large M at a value determined by the $O(h^2)$ contribution. The leveling off occurs, but the nice $O(k)$ decrease does not. What actually happens is that the error drops much faster than expected. This does not contradict our expectation that the error is $O(h^2) + O(k)$, but rather, it is an indication that for this particular problem our worst-case estimate is too pessimistic. The reason for this is that the $O(k)$ and $O(h^2)$ terms are of opposite signs and almost cancel each other when $k = h^2/6$ (see Exercise 3.19). The fact that you can find such an optimal grid spacing is interesting but secondary to our objective of developing the basic ideas. With this in mind, for another perspective, in the upper graph in Figure 3.10 the error is plotted as a function of M for fixed λ. Since λ is fixed and $h^2 = k/\lambda = O(k)$, the truncation error is $O(k)$. This indicates that for every power of ten increase in M the error should be reduced by a factor of ten, and this is clearly evident in both curves in Figure 3.10.

End Notes

We have completed the derivation and analysis of the explicit method. The approach we used is employed repeatedly in this and the following chapters, and so it is worth commenting on some of the steps taken. The stability argument, which is based on the test solution in (3.30), is called the Fourier, or von Neumann, stability method. It is relatively easy to use but it effectively ignores the boundary conditions. This is not an issue in the problem we studied, since the boundary conditions do not affect the finite difference equation other than as an inhomogeneous term on the right-hand side of the equation. In situations where approximations must be used on the boundary conditions, the requirement from the Fourier method is necessary but not sufficient to guarantee stability. If there is concern that the boundary conditions are affecting stability then another approach can be tried. One possibility is the matrix method, which uses the eigenvalues of the matrix $\mathbf{A}$ in (3.28) to determine stability. This is not pursued here, because the matrix approach is limited in its applicability to other types of partial differential equations, specifically, wave equations, and an interesting discussion of this can be found in Iserles [1996]. There is a third approach one can use for establishing stability, which involves a perturbation argument, and this is explored in Exercise 3.20.

The concept of stability and what it represents for the heat equation is similar to what we found for IVPs. However, the methods used to determine stability, and the subsequent conclusions that are made, differ between these two types of problems. For example, after determining that the trapezoidal method is A-stable we then expect it to work for a broad range of IVPs, such as problems involving $y' = -y$ or $y' = -y - y^3$. In contrast, the conclusions made about stability for the heat equation apply only to that particular equation. If the equation is changed, say to $u_{xx} = u_t + u$, then it is necessary to rederive the stability condition. In this sense stability for a finite difference approximation of a partial differential equation is a more specific property than it is for IVPs.

3.2.1 Implicit Method

In STEP 3, instead of a forward difference one can use a backward difference approximation in time. The stencil for this choice is shown in Figure 3.11. The result is that the equation in STEP 3 now takes the form

$$
\begin{aligned}
\frac{u(x_{i+1}, t_j) - 2u(x_i, t_j) + u(x_{i-1}, t_j)}{h^2} &+ \tau_{i,j} \\
&= \frac{u(x_i, t_j) - u(x_i, t_{j-1})}{k} + f(x_i, t_j),
\end{aligned}
\tag{3.40}
$$

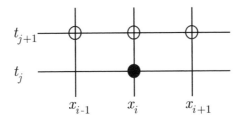

Figure 3.11. Stencil for the implicit method.

where the truncation error is $\tau_{i,j} = O(h^2) + O(k)$. This can be rewritten as

$$\lambda u(x_{i+1}, t_j) - (1 + 2\lambda)u(x_i, t_j) + \lambda u(x_{i-1}, t_j)$$
$$= -u(x_i, t_{j-1}) + kf(x_i, t_j) + k\tau_{i,j}, \tag{3.41}$$

where

$$\lambda = \frac{k}{h^2}. \tag{3.42}$$

Dropping the truncation error, we obtain

$$\lambda u_{i+1,j} - (1 + 2\lambda)u_{i,j} + \lambda u_{i-1,j} = -u_{i,j-1} + kf_{i,j}, \quad \text{for} \quad \begin{cases} i = 1, 2, \ldots, N, \\ j = 1, 2, \ldots, M. \end{cases} \tag{3.43}$$

The boundary and initial conditions are the same as for the explicit method, given in (3.23)–(3.25). In comparison to what occurred earlier, the unknowns, which are the values at time level t_j in the above equation, are coupled, and for this reason the method is implicit.

Collecting the unknowns at time level $t = t_j$ into a vector $\mathbf{u}_j$ we can rewrite (3.43) in matrix form as

$$\mathbf{B}\mathbf{u}_j = \mathbf{u}_{j-1} - k\mathbf{f}_j, \quad \text{for } j = 1, 2, \ldots, M, \tag{3.44}$$

where $\mathbf{u}_0 = \mathbf{g}$ and $\mathbf{B}$ is the $N \times N$ symmetric tridiagonal matrix given as

$$\mathbf{B} = \begin{pmatrix} 1 + 2\lambda & -\lambda & & & \\ -\lambda & 1 + 2\lambda & -\lambda & & 0 \\ & -\lambda & 1 + 2\lambda & -\lambda & \\ & & \ddots & \ddots & \ddots \\ 0 & & & & -\lambda \\ & & & -\lambda & 1 + 2\lambda \end{pmatrix}. \tag{3.45}$$

With (3.43), or (3.44), we have produced an implicit method with truncation error $\tau_{i,j} = O(h^2) + O(k)$. Moreover, using Theorem 2.1 it is not hard to prove that (3.44) has a unique solution. One can think of the method as backward Euler extended to the heat equation in the sense that a backward difference approximation is used for the time derivative. What about instant messaging? From (3.44), and a little algebra, one can show that $\mathbf{u}_j$ depends on every component of $\mathbf{u}_{j-1}$. In other words, this method has the instant messaging property. The question is, does this give it any significant advantage over the explicit method derived earlier? One way to answer this is to check the stability condition. As before, the start-off assumption is given in (3.30). Substituting this into (3.43), and setting $f(x,t) = 0$, one finds that $w_{j+1} = \kappa w_j$, where the amplification factor for this method is

$$\kappa = \frac{1}{1 + 4\lambda \sin^2(\frac{rh}{2})} . \tag{3.46}$$

The solution in this case is $w_j = \kappa^j w_0$. Therefore, the stability requirement is $|\kappa| \le 1$, and it is not hard to show that this always holds for the amplification factor in (3.46). In other words, this implicit method is stable.

Example
To begin we consider the example used for the explicit method, where $f(x,t) = 0$ and $g(x) = \sin(2\pi x)$. The exact solution is $u(x,t) = e^{-4\pi^2 t} \sin(2\pi x)$, and this function is plotted in Figure 3.2. Taking $N = 20$ and $T = 0.1$, the numerical solutions at $t = 0.02$, $t = 0.04$, and $t = 0.1$ are shown in Figure 3.12 using $M = 5, 10, 20$. It is seen that as the time step is reduced, the numerical solution at all three time levels approaches the exact solution. In other words, it is responding as we would expect a convergent method to behave.

Example
The second example involves the step function in (3.8) and $f(x,t) = 0$. We take $a = \frac{1}{4}$ and $b = \frac{3}{4}$, in which case the solution (3.5) is shown in Figure 3.3. Taking $N = 30$ and $T = 0.1$, the numerical solutions at $t = 0.02$, $t = 0.04$, and $t = 0.1$ are shown in Figure 3.13 using $M = 5, 20$. The numerical method does reasonably well in reproducing the exact solution, and this has been accomplished using far fewer time points than were used for the explicit method.

3.2.2 Theta Method

In looking at the numerical solutions in Figures 3.6 and 3.12 it would appear that the exact solution lies between the curves obtained for the explicit and implicit methods. This raises the question of whether we can combine the two to produce a method with a bit more accuracy. We can, and the starting point is to note that both can be rewritten as

$$\text{explicit: } u_{i,j+1} - u_{i,j} = H_{i,j}, \tag{3.47}$$
$$\text{implicit: } u_{i,j+1} - u_{i,j} = H_{i,j+1}, \tag{3.48}$$

where

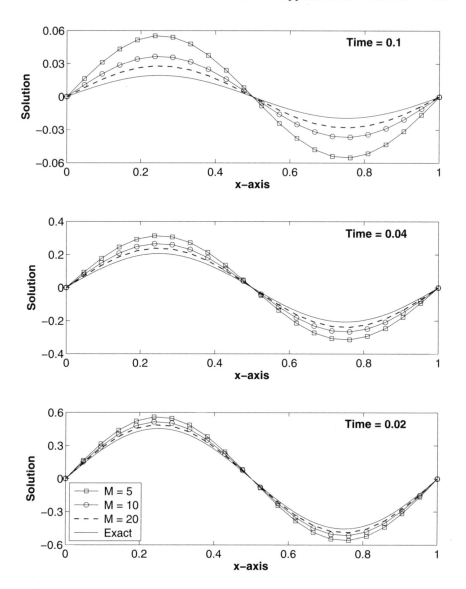

Figure 3.12. (**M**) Solution of the heat equation using the implicit method (3.44) in the case where the initial condition is $u(x, 0) = \sin(2\pi x)$. Also shown is the exact solution.

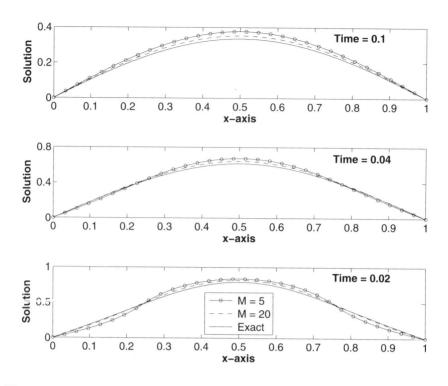

Figure 3.13. Solution of the heat equation using the implicit method (3.44) in the case where the initial condition is given in (3.5), with $a = \frac{1}{4}$ and $b = \frac{3}{4}$. Also shown is the exact solution.

$$H_{i,j} = \lambda \left(u_{i+1,j} - 2u_{i,j} + u_{i-1,j} \right) - k f_{i,j}. \tag{3.49}$$

Now, we could simply add the explicit and implicit terms together, but perhaps it would be better to take, say, $\frac{1}{3}$ of the explicit and $\frac{2}{3}$ of the implicit. Trying to keep our options open for the moment, we use a convex combination and do the following

$$
\begin{aligned}
u_{i,j+1} - u_{i,j} &= \theta(u_{i,j+1} - u_{i,j}) + (1-\theta)(u_{i,j+1} - u_{i,j}) \\
&= \theta H_{i,j} + (1-\theta) H_{i,j+1},
\end{aligned} \tag{3.50}
$$

or, more expansively,

$$
\begin{aligned}
u_{i,j+1} - u_{i,j} = (1-\theta)[&\lambda u_{i+1,j+1} - 2\lambda u_{i,j+1} + \lambda u_{i-1,j+1} - k f_{i,j+1}] \\
+ \theta[&\lambda u_{i+1,j} - 2\lambda u_{i,j} + \lambda u_{i-1,j} - k f_{i,j}],
\end{aligned} \tag{3.51}
$$

where $0 \le \theta \le 1$. This is called the θ-method. It is explicit only when $\theta = 1$, in which case it reduces to the explicit method introduced earlier. As for the truncation error, it should be at least $O(h^2) + O(k)$ since the two methods used to produce (3.51) have this as their truncation error. This assumption is correct, but the error is better than expected when $\theta = \frac{1}{2}$. This particular value of θ produces what is known as the Crank–Nicolson method, and the truncation error is $O(h^2) + O(k^2)$. The proof of this statement is presented in the next section.

Before blindly going out and coding the θ-method, it is worth our time to determine its stability properties. The procedure is the same as before, which means we set $f(x,t) = 0$ and assume $u_{i,j} = w_j e^{rx_i I}$. Substituting this into (3.51), and using the identities in Table 3.1, one finds that $w_{j+1} = \kappa w_j$, where the amplification factor is

$$\kappa = \frac{1 - 4\lambda\theta \sin^2\left(\frac{rh}{2}\right)}{1 + 4\lambda(1 - \theta)\sin^2\left(\frac{rh}{2}\right)}. \tag{3.52}$$

From this it follows that $w_j = \kappa^j w_0$. To obtain a bounded solution, the stability requirement is $|\kappa| \leq 1$ and it is not hard to show that this reduces to the single inequality $\lambda(2\theta - 1) \leq \frac{1}{2}$. The conclusion is therefore that if $0 \leq \theta \leq \frac{1}{2}$ then the resulting method is stable. If, however, $\frac{1}{2} < \theta \leq 1$, then the method is conditionally stable and the stability condition is

$$\lambda \leq \frac{1}{2(2\theta - 1)}. \tag{3.53}$$

It is interesting that $\theta = \frac{1}{2}$ separates stable from conditionally stable. This is the value used for the Crank–Nicolson method, and it produces the best truncation error among the θ-methods. Is it a problem that this value happens to be the stability boundary? We will look into this once we work out that the truncation error is actually as good as advertised, which is done in the next section.

As a final comment, from the above conclusions it is not uncommon to assume that the only stable methods are implicit. This is not true, and an example of a stable explicit method can be found in Exercise 3.8.

3.3 Methods Obtained from Numerical Quadrature

In Chapter 1 we found that numerical integration can be used to derive finite difference approximations for IVPs. This can be done with the heat equation in much the same way. To explain how, it is convenient to collect the terms without a time derivative and write the equation as

$$u_t = F(x,t), \tag{3.54}$$

where $F(x,t) = u_{xx} - f(x,t)$. The next step is to integrate the differential equation between two time points. We take t_j and t_{j+1}, and so from (3.54) we have

$$\int_{t_j}^{t_{j+1}} u_t \, dt = \int_{t_j}^{t_{j+1}} F(x,t) dt. \tag{3.55}$$

Using the Fundamental Theorem of Calculus yields

$$u(x, t_{j+1}) - u(x, t_j) = \int_{t_j}^{t_{j+1}} F(x, t)dt. \qquad (3.56)$$

3.3.1 Crank–Nicolson Method

The next choice to make is what integration method to use, and we take the trapezoidal rule (Table 1.4). This gives us

$$
\begin{aligned}
u(x, t_{j+1}) &- u(x, t_j) \\
&= \frac{k}{2}(F(x, t_{j+1}) + F(x, t_j)) + O(k^3) \\
&= \frac{k}{2}(u_{xx}(x, t_{j+1}) - f(x, t_{j+1}) + u_{xx}(x, t_j) - f(x, t_j)) + O(k^3).
\end{aligned}
$$
$$ (3.57) $$

We will replace the u_{xx} terms with centered difference approximations. Before doing this it is of interest to determine the truncation error for the time variable. From (3.57) one might expect that the error is $O(k^3)$. However, this is the error we generate at each time step. To get to T we take $M = T/k$ time steps, so the accumulated error we generate in getting to T is reduced by a factor of k. Therefore, the truncation error for the time approximation is $O(k^2)$.

We need to complete the derivation, and this means we need to take care of the u_{xx} terms in the equation. This will be accomplished by setting $x = x_i$ in (3.57) and then introducing centered difference approximations for the u_{xx} terms. Dropping the truncation error, and rearranging things slightly, one finds that

$$
\begin{aligned}
\lambda u_{i+1,j+1} &- 2(1 + \lambda)u_{i,j+1} + \lambda u_{i-1,j+1} \\
&= -\lambda u_{i+1,j} - 2(1-\lambda)u_{i,j} - \lambda u_{i-1,j} + k(f_{i,j} + f_{i,j+1}) \qquad (3.58)
\end{aligned}
$$
$$
\text{for } \begin{cases} i = 1, 2, \ldots, N, \\ j = 0, 1, \ldots, M - 1. \end{cases}
$$

The boundary and initial conditions are the same as for the explicit method, and are given in (3.23)–(3.25). Collecting everything together, this method can be expressed in matrix form as

$$(\mathbf{B} + \mathbf{I})\mathbf{u}_{j+1} = (\mathbf{A} + \mathbf{I})\mathbf{u}_j - k(\mathbf{f}_{j+1} + \mathbf{f}_j), \quad \text{for } j = 0, 1, 2, \ldots, M - 1, \quad (3.59)$$

where $\mathbf{B}$ is given in (3.45) and $\mathbf{A}$ is given in (3.28).

The difference equation (3.58), or its matrix version in (3.59), is the Crank–Nicolson method, and its truncation error is $O(h^2) + O(k^2)$. It is implicit, and as shown in the previous section, it is stable. The stencil for this method

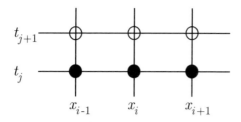

Figure 3.14. Stencil for the Crank–Nicolson method.

is shown in Figure 3.14. In terms of finding numerical solutions to diffusion equations, Crank–Nicolson is probably the most widely used finite difference method. Given its wide popularity it will be interesting to see how well it does on a couple of example problems.

Example
We first see how Crank–Nicolson does when $f(x,t) = 0$ and $g(x) = \sin(2\pi x)$. The exact solution is $u(x,t) = e^{-4\pi^2 t}\sin(2\pi x)$, and this function is plotted in Figure 3.2. Taking $N = 20$ and $T = 0.1$, the numerical solutions at $t = 0.02$, $t = 0.04$, and $t = 0.1$ are shown in Figure 3.15 using $M = 5, 20$. These values were used in Figures 3.6 and 3.12 for the explicit and implicit methods, respectively. In comparison to these other two methods, Crank–Nicolson produces a much more accurate solution. For example, the numerical solution in Figure 3.15 for $M = 20$ is not easily distinguished from the exact solution, something not occurring in Figures 3.6 and 3.12. To further support this conclusion, the error at $t = T$ is plotted in Figure 3.16. The results for k/h fixed, given in the upper plot, show the expected second-order convergence (i.e., increasing M by a factor of 10 decreases the error by a factor of 10^2). In the case N is fixed, which is the situation for the lower plot in Figure 3.16, we would expect the error to decrease as $O(k^2)$ and then plateau at a $O(h^2)$ level. This is seen in Figure 3.16, but there is also the appearance of an optimal M value where the error is better than predicted. For $N = 20$ it appears that $M = 12$ is the best choice, and for $N = 20$ it is $M = 25$. This is not a property of the method but, rather, something that happens for this particular problem. For example, as shown shortly, this does not happen when one uses an initial condition that contains one or more step functions.

That Crank–Nicolson does so much better than either the explicit or implicit method is not unexpected, given its better truncation error. It is, however, not perfect, and this is made evident in the next example.

Example
For a second example we use the step function in (3.8) and also set $f(x,t) = 0$. Taking $a = \frac{1}{4}$ and $b = \frac{3}{4}$, the solution calculated using the Fourier series (3.5) is shown in Figure 3.3. Taking $N = 30$ and $T = 0.1$, the numerical solutions

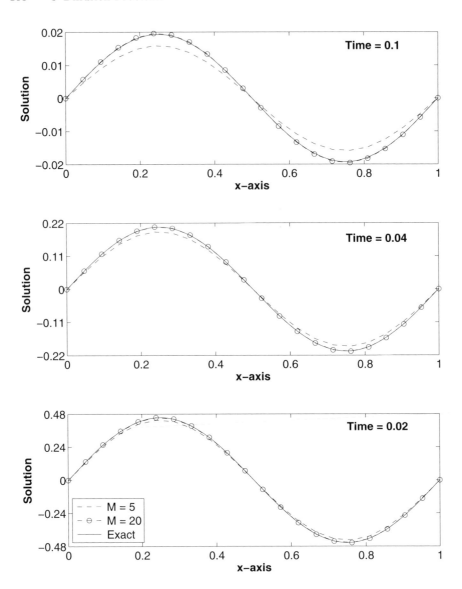

Figure 3.15. (**M**) Solution of the heat equation using the Crank–Nicolson method (3.58) in the case where the initial condition is $u(x, 0) = \sin(2\pi x)$. Also shown is the exact solution.

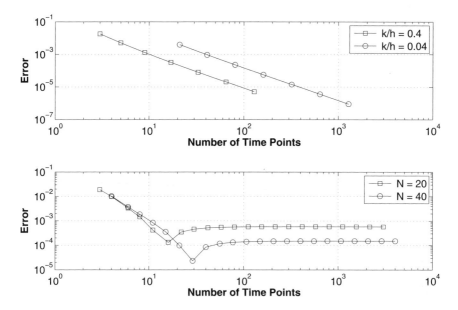

Figure 3.16. Error in the numerical solution at $t = T$ when using the Crank–Nicolson method (3.58) to solve the heat equation in the case where the initial condition is $u(x, 0) = \sin(2\pi x)$. In the lower plot the error is given for two values of N, and in the upper plot the error is given for two values of k/n.

at $t = 0.02$, $t = 0.04$, and $t = 0.1$ are shown in Figure 3.17 using $M = 5, 10, 20$. Clearly, something bad is happening with the numerical solution in a neighborhood of the jumps. Looking closely at these curves, it appears that the problem decreases with larger M. The strength of the jumps also appears to decrease, albeit slowly, as time increases. However, this still leaves the question of what is causing this to happen in the first place, and this is addressed next.

3.3.2 L-Stability

The last example indicates that the favorite finite difference method for solving diffusion problems needs to be thought about a bit more. What causes Crank–Nicolson to behave so badly near the jumps in Figure 3.17? One might argue that the good truncation error is actually a disadvantage for this example. The reason is that the higher the truncation error, the higher the derivative in the error term. In looking at the entries in Table 1.1 it is seen that the higher the power of h, the higher the derivative term in the error. At times this is an issue, and one should use a lower-order method to resolve jumps. However, the heat equation produces only smooth solutions, and as demonstrated in Figure 3.13, the implicit method has no difficulty with this example using the

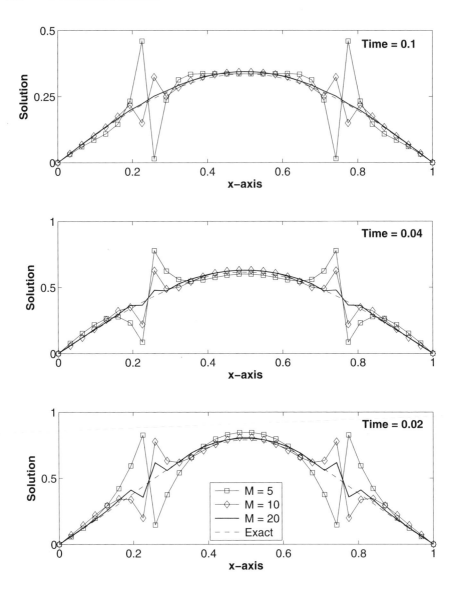

Figure 3.17. (**M**) Solution of the heat equation using the Crank–Nicolson method (3.58) in the case where the initial condition is given in (3.5), with $a = \frac{1}{4}$ and $b = \frac{3}{4}$. Also shown is the exact solution.

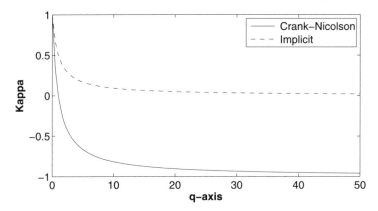

Figure 3.18. Sketch of the amplification factor for the Crank–Nicolson (3.60) and implicit (3.61) methods.

same spatial approximations Crank–Nicolson uses. Therefore, the problem is somewhere else. The place to look is at the stability analysis and the resulting amplification factor. It was shown earlier that for the Crank–Nicolson method the assumption $u_{i,j} = w_j e^{Ir x_i}$ leads to the conclusion that $w_j = w_0 \kappa^j$, where

$$\kappa = \frac{1 - q}{1 + q} \qquad (3.60)$$

and $q = 2\lambda \sin^2\left(\frac{rh}{2}\right)$. The values of κ are shown in Figure 3.18, and it is seen that for large values of q that κ is close to -1. This means that w_j decays with j, but it does so very slowly if q is large. This observation provides a hint as to why the problem in Figure 3.17 occurs, namely, the decay the Crank–Nicolson method is able to produce with jumps is not enough, and a residual of the jumps persists much longer than it should. To explain this in more depth, there are two aspects of this problem that come together to cause the problem. First, the Fourier series for the jumps converges slowly, and this means there are important terms in the series with large $\lambda_n = n\pi/\ell$ values. Given that $r = \lambda_n$ and

$$\frac{rh}{2} = \frac{\pi}{2} \frac{n}{n + 1},$$

then for these terms $\sin^2(rh/2) \approx 1$. Second, to obtain a reasonable numerical approximation of the jumps it is necessary to use a small value of h, and this produces a large value for λ unless k is also very small. Together these two properties cause the value of q to increase, and the consequence of this is that the decay is slower than expected near the jumps. This also explains why the persistence problem is reduced by increasing M in Figure 3.17.

Why didn't this happen with the implicit method, as demonstrated in Figure 3.13? Well, the amplification factor for it, in terms of q, is

$$\kappa = \frac{1}{1+q}. \tag{3.61}$$

In this case, κ decreases monotonically to zero as $q \to \infty$ (see Figure 3.18). The result is that there is no significant persistence in the jumps even if q is quite large. In fact, the larger q becomes, the faster the solution $w_j = w_0 \kappa^j$ decays as time increases.

This situation is what motivates the introduction of a stronger form of stability, something known as L-stability. The requirement is that if the method is stable and $w_j \to 0$ as $q \to \infty$, then the method is L-stable. With this, the implicit method is L-stable but Crank–Nicolson is not.

3.4 Methods of Lines

Another often-used method for solving diffusion equations is to reduce the problem to an IVP. This is done by introducing approximations for the x-derivatives, and then using an IVP method on the resulting problem. To illustrate the procedure we carry out the steps for the inhomogeneous heat equation, where the boundary conditions are $u(0,t) = u(1,t) = 0$. The first step is to evaluate the equation at $x = x_i$, giving us

$$u_{xx}(x_i, t) = u_t(x_i, t) + f(x_i, t). \tag{3.62}$$

Introducing the centered difference approximation for the spatial derivative, we obtain

$$u_t(x_i, t) = \frac{u(x_{i+1}, t) - 2u(x_i, t) + u(x_{i-1}, t)}{h^2} - f(x_i, t) + O(h^2). \tag{3.63}$$

Dropping the truncation error term yields

$$\frac{d}{dt} u_i(t) = \frac{u_{i+1}(t) - 2u_i(t) + u_{i-1}(t)}{h^2} - f_i(t), \tag{3.64}$$

where $u_i(t)$ is the resulting approximation for $u(x_i, t)$ and $f_i(t) = f(x_i, t)$. Collecting the u_i's together, this last equation can be written in vector form as

$$\frac{d}{dt} \mathbf{u}(t) = \mathbf{Cu} - \mathbf{f}(t), \quad \text{for } 0 < t, \tag{3.65}$$

where

$$\mathbf{u}(t) = \begin{pmatrix} u_1(t) \\ u_2(t) \\ \vdots \\ u_N(t) \end{pmatrix}, \tag{3.66}$$

$$
\mathbf{C} = \frac{1}{h^2}
\begin{pmatrix}
-2 & 1 & & & & \\
1 & -2 & 1 & & 0 & \\
& 1 & -2 & 1 & & \\
& & \ddots & \ddots & \ddots & \\
0 & & & 1 & -2 & 1 \\
& & & & 1 & -2
\end{pmatrix},
\quad \text{and} \quad
\mathbf{f}(t) =
\begin{pmatrix}
f_1(t) \\
f_2(t) \\
\vdots \\
f_N(t)
\end{pmatrix}.
\tag{3.67}
$$

The initial condition $u(x,0) = g(x)$ now takes the form

$$
\mathbf{u}(0) =
\begin{pmatrix}
g_1 \\
g_2 \\
\vdots \\
g_N
\end{pmatrix}.
\tag{3.68}
$$

With (3.65) and (3.68) we have a standard IVP, and this opens the door to using a wide variety of IVP solvers. Let's try something different from what we have done so far in this chapter and use RK4 (see Table 1.3). Calling this the L-RK4 method, then it is explicit with truncation error $O(h^2) + O(k^4)$. It is also conditionally stable, but determining the exact stability condition is not so easy with L-RK4, even on a simple heat equation problem. Nevertheless, it is interesting to see how well this method works, and this brings us to the next example.

Example
We solve the heat equation using L-RK4 in the case where the initial condition is $u(x,0) = \sin(2\pi x)$. Taking $N = 18$ and $T = 0.1$ the numerical solution at various time slices is shown in Figure 3.19 using $M = 5, 10, 20$. At the first time slice ($t = 0.02$), L-RK4 demonstrates its ability to describe the solution very accurately compared to the other methods considered in this chapter. It does so well that the three numerical solutions at this time level are indistinguishable from the exact result. However, at $t = 0.1$ it demonstrates its problems with stability. In fact, if you look closely, there is also evidence of instability when $t = 0.04$. To compare this method with Crank–Nicolson, the error at $t = 0.1$ is shown in Figure 3.20 for two values of N. It is seen that once M is large enough, L-RK4 is stable and the error very quickly plateaus to the level determined by the $O(h^2)$ term in the truncation error. This is very good, but the value of M needed to get into the stability region is rather large, and this is one of the criticisms of using L-RK4 on the heat equation.

3.5 Collocation

Another approach to solving the heat equation can be developed using collocation. The first step is to introduce an approximation function, and we will

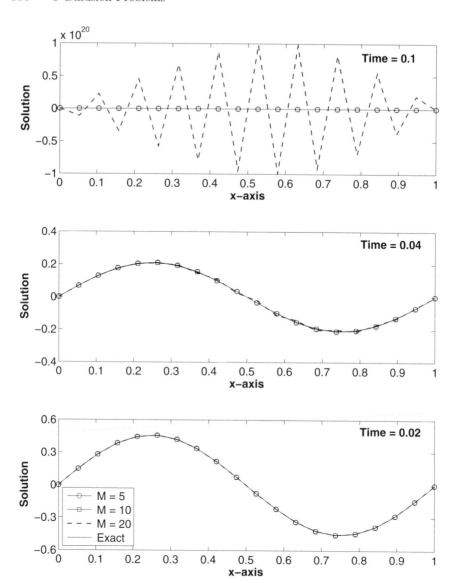

Figure 3.19. Solution of the heat equation using L-RK4, which is the method of lines using RK4, in the case where the initial condition is $u(x, 0) = \sin(2\pi x)$. The method is conditionally stable and problems with stability are evident at $t = 0.1$. For comparison, the exact solution is given at each time level.

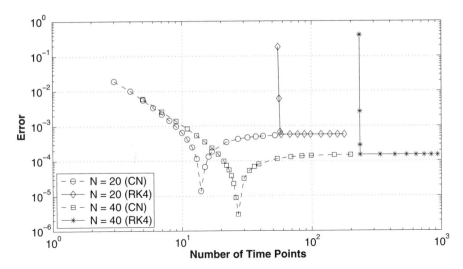

Figure 3.20. Error in the numerical solution at $t = T$ when using the L-RK4 and the Crank–Nicolson methods to solve the heat equation. In the calculation, $u(x,0) = \sin(2\pi x)$.

use B-splines. Adapting the formulas given in the last chapter, the solution of the heat equation is approximated with the function

$$U(x,t) = \sum_{k=-1}^{N+2} a_k(t)B_k(x), \quad \text{for } 0 \le x \le \ell, \tag{3.69}$$

where $B_k(x)$ is the cubic B-spline centered at $x = x_k$. The formula for B_k, and its properties, are given in Section 2.3.1. As illustrated in Figure 2.8, two properties that are particularly notable are that B_k has a continuous second derivative and that it is zero everywhere except if $x_{k-2} < x < x_{k+2}$. In fact, the only difference between what was done earlier for BVPs and (3.69) is that the $N + 4$ coefficients a_k now depend on t.

As usual in collocation, the approximation function in (3.69) is required to satisfy the heat equation at $N + 2$ points selected from the spatial interval, and it must also satisfy the two boundary conditions. The simplest choice is to select the collocation points to be the same as the grid points used in the construction of the B-splines. Also, to simplify things a bit it is assumed that $f = 0$. In this case, substituting (3.69) into the heat equation (3.10) and then evaluating the result at $x = x_i$ gives us the following

$$\sum_{k=-1}^{N+2} a'_k(t)B_k(x_i) = \sum_{k=-1}^{N+2} a_k(t)B_k''(x_i), \quad \text{for } i = 0, 1, 2, \ldots, N+1. \tag{3.70}$$

Using the properties of B-splines (see Table 2.2), this can be written as

$$a'_{i-1}(t) + 4a'_i(t) + a'_{i+1}(t) = \frac{6}{h^2}(a_{i-1}(t) - 2a_i(t) + a_{i+1}(t)),$$

$$\text{for } i = 0, 1, 2, \ldots, N+1. \tag{3.71}$$

From the boundary condition at $x = 0$ one finds that $a_{-1} = -4a_0 - a_1$. With this, and the $i = 0$ equation in (3.71), it follows that $a_0 = 0$ and $a_{-1} = -a_1$. Similarly, at $x = \ell$ one finds that $a_{N+1} = 0$ and $a_{N+2} = -a_N$.

To summarize what we have done, the problem that needs to be solved to find the coefficients in (3.69) is

$$\mathbf{A a'} = \mathbf{B a}, \tag{3.72}$$

where $\mathbf{A}$ and $\mathbf{B}$ are symmetric tridiagonal $N \times N$ matrices given as

$$\mathbf{A} = \begin{pmatrix} 4 & 1 & & & & \\ 1 & 4 & 1 & & 0 & \\ & 1 & 4 & 1 & & \\ & & \ddots & \ddots & \ddots & \\ 0 & & & 1 & 4 & 1 \\ & & & & 1 & 4 \end{pmatrix}, \quad \mathbf{B} = \frac{6}{h^2} \begin{pmatrix} -2 & 1 & & & & \\ 1 & -2 & 1 & & 0 & \\ & 1 & -2 & 1 & & \\ & & \ddots & \ddots & \ddots & \\ 0 & & & 1 & -2 & 1 \\ & & & & 1 & -2 \end{pmatrix}. \tag{3.73}$$

Also, $\mathbf{a} = (a_1(t), a_2(t), \ldots, a_N(t))^T$. From the initial condition it is required that $U(x_i, 0) = g(x_i)$, and so the initial condition for the collocation approximation is

$$\mathbf{A a}(0) = 6\mathbf{g}, \tag{3.74}$$

where $\mathbf{g} = (g(x_1), g(x_2), \ldots, g(x_N))^T$.

The problem that remains is to solve the IVP consisting of (3.72) and (3.74). We will, of course, solve this problem numerically. Using either forward or backward Euler would give us a method similar to the explicit and implicit methods studied earlier. It is more interesting to try something that can compete with Crank–Nicolson, which means we need a method with truncation error $O(k^2)$ and that is stable. The natural choice is the trapezoidal method. In regard to the formula for the trapezoidal method in Table 1.3, for this problem $\mathbf{f} = \mathbf{A}^{-1}\mathbf{B a}$. From this it follows that the finite difference equation to solve (3.72) is

T=40, Re=1000,   N=223481

0   10   20   30

Figure 3.21. Cubic B-splines are used to solve the diffusion problem that arises in the study of vortex transport. This figure shows the formation of eddies in the flow of fluid as it moves through a small narrow chamber (Kudela [1999]).

$$\left(\mathbf{A} - \frac{k}{2}\mathbf{B}\right)\mathbf{a}_{j+1} = \left(\mathbf{A} + \frac{k}{2}\mathbf{B}\right)\mathbf{a}_j, \quad \text{for } j = 0, 1, 2, \dots, M-1, \qquad (3.75)$$

where $\mathbf{Aa}_0 = 6\mathbf{g}$. At first sight this method might look to be more complicated than some of the other implicit methods we have used to solve the heat equation, but it really isn't. The equation in (3.75) involves tridiagonal matrices, so the computational effort is comparable to Crank–Nicolson. It does, however, have the somewhat unusual feature of requiring us to solve an equation to satisfy the initial condition. Anyway, let's take it for a test drive to see how well it does.

Example
Using one of our standard examples, we consider the case where $g(x)$ is the function given in (3.8). Taking $a = \frac{1}{4}$ and $b = \frac{3}{4}$, then the solution (3.5) is shown in Figure 3.3. We will determine the error using the B-spline method at time level $t = 0.1$. As discussed earlier, this depends on the number of points used in the space and time directions, and in Figure 3.22 it is plotted as a function of the number of time points used in the calculation. The B-spline curve behaves as expected in the sense that it drops like $O(k^2)$ and then levels off at a level determined by the $O(h^2)$ term in the error. Also, when the spatial points are doubled, the $O(h^2)$ component drops by about a factor of 4, as it should. From this calculation it appears that Crank–Nicolson does slightly better than the B-spline method with this problem, but they are comparable in terms of accuracy and convergence.

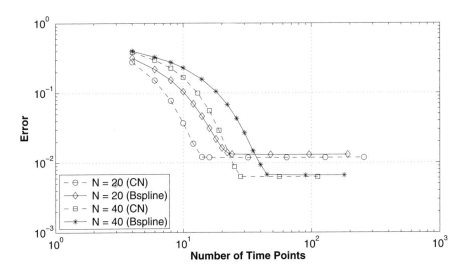

Figure 3.22. Error in the numerical solution at $t = 0.1$ using the B-spline and Crank–Nicolson methods to solve the heat equation in the case where the initial condition is given in (3.8), with $a = \frac{1}{4}$ and $b = \frac{3}{4}$.

3.6 Next Steps

The centerpiece for the theory of convergence of linear difference approximations of time-dependent partial differential equations is the Lax Equivalence
Theorem. To state this result suppose that to solve the heat equation prob-
lem (3.10)–(3.12), an approximation of the form $\mathbf{A}\mathbf{u}_{j+1} = \mathbf{B}\mathbf{u}_j + \mathbf{f}_j$ is used.
In this expression $\mathbf{u}_j$ is given in (3.26) and $\mathbf{f}_j$ contains the contributions of
the boundary conditions and forcing function. Using ideas from functional
analysis it is possible to establish the next result.

Theorem 3.1. *(Lax Equivalence Theorem) The finite difference approximation converges if and only if it is consistent and stable.*

A proof of this, along with an introduction to some of the background material necessary for the proof, can be found in Atkinson and Han [2005] and
Richtmyer and Morton [1994]. This result resembles the Dahlquist Equivalence
Theorem for IVPs as it reaffirms the necessity, and sufficiency, of stability and
consistency for convergence. What is different is the specific form for stability.
In both cases stability is a requirement of boundedness, but how this is expressed differs in the formulation of the two types of problems. With stability
playing such a prominent role one might ask how it arises when using collocation. Combining a finite difference approximation, in time, with a function
approximation, in space, requires a special set of functional analytic tools to
analyze stability. This is beyond the scope of this text and an introduction to
this can be found in Prenter [1989].

In terms of applications one of the more important generalizations of (3.1)
concerns problems involving multiple spatial dimensions. The complications
in such situations center on handling large matrix equations and how to develop approximations for complicated boundaries. These topics are taken up
in Chapter 6.

Another generalization that is of interest concerns nonlinear diffusion problems. Two well-studied examples are Burger's equation $u_t + u u_x = u_{xx}$ and the
porous media equation $\partial_x(D(u)u_x) = u_t$. The differences in these two equations are a small hint of the differences in their solutions, which are substantial.
Nevertheless, a few comments can be made. We found that stable methods
for the heat equation are implicit whereas the explicit methods are, at best,
conditionally stable. It is not unreasonable to assume this also holds for nonlinear heat equations. The computational difficulty with nonlinear problems
is that an implicit method does not result in a simple matrix equation but in
a nonlinear problem for $\mathbf{u}_j$ that must be solved at each time step. An explicit
method does not suffer this complication and this makes them a bit more
attractive than in the linear case, although they can still suffer from a severe step size requirement to achieve stability. What this means is that some
thought must be invested into deciding what method to use and how to solve
the resulting equation effectively. Another question is how to determine if a
method is stable as the methods normally used for linear problems are not

easily extendable to nonlinear problems. In such cases the usual approach is to use an energy method. The idea is to introduce an energy function in such a way that stability corresponds to the difference approximation keeping the energy bounded. The analysis can become quite involved and those interested in investigating this might start with Iserles [1996].

Exercises

3.1. Suppose it is proposed to use $u_{i,j+1} = \lambda u_{i+1,j} + (1 - 2\lambda - \beta k)u_{ij} + \lambda u_{i-1,j}$ to solve the heat equation (3.2) with $D = 1$. Assume that β is nonnegative.
(a) Is the method explicit or implicit?
(b) Is the method stable (or conditionally stable)?
(c) For what values of β, if any, is the method consistent?

3.2. This problem explores a consequence of consistency.
(a) If the constant function $u_{i,j} \equiv 1$ does not satisfy the finite difference approximation of the heat equation (3.10), then the approximation is not consistent. Explain why. Also explain why $u_{i,j} \equiv 1$ satisfying the finite difference equation does not necessarily mean that the approximation is consistent.
(b) Can the following be used to solve the heat equation?
   (i) $u_{i,j+1} = \lambda u_{i+1,j} + (1 - 2\lambda - k)u_{ij} + \lambda u_{i-1,j}$.
   (ii) $\lambda u_{i+1,j+1} - 2(1+\lambda)u_{ij+1} + \lambda u_{i-1,j+1} = -\lambda u_{i+1,j} - 2(1+\lambda)u_{ij} - \lambda u_{i-1,j}$.
(c) Will the observation in (a) work on the equation $Du_{xx} = u_t$, where D is a positive constant? What about $Du_{xx} = u_t + u$?

3.3. Assuming that the finite difference equation for solving the heat equation is linear (i.e., it depends linearly on the u_{ij}'s), answer the following.
(a) If the stencil is given in Figure 3.5 does the method have to be explicit?
(b) If the stencil is given in Figure 3.11 does the method have to be implicit?

3.4. This problem concerns the error curves in Figure 3.10.
(a) In regard to the lower graph, without solving the problem, sketch the error curve for $N = 80$. Explain your reasoning for the curve drawn.
(b) In regard to the upper graph, without solving the problem, sketch the error curve for $\lambda = 0.1225$. Explain your reasoning for the curve drawn.

3.5. This problem concerns the error curves in Figure 3.16.
(a) In regard to the lower graph, without solving the problem, sketch the error curve for $N = 80$. Explain your reasoning for the curve drawn.
(b) In regard to the upper graph, without solving the problem, sketch the error curve for $\frac{k}{h} = 0.004$. Explain your reasoning for the curve drawn.

3.6. Use centered difference approximations in space and time to derive a finite difference approximation of the heat equation (3.10) with truncation error that is $O(h^2) + O(k^2)$. Is the method explicit or implicit? Is the method stable (or conditionally stable)?

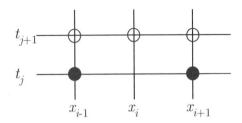

Figure 3.23. Stencil for solving the heat equation.

3.7. Consider the heat equation (3.10), with initial condition (3.12), and boundary conditions $u_x(0,t) = \alpha$ and $u(\ell,t) = 0$.

(a) Derive an explicit finite difference approximation to this problem that has truncation error $O(h^2) + O(k)$. Express the result in matrix form.

(b) Derive a finite difference approximation to this problem that has truncation error $O(h^2) + O(k^2)$. Express the result in matrix form.

3.8. The Dufort–Frankel method for solving the heat equation (3.10) is

$$\frac{u_{i+1,j} - u_{i,j+1} - u_{i,j-1} + u_{i-1,j}}{h^2} = \frac{u_{i,j+1} - u_{i,j-1}}{2k}.$$

(a) What is the stencil and what are the limits on i, j for this method?

(b) Show that the truncation error is $O(k^2) + O(h^2) + O(k^2/h^2)$. If you are unfamiliar with how to do this you might want to look at Exercise 3.19(a). Explain why this result shows that the method is only conditionally consistent.

(c) Is the method explicit or implicit?

(d) Show that the method is stable.

(e) Show that the method is not L-stable.

(f) Explain how to derive this method directly from the heat equation.

3.9. Consider the stencil shown in Figure 3.23.

(a) Derive a consistent finite difference approximation of the heat equation (3.10) that has this as its stencil.

(b) What is the truncation error?

(c) Is the method explicit or implicit?

(d) Is the method stable (or conditionally stable)?

3.10. Consider the stencil shown in Figure 3.24.

(a) Derive a consistent finite difference approximation of the heat equation (3.10) that has this as its stencil.

(b) What is the truncation error?

(c) Is the method explicit or implicit?

(d) Is the method stable (or conditionally stable)?

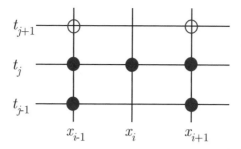

Figure 3.24. Stencil for solving the heat equation.

3.11. This problem considers whether there is any point to introducing the θ-method.
(a) For what values of θ is the θ-method L-stable?
(b) Assuming $0 \le \theta < \frac{1}{2}$, is there a value θ that is best? For example, is there any reason it would be better to take, say, $\theta = 0$ rather than $\theta = \frac{1}{10}$? Your answer should address stability, convergence, computational complexity, etc.
(c) Assuming $\frac{1}{2} \le \theta \le 1$, is there a value θ that is best?
(d) If you were writing a textbook chapter on solving the heat equation would you include a separate section on the θ-method? Explain your reasoning.

3.12. Using the numerical quadrature approach, derive the method stated below. Make sure to keep track of the truncation error in the derivation.
(a) The explicit method (3.22).
(b) The implicit method (3.43).
(c) The θ-method (3.51).

3.13. Using the numerical quadrature approach, and integrating from t_{j-1} to t_{j+1}, determine what finite difference approximation is obtained from (3.10) when using the stated integration rule. Make sure to keep track of the truncation error in the derivation. Also, determine whether the resulting method is stable.
(a) Simpson's rule.
(b) Midpoint rule.

3.14. This problem concerns the diffusion equation

$$Du_{xx} = u_t + bu, \quad \text{for} \quad \begin{cases} 0 < x < 1, \\ 0 < t, \end{cases}$$

where $u(0,t) = u(1,t) = 0$ and $u(x,0) = g(x)$. Also, D, b are positive constants.
(a) Show that the exact solution is

$$u(x,t) = \sum_{n=1}^{\infty} A_n e^{-(b+D\lambda_n^2)t} \sin(\lambda_n x),$$

where $\lambda_n = n\pi$ and $A_n = 2\int_0^1 g(x)\sin(\lambda_n x)dx$.

(b) What is the solution if $g(x) = \sin(m\pi x)$ where m is a positive integer?

(c) Derive an implicit finite difference approximation to this problem that has truncation error $O(h^2) + O(k)$. Express the result in matrix form.

(d) Determine if the method in (c) is stable. If it is conditionally stable make sure to state the stability condition.

(e) Assuming $g(x) = \sin(\pi x)$, $D = \frac{1}{10}$, $b = 1$, and $T = 1$, use the method in (c) to calculate the solution at $t = 1$ using $M = 4, 8, 16$. On the same axes plot the exact solution at $t = 1$ and the three numerical solutions. Make sure to state how you decide on how many points to use along the x-axis.

(f) Assuming $g(x) = \sin(\pi x)$, $D = \frac{1}{10}$, $b = 1$, plot the maximum error at $t = 1$ as a function of M for $M = 4, 8, 16, 32$ and $N = 10$. On the same axes also plot the maximum error at $t = 1$ for $M = 4, 8, 16, 32$ and $N = 20$. Explain the behavior of these two curves using the stated truncation error.

(g) Explain why the exact solution is bounded as $t \to \infty$, irrespective of the initial condition, as long as $b \geq -D\pi^2$. Is the method stable if $-D\pi^2 < b < 0$, where D is still assumed positive?

3.15. This problem concerns the advection–diffusion equation

$$Du_{xx} - au_x = u_t, \quad \text{for} \quad \begin{cases} 0 < x < 1, \\ 0 < t, \end{cases}$$

where $u(0,t) = u(1,t) = 0$ and $u(x,0) = g(x)$. Also, D, a are constant and D is positive.

(a) Show that the exact solution is

$$u(x,t) = \sum_{n=1}^{\infty} A_n e^{-\alpha_n t + \beta x} \sin(\lambda_n x),$$

where $A_n = 2\int_0^1 g(x)e^{-\beta x}\sin(\lambda_n x)dx$, $\lambda_n = n\pi$, $\alpha_n = D(\lambda_n^2 + \beta^2)$, and $\beta = \frac{a}{2D}$.

(b) Show that the function

$$u(x,t) = \frac{1}{\sqrt{1+4cDt}} e^{-c(x-at-b)^2/(1+4cDt)}$$

satisfies the differential equation, assuming that b, c are constant with c nonnegative. What initial and boundary conditions does this solution satisfy?

(c) Derive an implicit finite difference approximation to this problem that has truncation error $O(h^2) + O(k^2)$. Express the result in matrix form.

(d) Determine if the method in (c) is stable. If it is conditionally stable make sure to state the stability condition.

(e) Assume $g(x) = e^{-c(x-b)^2}$, where $c = 100$, $b = \frac{1}{3}$, $D = 0.01$, $a = 1$, and $T = 0.2$. Use the method in (c) to calculate the solution at $t = 0.2$ using $M = 4, 8, 16$. On the same axes plot the exact solution at $t = 0.2$ and the three numerical solutions. Make sure to state how you decide on how many points to use along the x-axis. Also, you can use the function from (b) in place of the exact solution but you need to explain why this is possible.

(f) Assuming $g(x) = e^{-c(x-b)^2}$, $c = 100$, $b = \frac{1}{3}$, $D = 0.01$, $a = 1$, and $T = 0.2$, plot the maximum error at $t = 0.2$ as a function of M for $M = 4, 8, 16, 32, 64, 128$ and $N = 30$. On the same axes also plot the maximum error at $t = 0.2$ for $M = 4, 8, 16, 32, 64, 128$ and $N = 60$. Explain the behavior of these two curves using the stated truncation error.

3.16. This problem concerns the advection–diffusion equation

$$u_{xx} + \mu(x)u_x = u_t, \quad \text{for} \quad \begin{cases} 0 < x < \ell, \\ 0 < t, \end{cases}$$

where $u(0, t) = u(\ell, t) = 0$ and $u(x, 0) = g(x)$.

(a) Derive an explicit finite difference approximation of this problem that has truncation error $O(h^2) + O(k)$. Express the result in matrix form.

(b) Assuming that μ is a positive constant, find the amplification factor κ obtained from the stability analysis for the method in (a). Use this to determine if the method is stable when μ is a positive constant (as a suggestion, you might see if $|\kappa|^2$ can be written as a quadratic function of s, where $s = \cos(rh)$). If it is conditionally stable make sure to state the stability condition.

(c) Suppose $\mu(x) = 1 + \cos(2\pi x)$, $\ell = 1$, and $g(x) = x(1 - x)^2$. Use the method in (a) to calculate and then plot the solution at $t = 1$ using $M = 60, 70, 80$ and $N = 20$. Explain the results using information from (a) and (b).

(d) The exact solution for the problem in (c) is unknown. Even so, give a convincing argument for why your computed solutions are correct. Your answer should address both the correctness of the algorithm and the computer program itself.

3.17. This problem concerns the diffusion equation

$$u_{xx} = u_t + b(x)u, \quad \text{for} \quad \begin{cases} 0 < x < 1, \\ 0 < t, \end{cases}$$

where $u(0, t) = h(t)$, $u(1, t) = 0$, and $u(x, 0) = 0$.

(a) Derive an implicit finite difference approximation of this problem that has truncation error $O(h^2) + O(k^2)$. Express the result in matrix form.

(b) Determine if the method in (a) is stable in the case where b is a positive constant. If it is conditionally stable make sure to state the stability condition.

(c) Suppose $h(t) = \sin(3\pi t)$ and $b(x) = x$. Use the method in (a) to plot the solution for $0 \le x \le 1, 0 \le t \le 1$. Make sure to pick a step size that produces a reasonably smooth surface. Does your result from part (b) help here?

(d) The exact solution for the problem in (c) is unknown. Even so give a convincing argument for why your computed solutions are correct. Your answer should address both the correctness of the algorithm and the computer program itself.

3.18. This problem concerns the diffusion equation

$$\frac{\partial}{\partial x}\left(D(x)\frac{\partial u}{\partial x}\right) = \frac{\partial u}{\partial t}, \quad \text{for} \quad \begin{cases} 0 < x < \ell \\ 0 < t, \end{cases}$$

where $u(0,t) = u(\ell, t) = 0$ and $u(x,0) = g(x)$.

(a) Derive an implicit finite difference approximation to this problem that has truncation error $O(h^2) + O(k)$. Express the result in matrix form.

(b) Determine if the method in (a) is stable in the case where D is a positive constant. If it is conditionally stable make sure to state the stability condition.

(c) Suppose $D(x) = \frac{1}{100}(2 + \sin(8\pi x))$, $\ell = 1$, and $g(x)$ is given in (3.8) with $a = 0.5$ and $b = 0.75$. Use the method from (a) to plot the solution for $0 \le x \le 1$, $0 \le t \le 1$. Make sure to pick a step size that produces a reasonably smooth surface. Does your result from part (b) help here?

3.19. This problem examines the better-than-expected error seen in Figure 3.10 for the explicit method.

(a) The truncation error can be determined by substituting the exact solution into the finite difference equation (3.22) and then using Taylor's theorem to express terms evaluated at $(x_{i\pm1}, t_j)$ or (x_i, t_{j+1}) in terms of the solution at (x_i, t_j). Doing this, show that the first term in the truncation error is $\tau_{i,j} = \frac{k}{2}u_{tt}(x_i, t_j) - \frac{h^2}{12}u_{xxxx}(x_i, t_j)$, which is the expected $O(h^2) + O(k)$ result.

(b) Show that by taking $6k = h^2$ the truncation error term in (a) is zero.

(c) If $6k = h^2$ then by examining the higher-order terms in the Taylor expansion show that $u(x_i, t_j) - u_{i,j} = O(h^4)$.

(d) Is it possible for this better-than-expected situation to arise with the θ-method?

(e) Because $\lambda = \frac{k}{h^2}$, the error for the explicit method can be written as $O(\frac{k}{\lambda}) + O(k)$. Consequently, for a given value of k, decreasing λ should increase the error. This does not happen in Figure 3.10. Why?

3.20. Another approach for determining stability is based on a perturbation argument. It starts with the homogeneous version of the problem, so $u = 0$ is the solution. With this, suppose $u_{i,j} = \epsilon$ at one grid point, where ϵ is small but nonzero. It is assumed that the solution is zero at all other grid points at this time level and earlier. However, it is not necessarily zero at $t = t_{j+1}, t_{j+2}, t_{j+3}, \ldots$, and the question is what happens to the solution at these later time levels. Answer this question for the explicit method by finding the solutions at $t = t_{j+1}, t_{j+2}, t_{j+3}$. Given that the stability requirement is that the solution not grow with j, what is your conclusion about the stability of the explicit method using the perturbation method?

3.21. Let L-Euler designate the method of lines when the Euler method is used to solve (3.65).
(a) In the case $f(x,t) = 0$ show that L-Euler reduces to an equation of the form $\mathbf{u}_{j+1} = \mathbf{A}\mathbf{u}_j$. What is the truncation error for L-Euler?
(b) Write the method as a finite difference equation involving the $u_{i,j}$'s. What is the stencil for L-Euler? How does this method differ from the explicit method?

3.22. Let L-Trap designate the method of lines when the trapezoidal method is used to solve (3.65).
(a) In the case $f(x,t) = 0$ show that L-Trap reduces to an equation of the form $\mathbf{A}\mathbf{u}_{j+1} = \mathbf{B}\mathbf{u}_j$. What is the truncation error for L-Trap?
(b) Write the method as a finite difference equation involving the $u_{i,j}$'s. What is the stencil for L-Trap? How does this method differ from the Crank–Nicolson method?

3.23. This exercise examines what happens to the method of lines if the problem is changed in various ways.
(a) Suppose the boundary conditions are changed to $u_x(0,t) = 0$ and $u(1,t) = t$. How is (3.65) affected? Your spatial truncation error should be $O(h^2)$.
(b) Suppose the problem is $u_t = u_{xx} + e^{-t}u_x$, where $u(0,t) = u(1,t) = 0$. How is (3.65) affected? Your spatial truncation error should be $O(h^2)$.
(c) Suppose the problem is $u_t = (Du_x)_x - u$, where $u(0,t) = u(1,t) = 0$ and $D = D(x)$ is a smooth positive function. How is (3.65) affected? Your spatial truncation error should be $O(h^2)$.

3.24. This exercise examines what happens to the collocation method using cubic B-splines if the problem is changed in various ways.
(a) Suppose the boundary conditions are changed to $u_x(0,t) = 0$ and $u(1,t) = t$. How is (3.72) affected? Your spatial truncation error should be $O(h^2)$.
(b) Suppose the problem is $u_t = u_{xx} + e^{-t}u_x$, where $u(0,t) = u(1,t) = 0$. How is (3.72) affected?
(c) Suppose the problem is $u_t = (Du_x)_x - u$, where $u(0,t) = u(1,t) = 0$ and $D = D(x)$ is a smooth positive function. How is (3.72) affected?
(d) Assuming that backward Euler is used to solve (3.72), how is (3.75) changed?

(e) Assuming that RK4 is used to solve (3.72), how is (3.75) changed?

3.25. Consider the nonlinear diffusion problem

$$u_{xx} = \frac{\partial}{\partial t}(u + u^3), \quad \text{for} \quad \begin{cases} 0 < x < 1, \\ 0 < t, \end{cases}$$

where

$$u(0,t) = \sin(t), \quad u(1,t) = 0, \quad \text{and} \quad u(x,0) = 0.$$

Derive a finite difference approximation for this equation that has a truncation error of $O(h^2) + O(k^2)$ and that uses values of the solution only at time levels t_j and t_{j+1}. Make sure to identify how the boundary and initial conditions are used in your approximation.

3.26. This problem investigates how fast the strength of the heat signal moves along the x-axis. The spatial region is assumed semi-infinite, so $\ell = \infty$ in (3.2). The imposed conditions are $u(x,0) = 0$, $u(0,t) = 1$, and $u(x,t) \to 0$ as $x \to \infty$.

(a) The solution has the form $u(x,t) = v(\eta)$, where $\eta = \frac{x}{\sqrt{t}}$. Rewrite the heat equation along with the boundary and initial conditions in terms of $v(\eta)$. You should find that $v(\eta)$ satisfies a BVP on a semi-infinite interval.

(b) Solve the BVP from (a) for $v(\eta)$ and express your answer in terms of the complementary error function, defined as

$$\text{erfc}(r) = \frac{2}{\sqrt{\pi}} \int_r^\infty e^{-s^2} ds.$$

(c) Show that the solution has the instant messaging property. In particular, even though the solution is zero at $t = 0$, it is nonzero for all x once $t > 0$.

(d) Taking $D = 1$, plot $v(\eta)$ for $0 \le \eta \le 10$. Also, evaluate $v(1)$, $v(3)$, and $v(10)$. Explain why this demonstrates that the strength of the signal far from the left end is rather weak and stays that way for some time.

(e) Suppose one introduces a threshold condition that states that at whatever x location is chosen, the solution must reach a particular level before it is considered to have effectively reached that location. If the threshold value is $u(x,t) = 0.31731050786291410283$, find t in terms of x for this to hold. For the mildly curious, this value is connected with the stability condition for the explicit method.

4

Advection Equation

4.1 Introduction

We now begin the study of numerical wave propagation. Everyone has experience with traveling waves, whether it is waves on a lake, sound waves, or perhaps an earthquake or two. It is not particularly difficult to write down a reasonable-looking finite difference approximation to a wave equation. Most of the effort is invested in trying to determine whether the method actually works. There are unique complications for numerical wave propagation, and so to introduce the ideas we use one of the simplest mathematical equations that produces traveling waves. This is the advection equation, given as

$$\frac{\partial u}{\partial t} + a\frac{\partial u}{\partial x} = 0, \quad \text{for} \quad \begin{cases} -\infty < x < \infty, \\ 0 < t, \end{cases} \tag{4.1}$$

where $u(x,0) = g(x)$. It is assumed that a is a positive constant.

The mathematical simplicity of (4.1) should not belie that fact that it plays an important role in a wide range of applications. For example, the momentum equation for the motion of a gas is an inhomogeneous version of (4.1) except that a depends on u. This is why a nonlinear version of the advection equation arises in the mathematical model of interstellar gas and its motion due to the solar wind (Figure 4.1). Another application arises in the study of the movement of traffic along a highway (Figure 4.1). In this case $u(x,t)$ represents the density of cars along the road and a is their speed.

4.1.1 Method of Characteristics

The solution of the advection equation can be found if one notes that it can be written as

$$\left(\frac{\partial}{\partial t} + a\frac{\partial}{\partial x}\right)u = 0. \tag{4.2}$$

The idea is to transform x, t to new variables r, s in such a way that the derivatives transform as

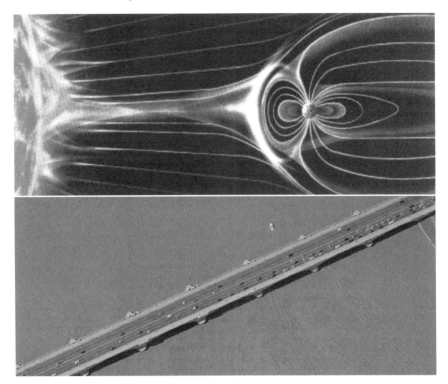

Figure 4.1. A nonlinear advection problem describes the Earth's bow shock associated with the solar wind (upper figure) as well as the flow of traffic along a highway (lower figure).

$$\frac{\partial}{\partial r} = \frac{\partial}{\partial t} + a\frac{\partial}{\partial x}. \tag{4.3}$$

If this is possible then (4.2) becomes $\frac{\partial u}{\partial r} = 0$, and this equation is very easy to solve. With this goal in mind let $x = x(r, s)$, $t = t(r, s)$, in which case using the chain rule, the r-derivative transforms as

$$\frac{\partial}{\partial r} = \frac{\partial x}{\partial r}\frac{\partial}{\partial x} + \frac{\partial t}{\partial r}\frac{\partial}{\partial t}. \tag{4.4}$$

Comparing this with (4.2), we require $\frac{\partial x}{\partial r} = a$ and $\frac{\partial t}{\partial r} = 1$. Integrating these equations yields $x = ar + q(s)$ and $t = r + p(s)$. To determine the s dependence recall that the initial condition specifies the solution along the x-axis. To make it easy to apply the initial condition in the transformed coordinates we ask that the x-axis $(t = 0)$ transform onto the s-axis $(r = 0)$. Specifically, we require that $r = 0$ imply that $t = 0$ and $x = s$. Setting $r = 0$ and $t = 0$, we conclude $q(s) = s$ and $p(s) = 0$, and so the change of variables we are looking for is

$$x = ar + s, \quad t = r. \tag{4.5}$$

Inverting this transformation, one obtains $r = t$ and $s = x - at$. We are now able to write (4.1) as $\frac{\partial u}{\partial r} = 0$, which means that $u = u(s) = u(x - at)$. With the initial condition we therefore conclude that the solution of the advection problem is

$$u(x, t) = g(x - at). \tag{4.6}$$

This is a traveling wave that moves with speed a. It is also seen that the solution is constant along lines of the form $x - at = constant$. These lines are called characteristics for the equation and the method we used to find the solution is known as the method of characteristics.

Example
Suppose the initial condition is the square bump shown in Figure 4.2. In mathematical terms,

$$g(x) = \begin{cases} 1 & \text{if } 0 \le x \le 1, \\ 0 & \text{otherwise}. \end{cases} \tag{4.7}$$

From (4.6) the solution is

$$u(x, t) = \begin{cases} 1 & \text{if } 0 \le x - at \le 1, \\ 0 & \text{otherwise}, \end{cases}$$

or equivalently,

$$u(x, t) = \begin{cases} 1 & \text{if } at \le x \le 1 + at, \\ 0 & \text{otherwise}. \end{cases} \tag{4.8}$$

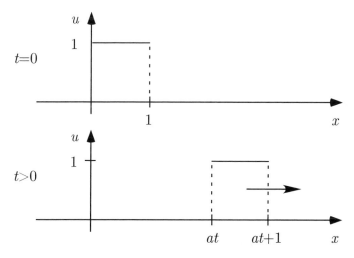

Figure 4.2. (M) Solution of the advection equation. The top figure is the initial condition, as given in (4.7). The bottom figure is the solution at a later time, as given in (4.8).

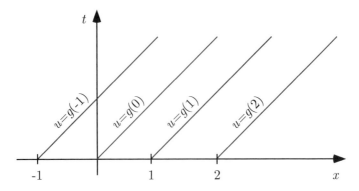

Figure 4.3. The solution of the advection equation, as given in (4.6), is constant along the characteristic lines $x - at = constant$. The value on each line is determined by the initial condition.

A typical solution profile is also shown in Figure 4.2, and it is apparent that at any given time t, the solution is simply the original square bump that has moved over to occupy the interval $at \leq x \leq 1 + at$. A different perspective of the solution is given in Figure 4.3, which shows a few of the characteristics $x - at = constant$ along with the value of the solution on each line. As can be seen, the solution is constant along each of these curves, and its value is determined by where the characteristic intersects the x-axis.

4.1.2 Solution Properties

The solution in Figure 4.2 is typical of what is seen in wave problems, and a few observations and comments that are helpful when finding the numerical solution appear below.

1. The initial shape is preserved, jumps and all. This is in marked contrast to the heat equation, where the solutions decay and any corners or jumps in the initial condition are immediately smoothed out. Because of this, one might question whether (4.8) is actually a solution since u_x is not defined at the jumps located at $x = at, 1 + at$. The short answer is that everything is fine. What is necessary is to introduce the concept of a weak solution, and the interested reader is referred to Evans [2002] for an extended discussion of this subject.
2. Information travels at speed a. This is discussed in detail later and it differs sharply with the heat equation, where information is instantly transmitted throughout the entire spatial domain. Also, waves travel in only one direction in this problem, and for this reason (4.1) is sometimes called a one-way wave equation.
3. The solution at a given spatial position $x = \bar{x}$ and time $t = \bar{t}$ is determined entirely by the value of the initial condition at $x = \bar{x} - a\bar{t}$. This situation

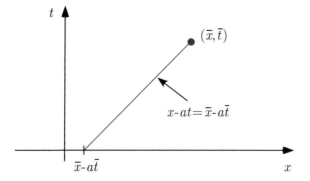

Figure 4.4. Domain of dependence. The solution at $(\bar{x}, \bar{t})$ is $u = g(\bar{x} - a\bar{t})$. Consequently, it is determined by the value of the initial condition at $x = \bar{x} - a\bar{t}$. For this reason $(\bar{x} - a\bar{t}, 0)$ is the domain of dependence for $u(\bar{x}, \bar{t})$.

is indicated schematically in Figure 4.4. This turns out to be a very important property for wave problems and it is the reason for introducing what is known as the domain of dependence. The domain of dependence for the solution at $(\bar{x}, \bar{t})$ consists of all the points on the x-axis ($t = 0$) that contribute to the solution at $(\bar{x}, \bar{t})$. For the advection equation, the domain of dependence for $u(\bar{x}, \bar{t})$ is the single point $(\bar{x} - a\bar{t}, 0)$.

4.1.3 Boundary Conditions

The unbounded spatial interval used in (4.1) is not going to make it into our numerical algorithms. To compute the solution we need to specify left (x_L) and right (x_R) endpoints for the spatial grid. In this case the grid points are $x_i = x_L + ih$, where $i = 0, 1, 2, \ldots, N + 1$ and $x_{N+1} = x_R$. This gives rise to the question as to what can, or should, be specified for boundary conditions at $x = x_L, x_R$. For example, suppose we wanted to compute the solution for $0 < t \leq 10$, where the initial condition is the square bump in Figure 4.2 with $a = 1$. The spatial interval that contains the nonzero portion of the solution during this time period is $0 \leq x \leq 11$. One option for our numerical algorithm is to choose the computational interval large enough so it includes this nonzero region, and a reasonable choice for the given time interval is $x_L = -1$ and $x_R = 12$. Because information in this problem is moving from left to right, a boundary condition consistent with the problem is $u(x_L, t) = 0$. For the same reason, we should not have to specify anything at $x = x_R$. The exact solution at this point is determined from the information flowing in from the left. Indeed, as shown in Exercise 4.19, attempting to specify a condition at $x = x_R$ almost certainly leads to a conflict with the initial condition. The unfortunate reality is, however, that some numerical methods for the advection equation require the boundary conditions at both ends to be specified. There are ways to deal with such situations and they will be discussed when the issue arises.

4.2 First-Order Methods

We have solved enough differential equations in this book that constructing a finite difference approximation is routine. We will compute the solution over the two-dimensional region $x_L < x < x_R$, $0 < t \leq T$, where x_L, x_R, T are specified at the time the computation is carried out. We confine our attention to a uniform grid and use the formulas introduced earlier, namely,

$$t_j = jk, \quad \text{for} \;\; j = 0, 1, 2, \ldots, M, \tag{4.9}$$

$$x_i = x_L + ih, \quad \text{for} \;\; i = 0, 1, 2, \ldots, N + 1, \tag{4.10}$$

where $k = T/M$ and $h = (x_R - x_L)/(N + 1)$. In preparation for introducing approximations for the derivatives we evaluate the differential equation at the grid point $(x, t) = (x_i, t_j)$ to obtain

$$u_t(x_i, t_j) + a u_x(x_i, t_j) = 0. \tag{4.11}$$

Now, what approximations should be used? For a little variety, we consider several, and two are listed below.

4.2.1 Upwind Scheme

Using a forward difference in time and a backward difference in space we have

$$\frac{u(x_i, t_{j+1}) - u(x_i, t_j)}{k} + a\frac{u(x_i, t_j) - u(x_{i-1}, t_j)}{h} + \tau_{ij} = 0, \tag{4.12}$$

where $\tau_{ij} = O(h) + O(k)$. The stencil for this choice is given in the upper grid in Figure 4.5. Dropping the truncation error and rearranging the terms we obtain what is known as the upwind scheme

$$u_{i,j+1} = (1 - \lambda)u_{ij} + \lambda u_{i-1,j}, \quad \text{for} \;\; \begin{cases} i = 1, 2, \ldots, N + 1, \\ j = 0, 1, 2, \ldots, M - 1, \end{cases} \tag{4.13}$$

where

$$\lambda = \frac{ak}{h}. \tag{4.14}$$

The initial condition is $u_{i,0} = g_i$. To use (4.13) we need $u_{0,j}$, which is the solution at $x = x_L$, for $0 < t$. There are various ways to handle this, depending on the exact formulation of the problem, and this will be discussed later when we work out a few examples.

4.2.2 Downwind Scheme

Using forward differences in space and time we have

$$\frac{u(x_i, t_{j+1}) - u(x_i, t_j)}{k} + a\frac{u(x_{i+1}, t_j) - u(x_i, t_j)}{h} + \tau_{ij} = 0, \tag{4.15}$$

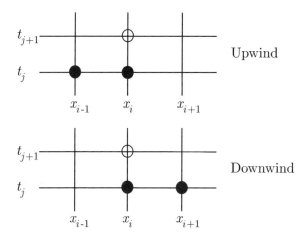

Figure 4.5. Stencils for the upwind (4.13) and downwind (4.16) schemes. The convention used here is that a solid dot indicates a point where the solution has already been determined, and a hollow dot indicates a point where the solution is going to be calculated using the given method.

where $\tau_{ij} = O(h) + O(k)$. The stencil for this choice is given in the lower grid in Figure 4.5. Dropping the truncation error and rearranging the terms, one obtains the downwind scheme, given as

$$u_{i,j+1} = (1+\lambda)u_{ij} - \lambda u_{i+1,j}, \quad \text{for} \quad \begin{cases} i = 0, 1, 2, \dots, N, \\ j = 0, 1, 2, \dots, M-1. \end{cases} \quad (4.16)$$

The initial condition is $u_{i,0} = g_i$. To use (4.16), $u_{N+1,j}$ must be specified, and how this might be done is discussed later.

The two methods, at first glance, seem to be more or less equal, because they are both explicit and have the same truncation error. As it turns out, they produce very different results. We will take the experimental approach and try them without determining ahead of time whether they actually work. Afterwards, we will do a little analysis to see if we can explain what happens.

Example
Suppose the initial condition is the square bump given in (4.7). The solution is given in (4.8) and it is sketched in Figure 4.2. We will compute the solution for $-10 \le x \le 10$ and $0 \le t \le 7$ with $a = 1$. The upwind scheme requires information, or a boundary condition, at $x = -10$ and we take $u_{0,j} = 0$. The downwind scheme requires a condition at $x = 10$ and in the calculations we take $u_{N+1,j} = 0$. To do a fair job in describing the initial condition we use 203 points along the x-axis. It is not clear how many time points to use, so we try a few values. The results are shown in Figure 4.6, and the outcomes are strikingly different. The solution computed using the downwind scheme in

(4.16) shows the overly oscillatory behavior associated with instability. More-over, the nonzero part of the solution is moving in the wrong direction! The latter is a particularly serious flaw and the reason we will not use this scheme to solve the advection equation (when $a > 0$). The upwind scheme, on the other hand, also shows instability when $M = 50$. It redeems itself, to a small degree, by at least sending the unstable solution in the right direction. If the number of time points is increased, to $M = 72$, the upwind method actually does a decent job solving the problem. In the hope of possibly improving the accuracy, the number of time points is increased to $M = 100$. Oddly, the com-puted solution gets worse. In particular, the method produces an amplitude that is too small, a bump that is too spread out, and a bump that looks to be traveling faster than it should. To its credit, however, the solution does not show unstable behavior.

Now comes the part where we explain what happened in the last example. For example, why is it that using fewer time points ($M = 72$ versus $M = 100$) actually produces a better numerical solution? This is different from the difficulty we had with the heat equation, where increasing the number of spatial points had the potential of causing the method to violate its stability condition. That is not happening here, because the $M = 100$ solution looks stable, it just isn't very accurate. To help with these questions we return to the idea of the domain of dependence.

4.2.3 Numerical Domain of Dependence

Given a spatial location $x = x_i$, and time $t = t_j$, the grid points along the x-axis that contribute to the solution at (x_i, t_j) form what is known as the numerical domain of dependence for the solution at (x_i, t_j). It is easy to de-termine this, because one simply has to identify what points are used in the previous time steps to calculate the solution. This is illustrated in Figures 4.7 and 4.8 for the downwind and upwind methods, respectively. Recall that for the advection equation the domain of dependence for (x_i, t_j) is the single point $(\overline{x}_0, 0)$, where $\overline{x}_0 = x_i - at_j$. In other words, the exact solution at (x_i, t_j) is determined solely by the value of the initial condition at $\overline{x}_0 = x_i - at_j$. If the computed solution is to have any hope of producing the correct answer, it is essential that the numerical domain of dependence bound the domain of dependence. This gives rise to the following principle.

CFL (COURANT–FRIEDRICHS–LEWY) CONDITION: The numerical domain of dependence must bound, or contain, the domain of dependence for the problem.

In looking at the situation in Figure 4.7 it is evident why the downwind scheme in (4.16) does so badly. For this method, information starting at, say,

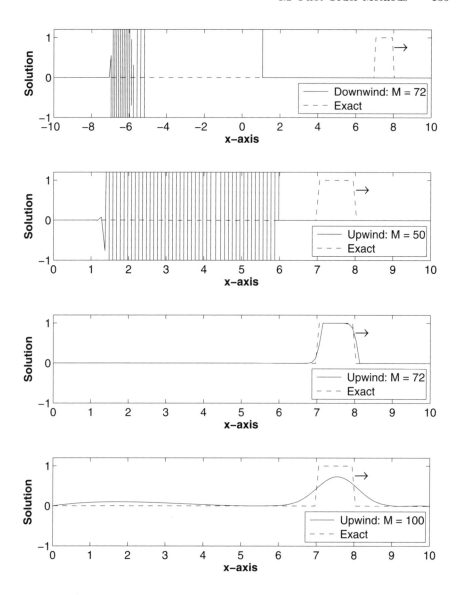

Figure 4.6. (**M**) Numerical solution of the advection equation at $t = 7$ when the initial condition is the square bump given in (4.7). The bottom three graphs use the upwind scheme (4.13), and the top one uses the downwind scheme (4.16). The number of time steps M used to reach $t = 7$ is given in each graph, as is the exact solution.

Methods for Solving

$$\frac{\partial u}{\partial t} + a\frac{\partial u}{\partial x} = 0$$

Using

$$u_{i,j+1} = Au_{i+1,j} + Bu_{i,j} + Cu_{i-1,j}$$

Method	Coefficients	Truncation Error	CFL Condition	Stability	Monotone
Upwind	$A = 0$ $B = 1 - \lambda$ $C = \lambda$	$O(h) + O(k)$	$\lambda \leq 1$	Conditional: $\lambda \leq 1$	Yes
Lax–Friedrichs	$A = \frac{1}{2}(1-\lambda)$ $B = 0$ $C = \frac{1}{2}(1+\lambda)$	$O(\frac{h^2}{k}) + O(k) + O(h^2)$	$\lambda \leq 1$	Conditional: $\lambda \leq 1$	Yes
Lax–Wendroff	$A = -\frac{\lambda}{2}(1-\lambda)$ $B = 1 - \lambda^2$ $C = \frac{\lambda}{2}(1+\lambda)$	$O(h^2) + O(k^2)$	$\lambda \leq 1$	Conditional: $\lambda \leq 1$	No
Centered	$A = -\frac{\lambda}{2}$ $B = 1$ $C = \frac{\lambda}{2}$	$O(h^2) + O(k)$	$\lambda \leq 1$	Unstable	No

Table 4.1. Explicit finite difference methods for solving the advection equation, assuming $a > 0$ and $\lambda = ak/h$.

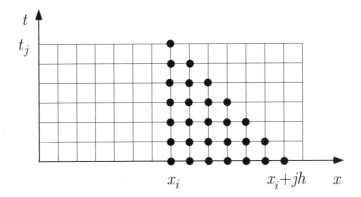

Figure 4.7. Grid points used to calculate the solution at (x_i, t_j) using the downwind scheme (4.16). The points on the x-axis, $\{x_i, x_i+h, \ldots, x_i+jh\}$, form the numerical domain of dependence for (x_i, t_j).

$x = x_i + jh$ affects points to the left as time increases. In other words, information moves only leftward and this explains why the wave goes in the wrong direction in Figure 4.6 for the downwind scheme. As for the CFL condition, with this method the numerical domain of dependence for (x_i, t_j) is $\{x_i, x_i+h, \ldots, x_i+jh\}$. It is impossible for $\bar{x}_0$ to fall within the region bounded by these points, that is, for $x_i \leq \bar{x}_0 \leq x_i + jh$. Therefore, this method does not satisfy the CFL condition.

As for the upwind scheme in (4.13), as seen in Figure 4.8, information moves to the right. All that is necessary is to make sure the numerical domain of dependence $\{x_i - jh, \ldots, x_i - h, x_i\}$ is large enough that it overlies the domain of dependence. This translates into the requirement that $x_i - jh \leq \bar{x}_0$.

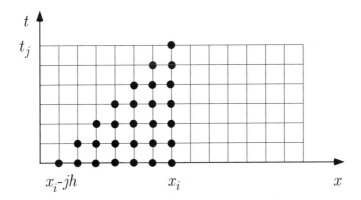

Figure 4.8. Grid points used to calculate the solution at (x_i, t_j) using the upwind scheme (4.13). The points on the x-axis, $\{x_i - jh, \ldots, x_i - h, x_i\}$, form the numerical domain of dependence for (x_i, t_j).

Since $t_j = jk$, the CFL condition for this method is

$$\lambda \leq 1, \tag{4.17}$$

where $\lambda = ak/h$. In Figure 4.6, when $M = 50$, $\lambda = 1.4$, and consequently the CFL condition is not satisfied. This correlates with the bad result, although a complete explanation will need to wait until the stability test is completed. The other two cases, $M = 72$ ($\lambda = 0.98$) and $M = 100$ ($\lambda = 0.7$), both satisfy the CFL condition. This still leaves the question as to why the one with the fewer time steps produces a better answer. The culprit is the truncation error. Numerical methods for the advection equation, like other wave equations, are sensitive to the form of the error term. We will investigate this shortly, and in the next chapter tools are developed that make it relatively easy to determine the wave properties of numerical methods.

The CFL condition is a valuable tool that we will routinely use for helping to develop effective numerical methods for solving wave problems. It is not perfect and comes with limitations and qualifications. These include the following:

- The CFL condition is, by itself, not enough to guarantee a stable method. An example of this will be given shortly. However, it does provide a quick and relatively painless way to determine whether there is any hope that such a method will work on a wave problem.
- The CFL condition is limited to wave problems. For example, with the heat equation the domain of dependence for any point is the entire x-axis. If the CFL condition were to apply, one would make the erroneous conclusion that the explicit method will fail for the heat equation.
- The CFL condition makes no statement about accuracy.

4.2.4 Stability

The stability analysis developed in the last chapter for the heat equation applies directly to the methods used to solve the advection equation. The start-off assumption is exactly the same, which is

$$u_{ij} = w_j e^{r x_i I}, \tag{4.18}$$

where $I = \sqrt{-1}$. The function w_j is determined from the difference equation, and the requirement for stability is that w_j remain bounded as j increases.

Example 1

For the upwind method in (4.13) one finds that $w_j = \kappa^j w_0$, where $\kappa = 1 - \lambda + \lambda \cos(rh) - I\lambda \sin(rh)$. In this case, $|\kappa|^2 = 1 - 4\lambda(1 - \lambda) \sin^2(rh/2)$. To guarantee that w_j remains bounded we require $|\kappa|^2 \leq 1$ (see Figure 4.9). From this we obtain the condition $0 \leq \lambda(1 - \lambda) \sin^2(rh/2)$. This holds, irrespective of the value of r, if $\lambda \leq 1$, and this is the stability condition for

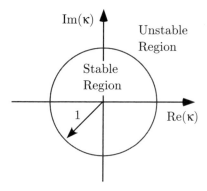

Figure 4.9. For stability of the upwind method it is required that the value of κ be on, or within, the unit circle.

the upwind method. Interestingly, this is exactly what was obtained from the CFL condition.

Example 2

Suppose one wants to do a better job with the x derivative and uses a $O(h^2)$ centered difference approximation. In this case the finite difference approximation is

$$u_{i,j+1} = -\frac{\lambda}{2}u_{i+1,j} + u_{i,j} + \frac{\lambda}{2}u_{i-1,j}. \qquad (4.19)$$

It is not hard to show that the CFL condition for this method is $\lambda \leq 1$. Carrying out the stability analysis, one finds that $w_j = \kappa^j w_0$, where $\kappa = 1 - \lambda I \sin(rh)$. From this we conclude that $|\kappa|^2 = 1 + \lambda^2 \sin^2(rh)$, which means that $|\kappa| > 1$. Therefore, the method is unstable. This is an example that demonstrates that the CFL condition, by itself, does not guarantee stability. It also reinforces something seen earlier. Namely, what appears to be a better approximation can actually lead to a much worse computational result.

4.3 Improvements

Although the upwind method does a reasonable job solving the advection equation, it is worth our time looking for a more accurate method. Our usual approach for this is to introduce higher-order difference approximations for the derivatives in the equation. This time, however, we try something different. In particular, given the form of the finite difference equations that have been obtained so far, we simply hypothesize that the form of the approximation is

$$u_{i,j+1} = Au_{i+1,j} + Bu_{ij} + Cu_{i-1,j}. \qquad (4.20)$$

The coefficients A, B, C are determined by requiring that the truncation error associated with this approximation be small. To calculate the error note that

$u_t = -au_x$, $u_{tt} = -au_{xt} = -a(u_t)_x = a^2 u_{xx}$, and $u_{ttt} = -a^3 u_{xxx}$. With this and Taylor's theorem, one finds that

$$u(x_i, t_j + k)$$
$$= u(x_i, t_j) + k u_t(x_i, t_j) + \frac{1}{2} k^2 u_{tt}(x_i, t_j) + \frac{1}{6} k^3 u_{ttt}(x_i, t_j) + \dots$$
$$= u(x_i, t_j) - ak u_x(x_i, t_j) + \frac{1}{2} a^2 k^2 u_{xx}(x_i, t_j) - \frac{1}{6} a^3 k^3 u_{xxx}(x_i, t_j) + \dots$$
$$(4.21)$$

Similarly,

$$u(x_i \pm h, t_j) = u(x_i, t_j) \pm h u_x(x_i, t_j) + \frac{1}{2} h^2 u_{xx}(x_i, t_j) \pm \frac{1}{6} h^3 u_{xxx}(x_i, t_j) + \dots$$
$$(4.22)$$

Now, as seen in (4.15) and (4.12), when $u(x_i, t_j)$ and the truncation error τ_{ij} are in the difference equation, the form of the equation is

$$u(x_i, t_j + k) = A u(x_i + h, t_j) + B u(x_i, t_j) + C u(x_i - h, t_j) + k\tau_{ij}. \quad (4.23)$$

The question is, how do we pick A, B, C such that τ_{ij} is small? Well, using (4.21) and (4.22) in (4.23) we have

$$\tau_{ij} = \frac{1}{k}(u(x_i, t_j + k) - A u(x_i + h, t_j) - B u(x_i, t_j) - C u(x_i - h, t_j))$$
$$= \frac{1}{k}(1 - A - B - C)u(x_i, t_j) - \frac{h}{k}(A - C + \lambda)u_x(x_i, t_j)$$
$$- \frac{1}{2}\frac{h^2}{k}(A + C - \lambda^2)u_{xx}(x_i, t_j) - \frac{1}{6}\frac{h^3}{k}(A - C + \lambda^3)u_{xxx}(x_i, t_j) + \dots$$
$$(4.24)$$

It is mandatory that A, B, C be selected so that the approximation is consistent (i.e., τ_{ij} must go to zero as h and k go to zero). This still leaves numerous ways to pick A, B, C, and this freedom gives rise to different methods.

4.3.1 Lax–Wendroff Method

Perhaps the most natural choice to make is simply to set as many coefficients to zero in (4.24) as possible. Working our way through the terms in (4.24) gives

$$u : 1 - A - B - C = 0,$$
$$u_x : \quad \lambda + A - C = 0,$$
$$u_{xx} : \quad -\lambda^2 + A + C = 0.$$

Solving these equations, we obtain what is known as the Lax–Wendroff method

$$u_{i,j+1} = -\frac{1}{2}\lambda(1 - \lambda)u_{i+1,j} + (1 - \lambda^2)u_{ij} + \frac{1}{2}\lambda(1 + \lambda)u_{i-1,j}. \quad (4.25)$$

The truncation error in this case is

$$
\begin{aligned}
\tau_{ij} &= -\frac{1}{6}\frac{h^3}{k}(A - C + \lambda^3)u_{xxx}(x_i, t_j) + \dots \\
&= \frac{1}{6}(-k^2 a^2 + ah^2)u_{xxx}(x_i, t_j) + \dots \\
&= O(k^2) + O(h^2).
\end{aligned}
\tag{4.26}
$$

This looks to be an improvement over the upwind scheme, but it remains to carry out the stability analysis. Before doing so, however, we determine the CFL condition for the method. The grid points that contribute to the solution at (x_i, t_j) are shown in Figure 4.10. From this it is seen that the numerical domain of dependence for Lax–Wendroff is $\{x_i - jh, x_i - (j-1)h, \dots, x_i + (j-1)h, x_i + jh\}$. For $\bar{x}_0 = x_i - at_j$ to be bounded by this set of points it is required that $\lambda \le 1$ and this is the CFL condition for Lax–Wendroff. There is another useful piece of information contained in Figure 4.10. Given that $x = x_i - jh$ $(t = 0)$ contributes to the solution at (x_i, t_j) then information moves to the right using Lax–Wendroff. By the same token, $x = x_i + jh$ $(t = 0)$ contributes to the solution at (x_i, t_j) and this means information moves to the left. In other words, the Lax–Wendroff method propagates information in both directions. This is a bit of a concern as the advection equation produces waves that travel in only one direction. As will be demonstrated below, however, this property of Lax–Wendroff will not affect its accuracy.

In remains to analyze the stability properties of the method. Assuming $u_{ij} = w_j \exp(Irx_i)$ one finds from (4.25) that $w_j = \kappa^j w_0$, where

$$
\kappa = 1 - 2\lambda^2 \sin^2\left(\frac{rh}{2}\right) - \lambda I \sin(rh).
\tag{4.27}
$$

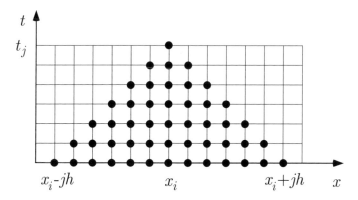

Figure 4.10. Grid points used to calculate the solution at (x_i, t_j) using the Lax–Wendroff method, (4.25). The points on the x-axis, $\{x_i - jh, x_i - (j-1)h, \dots, x_i + (j-1)h, x_i + jh\}$, form the numerical domain of dependence for (x_i, t_j).

Using the half-angle formula one can show that $\sin^2(rh) = 4\sin^2(rh/2) - 4\sin^4(rh/2)$. With this, we obtain

$$|\kappa|^2 = 1 - 4\lambda^2(1 - \lambda^2)\sin^4\left(\frac{rh}{2}\right). \tag{4.28}$$

For stability it is required that $|\kappa| \leq 1$. From (4.28) this happens, irrespective of the value of r, when $\lambda \leq 1$. Therefore, Lax–Wendroff is a conditionally stable second-order method. This has made it one of the more prominent methods for solving routine advection problems. However, it is not perfect, and we will explore this below.

There are two Lax methods listed in Table 4.1, and the derivation of the second, Lax–Friedrichs, is explored in the exercises. The method differs from the others listed since it is not consistent unless $h^2/k \to 0$ as $h, k \to 0$. This is not a limitation as this condition holds as long as the stability condition is satisfied.

The Lax methods introduce an interesting complication into the development. In coding these schemes one is faced with having to specify boundary conditions at both ends. As discussed earlier, this is not an issue at $x = x_L$ because the original advection problem on a finite interval also requires this information. Therefore the correct boundary condition should be part of the problem formulation. The question is what to do at $x = x_R$. One of the easiest ways to handle this is to use the upwind method to determine the solution at x_R. This is simple to implement, but the upwind scheme is first order, while Lax–Wendroff is a second-order method. This is an issue, since Lax–Wendroff allows information to propagate both left and right, so the inaccuracy at x_R has the potential to affect the solution over the entire interval. Fortunately, as shown in the next chapter, Lax–Wendroff forces the left-moving information to decay rapidly, so using a first-order approximation at x_R is not a particularly significant problem. It is possible to develop second-order-accurate conditions, but this is not pursued here. This subject, which falls into the more general topic of artificial boundary conditions, is discussed in depth in Durran [1998] and Colonius [2004].

Example
Returning to the problem solved in Figure 4.6, we use the Lax–Wendroff method to compute the solution for $0 \leq x \leq 10, 0 \leq t \leq 7$ with $a = 1$. The spatial grid is the same as used earlier. For boundary conditions we take $u_{0,j} = 0$ and $u_{N+1,j} = 0$. For the number of time points we will try $M = 70$ ($\lambda = 1.01$), $M = 72$ ($\lambda = 0.99$), $M = 74$ ($\lambda = 0.96$), and $M = 100$ ($\lambda = 0.7$). The results are shown in Figure 4.11. Not unexpectedly, the method has serious problems when 70 time steps are used, but not for 72. In comparison to the upwind results in Figure 4.6, the two methods appear comparable when $M = 72$. This is consistent with the fact that both methods produce the exact solution when $\lambda = 1$ (Exercise 4.2). Also, the Lax–Wendroff method does a better job when $M = 100$, and this is a consequence of its better

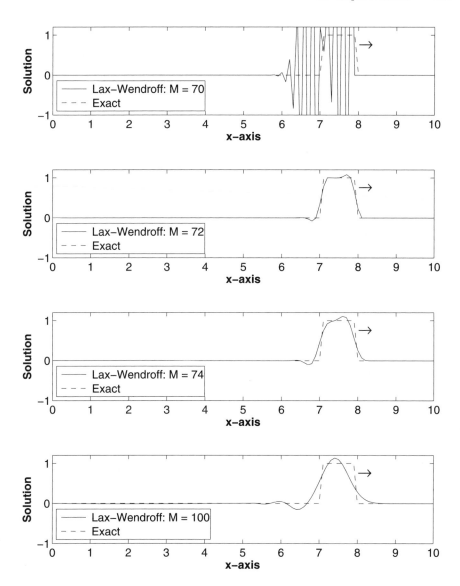

Figure 4.11. (**M**) Solution of the advection equation at $t = 7$ using the Lax–Wendroff method with various values of the time step. The initial condition is the square bump given in (4.7). The number of time steps M used to reach $t = 7$ is given in each graph, as is the exact solution.

truncation error. However, the Lax–Wendroff solution overshoots ($u > 1$) the transition from $u = 0$ to $u = 1$ and undershoots ($u < 0$) it when going from $u = 1$ to $u = 0$. This behavior is due to the method not being monotone, and this observation brings us to our next topic.

4.3.2 Monotone Methods

A numerical scheme is said to be monotone if it preserves monotonicity in the solution. For example, if $g(x) = \tanh(10x)$ then the solution $u(x, t) = \tanh(10(x - at))$ is a monotonically increasing function of x at any time t. A monotone scheme, in this case, will produce a solution that is a monotonically increasing function of x_i at any time step t_j. The precise definition is given below.

Definition 4.1. *A finite difference approximation is monotone if $u_{ij} \geq u_{i+1,j}$, $\forall i$ implies that $u_{i,j+1} \geq u_{i+1,j+1}$, $\forall i$.*

One benefit of a monotone scheme is that it will not create new max/min points. Because of this the over- and undershoots such as those in Figure 4.11 will not occur when a monotone scheme is used. Even though these max/min points can have small amplitude, they do affect the accuracy and interpretation of the solution. For example, if $u(x, t)$ represents a mass variable then having negative undershoots is not particularly desirable.

It is not difficult to determine if an explicit scheme is monotone when the method involves a single time step like those in Table 4.1. To state the condition, suppose $u_{i,j+1}$ is determined using a linear combination of the values of the solution at t_j. Writing $u_{i,j+1} = \sum_p A_p u_{p,j}$, then the method is monotone if all the coefficients A_p are nonnegative (see Exercise 4.22). This makes checking for monotonicity very easy, since one simply looks to see if the coefficients are nonnegative. So, in the stability region, the upwind scheme is monotone, and for this reason you do not see over- or undershoots in Figure 4.6. For the Lax–Wendroff method, $A < 0$ for $0 < \lambda < 1$. Therefore, unless $\lambda = 1$, the method is not monotone. For this reason it is not surprising there are over- and undershoots in Figure 4.11.

In looking at Table 4.1 it is natural to ask whether it is possible to have a monotone scheme that is more accurate than the upwind method. The answer to this is contained in Godunov's theorem, which states that except for isolated values of λ, a linear monotonic scheme is no greater than first-order accurate. In other words, there is no easy fix or adjustment to get Lax–Wendroff to be monotone. This is unfortunate and has repercussions beyond the simple advection equation we are studying. In many applications the equations are nonlinear, and an example that often arises is $u_t + q(u)u_x = 0$. Deriving a finite difference approximation for this is relatively easy. The hard part is proving stability, since the methods we have been using no longer work. Consequently, some other measure of stability must be introduced and it has been found that monotonicity can be used. This observation, in conjunction with Godunov's

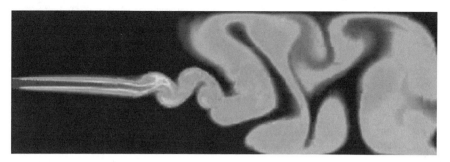

Figure 4.12. (M) Simulation of advection-dominated transport in an unsteady flow obtained by solving the Navier-Stokes equations using a monotone scheme (Denaro et al. [1995]).

theorem, has generated a good deal of research on nonlinear schemes, with one well-known example being the total variation diminishing (TVD) family of methods. This is beyond the scope of this text, and an extended discussion of this topic can be found in Toro [1999] and LeVeque [2003].

4.3.3 Upwind Revisited

We still have not explained why, in Figure 4.6, increasing the number of time points to $M = 100$ causes the results actually to get worse. The answer is contained in (4.24). Using the A, B, C values for the upwind method, the first term in the truncation error is

$$\tau_{ij} = -\frac{1}{2}ah(1 - \lambda)u_{xx}(x_i, t_j) + \dots . \tag{4.29}$$

This shows that the strength of the error is determined by the value of $h(1-\lambda)$. In Figure 4.6, when going from $M = 72$ to $M = 100$, the value of λ changed from 0.99 to 0.7, while h stayed fixed. Consequently, the error is increased when reducing k, and this is the reason for the relatively poor $M = 100$ result. Another worthwhile observation to make is that the error involves the spatial term found in the heat equation. One way to look at this is that the method is actually solving $u_t + au_x = \epsilon u_{xx}$, where $\epsilon = \frac{1}{2}ah(1 - \lambda)$. Now, $h(1 - \lambda)$ is supposed to be small and positive, so the contribution of the u_{xx} term is relatively small. However, over time this term does make itself felt, and it does what all good diffusion equations do: it causes the solution to decay. This behavior is evident in the $M = 100$ curve in Figure 4.6. Interestingly, this is not seen in the $M = 100$ curve for Lax–Wendroff given in Figure 4.11. The reason is that the error in this case involves u_{xxx}, and this term introduces an effect known as dispersion. In the next chapter the method of plane waves is developed to help explain what is happening in situations such as this.

4.4 Implicit Methods

Implicit methods played a central role in solving the diffusion problems of the last chapter. They were unconditionally stable, and this gave them an advantage over the conditionally stable explicit methods because it was not necessary to use an unreasonably small time step to compute the solution. This is not the situation with wave problems, because the stability requirement is not as severe. For the advection equation, with $a = 1$, it is required only that $k \leq h$, while for the heat equation we needed $2k \leq h^2$. So for the advection equation, doubling the number of spatial points results in having to double the number of time points. For the heat equation one would have to increase the time points by a factor of four. Consequently, for the heat equation the computational overhead of using an implicit method is worth the investment, but not so with the advection equation.

To illustrate some of the differences, suppose in (4.11) one uses a backward difference in time and a centered difference in space. The resulting finite difference equation is

$$\lambda u_{i+1,j} + 2u_{i,j} - \lambda u_{i-1,j} = 2u_{i,j-1}. \tag{4.30}$$

This method is stable. Also, the implicit nature of the approximation means that the numerical solutions at time level $t = t_j$ depend on all values at $t = t_{j-1}$. Consequently, the numerical domain of dependence of (x_i, t_j) is the entire x-axis, and this clearly satisfies the CFL condition. However, a larger numerical domain of dependence is not necessarily a good thing, and this becomes crystal clear from numerical experiments.

Example
Returning to the problem solved in Figure 4.6, we use the implicit method to compute the solution for $0 \leq x \leq 10$, $0 \leq t \leq 7$ with $a = 1$. As before, we use 100 points along the x-axis. For the number of time points we try $M = 72$ ($\lambda = 0.98$), $M = 100$ ($\lambda = 0.7$), $M = 200$, and $M = 400$. The results are shown in Figure 4.13, and a reasonable description is that the method does a mediocre job solving the problem. For example, the $M = 72$ solution in Figure 4.13 in nowhere near as accurate as the corresponding result in Figure 4.6. This can be explained, in part, by looking at the truncation error, which for this method is $\tau_{ij} = -ah\lambda u_{xx}(x_i, t_j) + \dots$ The error for the upwind scheme is given in (4.29). Comparing these expressions, it is seen that unlike the case for the upwind scheme, this implicit scheme does not benefit from taking λ close to one.

There are implicit methods that are competitive with their explicit counterparts, although more care is needed in their derivation. One example is the box scheme in Exercise 4.5. However, overall implicit methods have not found widespread use in numerical wave problems.

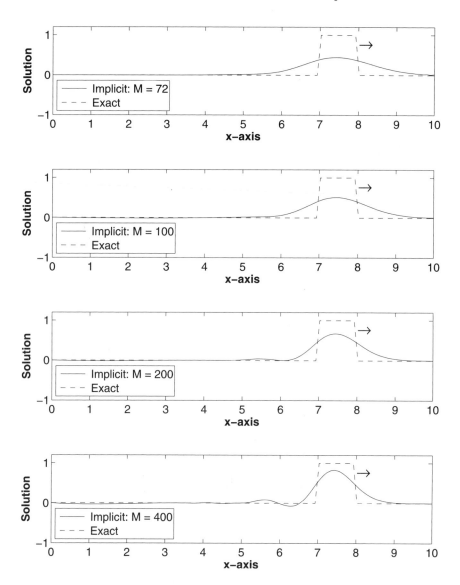

Figure 4.13. Solution of the advection equation at $t = 7$ obtained using the implicit method (4.30) for various values of the time step. The initial condition is the square bump given in (4.7). The exact solution is shown in each case.

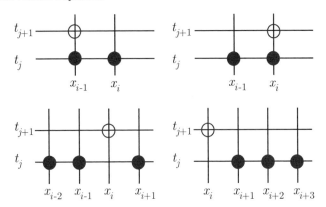

Figure 4.14. Stencils for Exercise 4.4.

Exercises

4.1. Suppose $a < 0$ in (4.1). The schemes in (4.13) and (4.16) are still valid although they can no longer be called the upwind and downwind schemes. What happens to the CFL and stability conditions in this case?

4.2. Show that the upwind and Lax–Wendroff methods give the exact solution if $\lambda = 1$.

4.3. This problem explores a consequence of consistency.
(a) If the constant function $u_{i,j} \equiv 1$ does not satisfy the finite difference approximation of the advection equation (4.1) then the approximation is not consistent. Explain why. Also explain why $u_{i,j} \equiv 1$ satisfying the finite difference equation does not necessarily mean the approximation is consistent.
(b) Can either of the following be used to solve the advection equation?
   (i) $u_{i,j+1} = u_{i,j-1} - \frac{1}{2}\lambda(u_{i+1,j} + u_{i-1,j})$
   (ii) $(1+\lambda)u_{i,j+1} + (1-\lambda)u_{i-1,j+1} = 2(1-\lambda)u_{i,j} + (1+\lambda)u_{i-1,j}$
(c) Will the observation in (a) work on the equation $u_t + u_x + u = 0$?

4.4. Suppose that for a wave problem the domain of dependence of the solution at $(\bar{x}, \bar{t})$ is the interval $\bar{x} \le x \le \bar{x} + 2\bar{t}$. The stencils for various finite difference approximations are given in Figure 4.14. Identify which should not be used and explain why. For those that can be used, determine the CFL condition.

4.5. The box method for solving the advection equation (4.1) is

$$(1+\lambda)u_{i,j+1} + (1-\lambda)u_{i-1,j+1} = (1-\lambda)u_{i,j} + (1+\lambda)u_{i-1,j}.$$

(a) What is the stencil for the method?

(b) Derive this method by first integrating the advection equation in both x and t and then using the appropriate quadrature rules. In the process show that the truncation error is $O(h^2) + O(k^2)$.

(c) What is the CFL condition? Is the method stable?

(d) Suppose the solution at the left endpoint is specified, so $u_{0,j+1}$ is known. Explain why the scheme is implicit but $u_{i,j+1}$ can be computed as if it were an explicit method.

4.6. The leapfrog method for solving the advection equation (4.1) is

$$u_{i,j+1} = u_{i,j-1} - \lambda(u_{i+1,j} - u_{i-1,j}).$$

(a) Derive this method using the derivative approximation method and in the process show that the truncation error is $O(h^2) + O(k^2)$.

(b) What is the stencil for the method?

(c) What is the CFL condition?

(d) Show that the method is stable if $\lambda < 1$ and unstable if $1 \le \lambda$.

(e) Explain why the odd and even numbered grid points along the x-axis do not communicate with each other, depending on the time level. This can lead to problems computationally.

4.7. The Fromm method for solving the advection equation (4.1) is

$$u_{i,j+1} = \frac{1}{4}\lambda(\lambda-1)u_{i-2,j} + \frac{1}{4}\lambda(5-\lambda)u_{i-1,j} - \frac{1}{4}(\lambda-1)(\lambda+4)u_{i,j} + \frac{1}{4}\lambda(\lambda-1)u_{i+1,j}.$$

(a) What is the stencil for the method?

(b) What is the CFL condition? Is the method stable? Is it monotone?

(c) What is the truncation error?

4.8. This problem explores using Runge–Kutta methods to solve the advection equation using the method of lines.

(a) Reduce the advection equation to an IVP by using a $O(h^2)$ approximation of the spatial derivative. With this use Heun's method (Table 1.3) to derive a finite difference equation for solving the problem.

(b) What is the stencil and CFL condition of the method in (a)? Is the method monotone?

(c) Is the method stable?

(d) Based on what occurred in (a), what will be the stencil if RK4 is used in place of Heun?

4.9. Consider the stencil shown in Figure 4.15.

(a) Assuming $\lambda \le 1$, derive a consistent finite difference approximation of the advection equation (4.1) that has this as its stencil and a truncation error that is $O(h^2)$.

(b) Is the method explicit or implicit?

(c) What is the CFL condition?

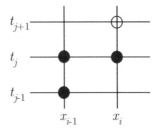

Figure 4.15. Stencil for solving the advection equation.

(d) Is the method stable?

4.10. Consider the stencil shown in Figure 4.16.
(a) Derive a consistent finite difference approximation of the advection equation (4.1) that has this as its stencil and a truncation error that is $O(h^2) + O(k^2)$.
(b) Is the method explicit or implicit?
(c) What is the CFL condition?
(d) Is the method stable?

4.11. Assume $a = 1$ and

$$g(x) = \begin{cases} -4(x+8)(x+9) & \text{if } -9 \le x \le -8, \\ 0 & \text{otherwise.} \end{cases}$$

(a) What is the exact solution of the resulting advection problem?
(b) Use the upwind and Lax–Wendroff methods to solve the problem for $-10 \le x \le 10$ using a grid where λ is approximately 0.95. With this plot, on the same axes, the numerical and exact solutions at $t = 6$.
(c) Using the grid from (b), plot, on the same axes, the solution at $t = 18$.
(d) Redo (b) and (c) using a grid where λ is approximately 0.5.
(e) Comment on the effectiveness of the two methods, including how they do with amplitude decay, spreading of the bump, speed of the bump, and smoothness of the bump in comparison to the exact solution.

4.12. Consider the equation $u_t + au_x + bu = 0$, where a, b are positive constants and $u(x,0) = g(x)$.

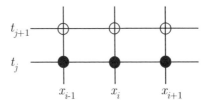

Figure 4.16. Stencil for solving the advection equation.

(a) By making the change of variables $u(x,t) = v(x,t)e^{-bt}$ find the solution of this problem. What is the domain of dependence for a point $(\overline{x}, \overline{t})$?

(b) Derive an upwind scheme for this equation (it should have the same stencil as the upwind scheme).

(c) Determine the CFL condition, numerical domain of dependence, and stability condition for the method in part (b).

(d) Use the method from (b) to solve the problem for $0 \le t \le 7$ in the case $a = 1$, $b = 1/7$, and

$$
g(x) = \begin{cases} \frac{1}{2}(1 - \cos(2\pi x)) & \text{if } 0 \le x \le 1, \\ 0 & \text{otherwise.} \end{cases}
$$

Use a grid where λ is approximately 0.99 (but $\lambda < 1$), and another where λ is about 0.5. Comment on the effectiveness of the method, including how it does with amplitude decay, spreading of the bump, and the speed of the bump in comparison to the exact solution.

4.13. Consider the equation $u_t + au_x + bu = 0$, where a, b are positive constants and $u(x, 0) = g(x)$.

(a) Derive a Lax–Wendroff scheme for this equation (it should reduce to the Lax–Wendroff scheme when $b = 0$).

(b) Determine the CFL condition, numerical domain of dependence, and stability condition for the method in part (b).

(c) Use the method from (b) to solve the problem for $0 \le t \le 7$ in the case $a = 1$, $b = 1/7$, and

$$
g(x) = \begin{cases} \frac{1}{2}(1 - \cos(2\pi x)) & \text{if } 0 \le x \le 1, \\ 0 & \text{otherwise.} \end{cases}
$$

Use a grid where λ is approximately 0.99 (but $\lambda < 1$), and another where λ is about 0.5. Comment on the effectiveness of the method, including how it does with amplitude decay, spreading of the bump, and the speed of the bump in comparison to the exact solution.

4.14. Consider the problem of solving $xu_t + u_x = x$, for $x > 0$ and $t > 0$, where $u(0, t) = 0$ and $u(x, 0) = 0$.

(a) Use the method of characteristics to show that $u(x, t) = t - (t - \frac{1}{2}x^2)H(t - \frac{1}{2}x^2)$, where H is the Heaviside step function.

(b) Derive an upwind scheme for this equation (it should have the same stencil as the upwind scheme).

(c) Using the method from (b), plot the solution at $t = 10$ for $0 \le x \le 10$. Also plot the exact solution on the same axes. In the calculation use 50 points along the x-axis, and state how many time points you used (and why).

(d) Redo (c) but use 100 points along the x-axis.

(e) Is there any truth to the rumor that for stability one must require $k < 2h^2$? Explain clearly how you come to your conclusion.

4.15. Consider the problem of solving $u_t + (x + t)u_x = 0$, where the initial condition is $u(x, 0) = g(x)$.
(a) Use the method of characteristics to show that $u(x, t) = g((1 + x + t)e^{-t} - 1)$.
(b) Derive an upwind scheme for this equation (it should have the same stencil as the upwind scheme).
(c) Assuming that $g(x)$ is the square bump in (4.7), use the method from (b) to plot the numerical solution at $t = 10$ for $0 \le x \le 10$. Also plot the exact solution on the same axes. In the calculation use 50 points along the x-axis, and state how many time points you used (and why).
(d) Redo (c) but use 100 points along the x-axis.

4.16. This problem concerns the derivation of the Lax–Friedrichs method (Table 4.1).
(a) Show that $y(x_i) = \frac{1}{2}(y(x_{i+1}) + y(x_{i-1})) + O(h^2)$.
(b) Using a forward difference in time, a centered difference in space, and the result from part (a), derive the Lax–Friedrichs method. Make sure to include the truncation error in the derivation.
(c) Derive the stability condition for this method.

4.17. One might argue that the wrong three spatial points are used in (4.20) to solve the advection equation. If $a > 0$ then it would be better to hypothesize that
$$u_{i,j+1} = Au_{ij} + Bu_{i-1,j} + Cu_{i-2,j}.$$

(a) Using an argument similar to what was used to derive the Lax–Wendroff scheme, find the coefficients for this method. Make sure to state the resulting truncation error. This is known as the Beam–Warming method.
(b) What is the CFL condition? Is the method stable? Is it monotone?
(c) Solve the same problem that produced Figures 4.6, 4.11.
(d) Based on your results from (a)–(c), discuss the benefits and drawbacks of this method in comparison to the upwind and Lax–Wendroff methods.

4.18. In this problem assume that $u(x, t)$ is a solution of the advection problem with $u = 0$ at $x = x_L, x_R$.
(a) Show that $M(t) = \int_{x_L}^{x_R} u(x, t)dx$ is conserved, that is, $\frac{d}{dt}M = 0$.
(b) Show that the upwind method conserves $M_j = \sum_{i=0}^{N+1} u_{ij}$, that is, $M_j = M_{j+1}$.
(c) Show that the Lax–Wendroff method conserves M_j.
(d) The sum in (b) is obtained from the trapezoidal approximation of the integral in (a). Does the conclusion in (b) hold if Simpson's rule is used for the integral?

4.19. This exercise examines the effects of boundary conditions when the advection equation is solved over a finite interval. The problem to solve is

$$\frac{\partial u}{\partial t} + \frac{\partial u}{\partial x} = 0, \quad \text{for} \quad \begin{cases} 0 < x < 1, \\ 0 < t, \end{cases}$$

where $u(x,0) = x(1-x)$. The key to this exercise is the observation that the solution is constant along any line of the form $x - t = constant$.

(a) Explain why the equation and initial condition determine only the solution in a triangular region in the xt-plane, given as $0 \le t \le x$ for $0 \le x \le 1$.

(b) What is the solution of the problem if the boundary condition is $u(0,t) = \sin(\pi t)$?

(c) Instead of the boundary condition in (b), suppose one tried to use $u(1,t) = \sin(\pi t)$. Explain why there is no solution of this problem.

(d) Another approach is to use a periodic boundary condition, which for this problem is $u(0,t) = u(1,t)$. This uses the known value at $x = 1$ as the boundary condition at $x = 0$. What is the resulting solution?

4.20. This problem concerns using the method of characteristics to solve first-order wave equations.

(a) Using the method of characteristics to solve

$$u_t + au_x + bu = f(x,t),$$

where $u(x,0) = g(x)$, show that

$$u(x,t) = g(x - at)e^{-bt} + \int_0^t f(x - a(t-s), s)e^{-b(t-s)}ds.$$

(b) Use the method of characteristics to solve $u_t + e^{\alpha x}u_x = 0$, where $u(x,0) = g(x)$. Explain why the solution is well defined if $\alpha \ge 0$ but is not if $\alpha < 0$.

4.21. This problem explores how the method of characteristics can be used to derive finite difference approximations.

(a) Show that the exact solution of the advection equation satisfies $u(x_i + ak, t_{j+1}) = u(x_i, t_j)$.

(b) Using the result from part (a), the solution at (x_{i-1}, t_j) determines the solution at a point $(\bar{x}, t_{j+1})$. Similarly, the solution at (x_{i+1}, t_j) determines the solution at $(\bar{\bar{x}}, t_{j+1})$. Find $\bar{x}$ and $\bar{\bar{x}}$.

(c) Find an approximation of the solution at $x = x_i$ ($t = t_{j+1}$) using linear interpolation on the solutions found in part (b) at the points $\bar{x}$ and $\bar{\bar{x}}$. Explain why this approximation for $u(x_i, t_{j+1})$ is equivalent to one listed in Table 4.1.

(d) Redo (b) and (c), but use three points (x_{i-1}, t_j), (x_i, t_j), and (x_{i+1}, t_j).

4.22. This exercise examines the importance of a scheme being monotone.

(a) Suppose that for a given value of j, u_{ij} is monotone increasing in space (i.e., $u_{i+1,j} \geq u_{ij}, \forall i$). Using the general formula in (4.20) show that $u_{i,j+1}$ will also be monotone increasing in space if A, B, C are nonnegative.

(b) In part (a) suppose one of the coefficients A, B, C is negative. Explain why an initial condition that contains a jump is almost certain to result in loss of monotonicity.

(c) Suppose the initial condition is $g(x) = \tanh(10(x + 8))$. Solve the resulting advection problem for $-10 \leq x \leq 10$, $0 \leq t \leq 15$ using the first three methods listed in Table 4.1. The boundary condition is $u(-10, t) = \tanh(10(-t + 8))$ and $a = 1$. Use a grid where λ is approximately 0.99.

(d) Redo (b) but use a λ that is approximately 0.8, and another that is about 0.5.

(e) Based on your results from (b) and (c), are there any advantages to having a monotone scheme? Does it make any difference on what the value of λ is?

4.23. It might be argued that the conclusion about implicit methods coming from Figure 4.13 is unfair and that one should really use another approximation. This exercise explores that option.

(a) What implicit method is obtained using first-order backward differences in space and time? Make sure to state the truncation error.

(b) Use quadrature in time to derive a second-order implicit method for the advection equation.

(c) Solve the same problem that produced Figures 4.6, 4.11 using the methods from (a) and (b).

(d) Based on your results from (c), what is your opinion of implicit methods for the advection equation? Make sure to explain why you come to the conclusion you do.

4.24. This problem considers how to use the Lax–Wendroff method to solve the inhomogeneous advection equation $u_t + au_x = f(x,t)$.

(a) Explain how the hypothesis in (4.20) should be modified for this equation.

(b) Use your result from (a) to derive a Lax–Wendroff method for this equation.

5

Numerical Wave Propagation

5.1 Introduction

In studying phenomena in such diverse areas as electrodynamics, fluid dynamics, and acoustics, it is almost inevitable to come across what is known as the wave equation. This ubiquitous equation is a prototype for many of the waves seen in nature, and it is the subject of this chapter. The specific problem we start with is the wave equation

$$c^2 \frac{\partial^2 u}{\partial x^2} = \frac{\partial^2 u}{\partial t^2}, \quad \text{for} \quad \begin{cases} 0 < x < \ell, \\ 0 < t, \end{cases} \tag{5.1}$$

where c is a positive constant. The boundary conditions are

$$u(0, t) = u(\ell, t) = 0, \tag{5.2}$$

and the initial conditions are

$$u(x, 0) = f(x), \quad u_t(x, 0) = g(x). \tag{5.3}$$

A simple example of where this problem arises is in the study of the motion of an elastic string. An everyday instance of this is a string on a guitar. In this case $u(x, t)$ represents the vertical deflection of the string where, at rest, the string occupies the interval $0 \le x \le \ell$. Assuming that the string is held fixed at both ends, we get the boundary conditions in (5.2). The initial conditions correspond to specifying the initial deflection, $f(x)$, and the initial velocity, $g(x)$.

5.1.1 Solution Methods

There are various ways to express the solution of the wave problem, and those that we will find useful in this chapter are described below.

Fourier Series Representation

Just as with the heat equation, it is possible to solve the wave equation problem using separation of variables. One finds that

$$u(x,t) = \sum_{n=1}^{\infty} [a_n \cos(c\lambda_n t) + b_n \sin(c\lambda_n t)] \sin(\lambda_n x), \tag{5.4}$$

where $\lambda_n = n\pi/\ell$, and the Fourier coefficients are

$$a_n = \frac{2}{\ell} \int_0^\ell f(x) \sin(\lambda_n x) dx, \tag{5.5}$$

$$b_n = \frac{2}{c\pi n} \int_0^\ell g(x) \sin(\lambda_n x) dx. \tag{5.6}$$

Comparing this with the series solution (3.5) for the heat equation, it is seen that both consist of the superposition of terms of the form $w_n(t) \sin(\lambda_n x)$. The difference, however, is that for the heat equation $w_n(t)$ is a decaying exponential, whereas for the wave equation it is oscillatory. The latter produce standing waves and in this sense (5.4) consists of the superposition of standing waves. As a simple example, suppose $f(x) = \sin(2\pi x)$ and $g(x) = 0$. Assuming that $c = \ell = 1$, then (5.4) reduces to the standing wave solution $u(x,t) = \cos(2\pi t) \sin(2\pi x)$.

d'Alembert Representation

The Fourier series representation of the solution (5.4) is an important result, but it is rather difficult to extract information about the traveling wave properties of the solution from this expression. Fortunately, there is another form that is particularly useful, and to derive it assume for the moment that the spatial interval is infinite, that is, $-\infty < x < \infty$. The first step in the derivation is to note that the wave equation can be written as

$$\left(c\frac{\partial}{\partial x} + \frac{\partial}{\partial t} \right) \left(c\frac{\partial}{\partial x} - \frac{\partial}{\partial t} \right) u = 0. \tag{5.7}$$

This type of factorization was used to solve the advection equation. As before, we change variables from x, t to r, s in such a way that the derivatives transform as

$$c\frac{\partial}{\partial x} + \frac{\partial}{\partial t} = \frac{\partial}{\partial r}, \tag{5.8}$$

$$c\frac{\partial}{\partial x} - \frac{\partial}{\partial t} = \frac{\partial}{\partial s}. \tag{5.9}$$

A change of variables that will accomplish this, up to a constant multiplicative factor, is $r = x + ct, s = x - ct$. In this case, from (5.7), the wave equation transforms into

$$\frac{\partial}{\partial r}\left(\frac{\partial}{\partial s}u\right) = 0. \tag{5.10}$$

Integrating this, we obtain

$$u(x,t) = F(x + ct) + G(x - ct), \tag{5.11}$$

where the functions F, G are determined from the initial conditions. This form of the solution is interesting, because it clearly shows the traveling waves forming the solution. One, with profile F, moves to the left with speed c, and the other, with profile G, moves to the right with speed c. It remains to satisfy the initial conditions, and in doing so one finds that the solution of the wave equation is

$$u(x,t) = \frac{1}{2}f(x - ct) + \frac{1}{2}f(x + ct) + \frac{1}{2c}\int_{x-ct}^{x+ct} g(z)dz. \tag{5.12}$$

This is the d'Alembert form of the solution. The procedure used to obtain this result is known as the method of characteristics, and the lines $x \pm ct = const$ are the characteristics. In the derivation it was assumed that $-\infty < x < \infty$, but as will be seen shortly, in certain circumstances it is possible to adapt the solution to a finite interval.

Example
As an example, suppose the initial conditions are $g(x) = 0$ and $f(x)$ is the rectangular bump

$$f(x) = \begin{cases} 1 & \text{if } -1 \le x \le 1, \\ 0 & \text{otherwise.} \end{cases} \tag{5.13}$$

From (5.12) one finds that the solution in this case is

$$u(x,t) = \frac{1}{2}f(x - ct) + \frac{1}{2}f(x + ct). \tag{5.14}$$

This is shown in Figure 5.1, and it is seen that the solution consists of two rectangular bumps, half the height of the original, traveling to the left and right with speed c.

In certain cases it is possible to adjust the d'Alembert solution so that it works with problems that have boundary conditions. To explain how this is done, suppose that in the above example the interval is $-10 \le x \le 10$ and the boundary conditions are $u = 0$ at $x = \pm 10$. In this case the solution in (5.14) applies up until the bumps reach the boundaries. When they arrive at $x = \pm 10$ the boundary conditions are no longer satisfied, and the formula for the solution must be modified. To correct the situation at $x = 10$ we add a left-traveling wave that exactly cancels $\frac{1}{2}f(x - ct)$ as it crosses $x = 10$. This is accomplished using symmetry (see Figure 5.2). In particular, a rectangular

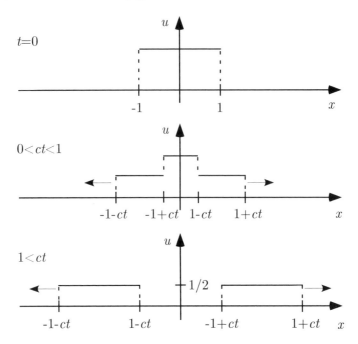

Figure 5.1. (**M**) Schematic of the solution of the wave equation in the case where the initial profile is a rectangular bump and $g(x) = 0$.

bump is added in that starts out centered at $x = 20$ and has height $-\frac{1}{2}$. Using (5.13) this is equivalent to adding in a term of the form $-\frac{1}{2}f(20 - x - ct)$. A similar correction is needed for the boundary condition at $x = -10$. This fix works until the new bumps cross the interval and reach the opposite boundary, and this occurs when $ct = 29$. Up to that point, the resulting solution can be written as

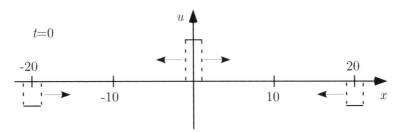

Figure 5.2. To extend the d'Alembert representation so that it satisfies the boundary conditions one can use reflections. As shown, the original interval is $-10 \leq x \leq 10$, and so virtual waves are placed at $x = \pm 20$.

$$u(x,t) = \begin{cases} \frac{1}{2}f(x-ct) + \frac{1}{2}f(x+ct) & \text{if } 0 \le t < 9/c, \\ \frac{1}{2}f(x-ct) - \frac{1}{2}f(20-x-ct) & \\ \quad + \frac{1}{2}f(x+ct) - \frac{1}{2}f(-20-x+ct) & \text{if } 9/c \le t < 29/c. \end{cases} \qquad (5.15)$$

This process of adding in reflected bumps can be continued indefinitely, although we do not have need of such situations in this book.

The properties of the solution discussed in the above example are reminiscent of those obtained for the advection equation. For example, the initial jumps in the solution result in jumps in the solution for all time and information travels at finite speed. There are also differences with the advection equation, and the most obvious one is that waves are now able to propagate in both directions. This has a consequence for the domain of dependence. As seen in (5.12), the solution at a given spatial position $x = \bar{x}$ and time $t = \bar{t}$ is determined by the values of the initial conditions over the interval $\bar{x} - c\bar{t} \le x \le \bar{x} + c\bar{t}$. Consequently, this interval is the domain of dependence for $(\bar{x}, \bar{t})$, and this situation is indicated schematically in Figure 5.3.

The domain of dependence (DoD) played an important role in developing numerical methods for the advection equation, and the same will be true for higher-order wave equations. However, to be able to use the CFL condition we need to know what points make up the DoD. This has been determined for the wave question, but what if we want to solve the Klein–Gordon equation $u_{xx} = u_{tt} + u$? To answer this, consider the equation $c^2 u_{xx} = u_{tt} + \alpha(x,t)u_t + \beta(x,t)u_x + \gamma(x,t)u$, where α, β, γ are continuous functions. One can prove that this has the same DoD as the wave equation, which is given in Figure 5.3 (see Courant and Hilbert [1989] for a proof of this statement). An equation with a different DoD is $\alpha u_{xx} + \beta u_{xt} + u_{tt} = 0$, and this is investigated in Exercise 5.2.

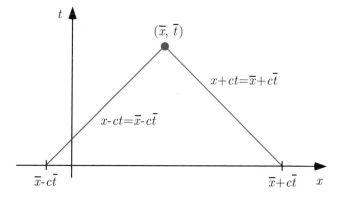

Figure 5.3. Domain of dependence. As shown in (5.12), the solution of the wave equation at $(\bar{x}, \bar{t})$ depends of the values of the initial conditions from the interval $\bar{x} - c\bar{t} \le x \le \bar{x} + c\bar{t}$. This interval forms the domain of dependence for $(\bar{x}, \bar{t})$.

5.1.2 Plane Wave Solutions

In developing and testing numerical methods it is very helpful to have relatively simple solutions that can be used to check on how well the methods work. An important tool for analyzing wave problems involves the use of plane wave solutions, and these have the form

$$u(x,t) = e^{I(\bar{k}x - \bar{\omega}t)}, \tag{5.16}$$

where $I = \sqrt{-1}$. In this expression $\bar{k}$ is the wave number and it is related to the wavelength $\bar{\lambda}$ of the wave through the equation $\bar{\lambda} = 2\pi/\bar{k}$. In what follows it is assumed that $0 < \bar{k} < \infty$. To determine the plane wave solutions we substitute (5.16) into (5.1), and the result is the algebraic equation

$$\bar{\omega}^2 = c^2 \bar{k}^2. \tag{5.17}$$

This is the dispersion relation for the wave equation, and it determines the frequency $\bar{\omega}$ in terms of $\bar{k}$ for there to be plane wave solutions. Solving this quadratic equation, one obtains $\bar{\omega} = \pm c\bar{k}$. Consequently, for each wave number there are two plane waves, traveling in opposite directions. The directions and speed of these waves agree with what we found earlier using the d'Alembert solution.

Although it did not happen above, it is possible that the solutions of the dispersion relation are complex-valued. In such situations the general solution will have the form $\bar{\omega} = \bar{\omega}_r + I\bar{\omega}_i$, where $\bar{\omega}_r, \bar{\omega}_i$ are the real and imaginary components of the (complex) frequency. With this, after substituting $\bar{\omega}$ back into (5.16) we obtain the solution

$$u(x,t) = e^{\bar{\omega}_i t} e^{I(\bar{k}x - \bar{\omega}_r t)}. \tag{5.18}$$

Written this way, there is a traveling wave component (the second exponential term) and an amplitude that is time-dependent if $\bar{\omega}_i \neq 0$. This solution is used to define various properties of the problem, and in preparation for this we introduce the phase velocity v_{ph}, defined as

$$v_{ph} \equiv \frac{\bar{\omega}_r}{\bar{k}}. \tag{5.19}$$

With this we have the following plane wave properties:

- An equation is stable if $\bar{\omega}_i \leq 0 \ \forall \bar{k}$; otherwise, it is unstable. It is assumed in what follows that the equation is stable.
- An equation is dispersive if v_{ph} depends on $\bar{k}$; otherwise, it is nondispersive.
- An equation is dissipative if $\bar{\omega}_i$ is not identically zero; otherwise, it is nondissipative.

For the wave equation we found that $v_{ph} = \pm c$. Because v_{ph} does not depend on $\bar{k}$, it follows that the equation is nondispersive. Consequently, long waves

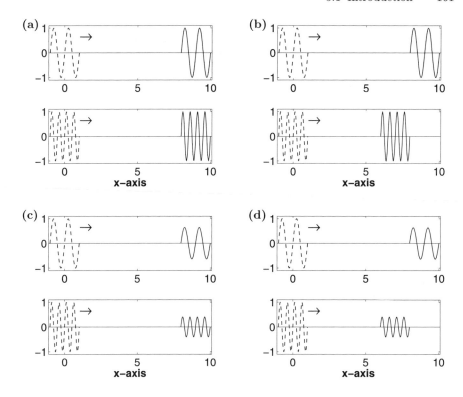

Figure 5.4. The dashed curves are the initial profiles and the solid curves are the profiles of the wave after a fixed amount of time. The situation is shown where the equation is (a) nondispersive and nondissipative, (b) dispersive and nondissipative, (c) nondispersive and dissipative, and (d) dispersive and dissipative.

($\overline{k}$ small) and short waves ($\overline{k}$ large) all travel at the same speed. Moreover, since $\overline{\omega}_i = 0$, the equation is nondissipative. This means that all waves travel without decaying. An example of this situation is illustrated in Figure 5.4(a). In this figure a portion of the initial profiles of both a short and long wave are shown with dashed curves. The solid curves on the same axes are the respective positions of the waves after a given amount of time has passed. As shown, the short and long waves travel the same distance (because the equation is nondispersive) and neither decays (because the equation is nondissipative).

An example of a dispersive, and nondissipative, equation is the Klein–Gordon equation (see Table 5.1). From the phase velocity it is seen that the shorter the wavelength, the slower the wave moves. This situation is illustrated in Figure 5.4(b), where the long wave travels farther than the shorter wave over the given time interval and neither decays (because the equation is nondissipative). A consequence of this is that a waveform composed of multiple plane waves will spread out as it moves, with the longer waves out in front. A somewhat different situation arises with the beam equation, which is also dispersive, but the shorter waves travel faster than longer ones.

An example of a dissipative equation is the advection–diffusion equation (see Table 5.1). Because it is nondispersive, one would expect that a waveform made up of multiple waves would keep its overall shape as it moves but that it would decay as it proceeds. In fact, the shorter the waves, the faster the decay. An example of this situation is illustrated in Figure 5.4(c).

Group Velocity

For nondissipative equations there is a second velocity that is derived from the dispersion relation that will prove useful when studying waveforms made up of multiple plane waves. This is the group velocity, and it is defined as

$$v_g \equiv \frac{d\overline{\omega}}{d\overline{k}}. \tag{5.20}$$

To explain why v_g is important, consider the situation in which two waves of slightly different wavelengths are added together. Let the wave numbers of the two waves be $\overline{k}_1$ and $\overline{k}_2 = \overline{k}_1 + \Delta\overline{k}$. In this case, from the dispersion relation the corresponding frequencies are $\overline{\omega}_1 = \overline{\omega}(\overline{k}_1)$ and

$$
\begin{aligned}
\overline{\omega}_2 &= \overline{\omega}(\overline{k}_2) \\
&= \overline{\omega}(\overline{k}_1 + \Delta\overline{k}) \\
&\approx \overline{\omega}(\overline{k}_1) + \Delta\overline{k}\,\frac{d\overline{\omega}}{d\overline{k}}(\overline{k}_1) \\
&= \overline{\omega}_1 + \Delta\overline{\omega}, \tag{5.21}
\end{aligned}
$$

where

$$\Delta\overline{\omega} = \Delta\overline{k}\,\frac{d\overline{\omega}}{d\overline{k}}. \tag{5.22}$$

Adding the waves together, we obtain

$$
\begin{aligned}
u &= \cos(\overline{k}_1 x - \overline{\omega}_1 t) + \cos(\overline{k}_2 x - \overline{\omega}_2 t) \\
&= A(x,t)\cos(\overline{k}_a x - \overline{\omega}_a t), \tag{5.23}
\end{aligned}
$$

where

$$A(x,t) = 2\cos\left(\frac{1}{2}\Delta\overline{k}\,x - \frac{1}{2}\Delta\overline{\omega}\,t\right), \tag{5.24}$$

$\overline{k}_a = \overline{k}_1 + \frac{1}{2}\Delta\overline{k}$, and $\overline{\omega}_a = \overline{\omega}_1 + \frac{1}{2}\Delta\overline{\omega}$. As given in (5.23), the two waves combine to form a single wave with wave number k_a, frequency ω_a, and (signed) amplitude $A(x,t)$. A typical example is shown in Figure 5.5. Now, the energy in a wave is determined by its envelope, indicated with the dashed curves in Figure 5.5. In looking at the formula for $A(x,t)$ in (5.24), it is seen that for this example the amplitude is moving with velocity $\Delta\overline{\omega}/\Delta\overline{k}$. From (5.20) and

PDE	Dispersion Relation	v_{ph}	v_g	Dispersive	Dissipative
Wave Equation $c^2 u_{xx} = u_{tt}$	$\bar{\omega} = \pm c\bar{k}$	$\bar{\omega} = \pm c$	$\bar{\omega} = \pm c$	No	No
Klein–Gordon Equation $c^2 u_{xx} = u_{tt} + bu$	$\bar{\omega} = \pm\sqrt{c^2\bar{k}^2 + b}$	$\pm c\sqrt{1 + \frac{b}{\bar{k}^2}}$	$\pm\dfrac{c^2\bar{k}}{\sqrt{c^2\bar{k}^2 + b}}$	Yes	No
Modified Advection Equation $u_t + au_x + bu = 0$	$\bar{\omega} = a\bar{k} - Ib$	a	a	No	Yes
Beam Equation $u_{xxxx} + u_{tt} = 0$	$\bar{\omega} = \pm\bar{k}^2$	$\bar{\omega} = \pm\bar{k}$	$\bar{\omega} = \pm 2\bar{k}$	Yes	No
Advection–Diffusion Equation $D u_{xx} = u_t + au_x$	$\bar{\omega} = a\bar{k} - ID\bar{k}^2$	a	a	No	Yes

Table 5.1. Plane wave characteristics of various partial differential equations. It is assumed that $a, b, c > 0$.

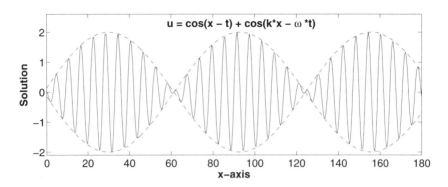

Figure 5.5. (**M**) Adding two plane waves with slightly different frequencies and wave numbers produces a single wave with an envelope that moves with the group velocity. The two dashed curves are the functions $\pm A(x,t)$, where $A(x,t)$ is given in (5.24). Also, $k^* = 1.1$ and $\omega^* = 1.05$.

(5.22) it follows that the envelope is moving at the group velocity. This differs from the phase velocity $\overline{\omega}_a/\overline{k}_a$, which is the speed of the wave formed by the solid curve in Figure 5.5. For example, depending on the slope of $\overline{\omega}(\overline{k})$ the group velocity can be greater than, equal to, or less than the phase velocity.

It is possible to generalize this example and show that when waveforms are localized disturbances, such as a wave packet, that the energy moves with the group velocity. This is investigated in more detail later in the chapter. It can also be used at times to estimate the domain of dependence and an example of this is examined in Exercise 5.11.

5.2 Explicit Method

As with most problems, there is a variety of ways to find the numerical solution of a second-order equation such as (5.1). In many situations the easiest approach is to reduce the problem to one that has already been solved. Based on this, given our work on the advection equation in the last chapter, one might try rewriting (5.1) as a first-order system and then use one or more of the methods discussed earlier. It is certainly easy to convert the equation to system form, and an example is to let $\mathbf{u} = (u, v)^T$, where $u_t = cv_x$ and $v_t = cu_x$. In this case one produces an advection equation of the form $\mathbf{u}_t + \mathbf{A}\mathbf{u}_x = \mathbf{0}$. Although this approach works, there are reasons not to do it. For example, the boundary conditions complicate the development. There are ways around these difficulties, but instead we keep the problem as it is and simply introduce finite difference approximations directly into the original problem. This is undertaken using the same five steps employed in the earlier chapters.

STEP 1. We will compute the solution of the wave equation problem for $0 \leq x \leq \ell, 0 \leq t \leq T$ using a uniform grid. The usual formulas are used, namely,

$$t_j = jk, \quad \text{for } j = 0, 1, 2, \ldots, M, \tag{5.25}$$
$$x_i = ih, \quad \text{for } i = 0, 1, 2, \ldots, N+1, \tag{5.26}$$

where $k = T/M$ and $h = \ell/(N+1)$.

STEP 2. In preparation for introducing finite difference approximations we evaluate the differential equation at the grid point $(x, t) = (x_i, t_j)$ to obtain

$$c^2 u_{xx}(x_i, t_j) = u_{tt}(x_i, t_j). \tag{5.27}$$

STEP 3. Using centered differences to approximate the derivatives gives us

$$c^2 \frac{u(x_{i+1}, t_j) - 2u(x_i, t_j) + u(x_{i-1}, t_j)}{h^2} + O(h^2)$$
$$= \frac{u(x_i, t_{j+1}) - 2u(x_i, t_j) + u(x_i, t_{j-1})}{k^2} + O(k^2). \tag{5.28}$$

This can be rewritten as

$$u(x_i, t_{j+1}) = \lambda^2 u(x_{i+1}, t_j) + 2(1 - \lambda^2) u(x_i, t_j)$$
$$+ \lambda^2 u(x_{i-1}, t_j) - u(x_i, t_{j-1}) + k^2 \tau_{ij}, \tag{5.29}$$

where $\tau_{ij} = O(h^2) + O(k^2)$ is the truncation error and

$$\lambda = \frac{ck}{h}. \tag{5.30}$$

STEP 4. Dropping the truncation error gives us the following finite difference approximation to the wave equation

$$u_{i,j+1} = \lambda^2 u_{i+1,j} + 2(1 - \lambda^2) u_{ij} + \lambda^2 u_{i-1,j} - u_{i,j-1}, \quad \text{for } \begin{array}{l} i = 1, \ldots, N, \\ j = 1, \ldots, M-1. \end{array} \tag{5.31}$$

The stencil for this method is given in Figure 5.6. From the boundary conditions we get that $u_{0,j} = u_{N+1,j} = 0$, and the first initial condition in (5.3) translates into $u_{i,0} = f_i$. To be able to use (5.31) also requires $u_{i,1}$, and this can be determined using the second initial condition. It is important to preserve the quadratic truncation error, and possibilities include introducing a ghost point or to use a higher-order one-sided difference (see Exercise 5.3). However, for this particular problem there is a more direct approach using Taylor's theorem. Keeping in mind that $u_{tt} = c^2 u_{xx}$ then, for k small,

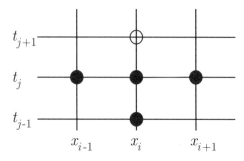

Figure 5.6. Stencil for the explicit method in (5.31).

$$u(x_i, t_1) = u(x_i, k)$$

$$= u(x_i, 0) + ku_t(x_i, 0) + \frac{1}{2}k^2 u_{tt}(x_i, 0) + O(k^3)$$

$$= u(x_i, 0) + ku_t(x_i, 0) + \frac{1}{2}k^2 c^2 u_{xx}(x_i, 0) + O(k^3)$$

$$= f(x_i) + kg(x_i) + \frac{1}{2}k^2 c^2 f''(x_i) + O(k^3). \tag{5.32}$$

With this, we have that

$$u_{i,1} = f_i + kg_i + \frac{\lambda^2}{2}(f_{i+1} - 2f_i + f_{i-1}), \quad \text{for } i = 1, \dots, N. \tag{5.33}$$

Together (5.31) and (5.33) produce a $O(k^2) + O(h^2)$ approximation of the wave equation problem. There are, however, a couple of oddities in (5.32) and (5.33). One question is, why is the Taylor series not stopped at the $O(k^2)$ term since that is the overall order of truncation error we are trying to obtain? The answer is that (5.32) is being used to construct an approximation of the second initial condition. This means that we should solve (5.32) for $u_t(x_i, 0)$, and that requires a division by k. This is the reason for carrying the expansion up to $O(k^3)$. The second oddity is that $f''(x_i)$ has been replaced with a finite difference approximation. This is odd, because we can calculate $f''(x_i)$ directly, so it would seem to be unnecessary to introduce another approximation. The reason has to do with the domain of dependence requirement. As will be shown below, without using the approximation for $f''(x_i)$ it is not possible to satisfy the CFL condition.

This completes the derivation of the finite difference approximation. In the previous chapter, when studying explicit methods for the advection equation, we found that the stability requirement limited what could be used for the step sizes. It is worth knowing this before running any numerical experiments, and this brings us to the next topic.

5.2.1 Diagnostics

With the development of a new finite difference approximation we apply our diagnostic tests to see whether it actually might work. This is, in effect, STEP 5 of our derivation.

CFL Condition and Grid Velocity

The first, and easiest, test is to see if the numerical domain of dependence contains, or bounds, the domain of dependence of the wave equation. Recall that given a spatial location $x = x_i$, and time $t = t_j$, the grid points along the x-axis that potentially contribute to the solution at (x_i, t_j) form the numerical domain of dependence for (x_i, t_j). Given the stencil in Figure 5.6, along with the initial condition approximation in (5.33), the numerical domain of dependence consists of the points $\{x_i - jh, x_i - (j-1)h, \ldots, x_i + (j-1)h, x_i + jh\}$. Recalling that the domain of dependence for the wave equation in this case is $x_i - ct_j \leq x \leq x_i + ct_j$, then to guarantee that the numerical points bound this interval we need $x_i - jh \leq x_i - ct_j$ and $x_i + ct_j \leq x_i + jh$. Both of these reduce to

$$\frac{ck}{h} \leq 1, \tag{5.34}$$

and this is the CFL condition for this method. One can look at this as simply a requirement that if one picks a certain h value then it is necessary to use a time step k small enough to satisfy the given inequality. However, one can also think of the CFL as a requirement on the grid velocity h/k. As seen in the stencil in Figure 5.6, with each time step the solution at any spatial grid point x_i can affect the solution at only those grid points to the immediate left and right. In this sense, information for the explicit method moves through the grid with speed h/k. Consequently, the CFL condition means that the grid speed cannot be smaller than the speed of the waves for the equation. This connection between the wave speed and the CFL is investigated further in Exercise 5.11.

Stability

As usual, to start the stability analysis we assume

$$u_{i,j} = w_j e^{rx_i I}, \tag{5.35}$$

where $I = \sqrt{-1}$. The function w_j is determined from the difference equation. Plugging (5.35) into (5.31) and then simplifying the resulting expression, one finds that the problem reduces to

$$w_{j+1} - 2sw_j + w_{j-1} = 0, \tag{5.36}$$

where $s = 1 - 2\lambda^2 \sin^2(rh/2)$. This second-order difference equation can be solved by assuming $w_j = \kappa^j$. Substituting this into (5.36) and simplifying a bit one finds that $\kappa^2 - 2s\kappa + 1 = 0$. Assuming for the moment that $s \neq \pm 1$, the two solutions are

$$\kappa_\pm = s \pm \sqrt{s^2 - 1}. \tag{5.37}$$

Therefore, the general solution of (5.36) is

$$w_j = \alpha \kappa_+^j + \beta \kappa_-^j, \tag{5.38}$$

where α, β are constants. For stability we require that w_j remain bounded, and this means we must have $|\kappa_\pm| \leq 1$. To determine when and where this occurs note that $s \leq 1$. Now, if $s < -1$ then $\kappa_- = s - \sqrt{s^2 - 1} < -1$, and in this case the solution is unbounded. On the other hand, if $-1 < s < 1$ then $\kappa_\pm = s \pm I\sqrt{1 - s^2}$. With this, $|\kappa_\pm|^2 = 1$, and so for these s values w_j remains bounded. This leaves us with having to determine what happens when $s = \pm 1$. For these values there is a repeated root of (5.36), and the resulting general solution has the form $w_j = (\alpha + \beta j)s^j$. Although the linear growth of js^j, for $s = \pm 1$, is nowhere near as significant as the geometric growth obtained when we get roots that satisfy $|\kappa| > 1$, it does nevertheless result in an unbounded solution.

Based on the above analysis we have that the method is stable if $-1 < s < 1$, or equivalently, if the following holds

$$0 < \lambda^2 \sin^2\left(\frac{rh}{2}\right) < 1. \tag{5.39}$$

To translate this into a requirement independent of r we need to look a little closer at the stability test. The assumption in (5.35) is based on the fact that the general solution (5.4) is composed of such functions with $r = \lambda_n$,

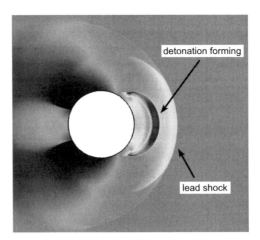

Figure 5.7. To study the detonation of a gas due to the impulsive motion of a cylinder, finite difference methods that conserve energy are used to solve the reactive Euler equations (Henshaw and Schwendeman [2007]).

where $\lambda_n = n\pi/\ell$. When carrying out the stability test we make it easy on ourselves and simply assume that r is an arbitrary positive constant. Normally, extending the interval for r in this way is not an issue, but this opens up the possibility that $\sin^2(rh/2) = 0$ in (5.39). To determine whether this happens, for the given r values in this problem,

$$
\begin{aligned}
\frac{hr}{2} &= \frac{h}{2}\lambda_n \\
&= \frac{h}{2}\frac{n\pi}{\ell} \\
&= \frac{\pi}{2}\frac{n}{n+1}.
\end{aligned}
$$

Consequently, for each value of r we have $0 < \sin^2(rh/2) < 1$. Given this, then (5.39) will hold if $\lambda \leq 1$, or equivalently,

$$
\frac{ck}{h} \leq 1. \tag{5.40}
$$

This is the sought-after stability condition for the method.

5.2.2 Numerical Experiments

To see how well the explicit method does we solve the problem using the initial conditions

$$
f(x) = \begin{cases} \frac{1}{2}(1 - \cos(\frac{2\pi x}{a})) & \text{if } 0 \leq x \leq a, \\ 0 & \text{otherwise,} \end{cases} \tag{5.41}
$$

and $g(x) = -cf'(x)$. It is assumed that $0 < a < 1$. These initial conditions give rise to what we will affectionately refer to as the cosine bump. Note that they have been rigged so the solution starts out as a right-traveling wave $f(x - ct)$. However, eventually the bump reaches $x = 1$, in which case it is necessary to include its reflection, as illustrated in Figure 5.2. Up to the second reflection the formula for the solution is

$$
u(x,t) = \begin{cases} f(x - ct) & \text{if } 0 \leq t < \frac{1}{c}(1 - a), \\ f(x - ct) - f(2 - x - ct) & \text{if } \frac{1}{c}(1 - a) \leq t < \frac{1}{c}(2 - a). \end{cases} \tag{5.42}
$$

We will compute the solution up to $T = 1.8$, assuming $c = 1$ and $a = 0.09$. To decide on what to use for the spatial step size, 15 or so points will do a reasonable job resolving one period of the cosine. Based on this we take $N = 150$. With this, to satisfy the stability requirement, we need $M \geq 272$. We will take $M = 272$ ($\lambda = 0.999$) and $M = 546$ ($\lambda = 0.498$). The results from the computations are shown in Figure 5.8. Some observations about the computed solutions are these:

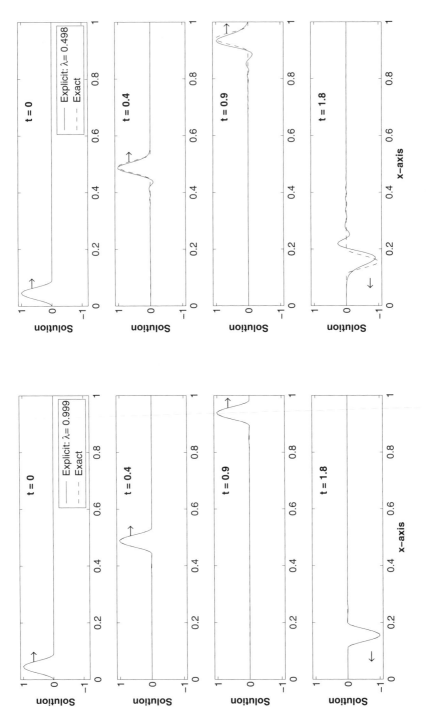

Figure 5.8. (M) Does using a smaller time step improve the accuracy of the explicit method? On the left, $M = 272$ time steps are used to reach $t = 1.8$, and on the right, $M = 546$ time steps are used. In the calculation $N = 150$.

1. The $M = 272$ results are quite good over the entire time interval. The amplitude, speed, and spread of the bump are computed very accurately, because there are no visible differences between the numerical and exact solutions. If we were to let the computation continue, generating multiple reflections of the bump off the boundaries, whatever differences there are between the exact and computed solutions would become more evident. However, over the time interval used in Figure 5.8 the two are in close agreement.

2. If the time step is reduced ($M = 546$) then the computed results are rather poor. The computed bump is too small, a little too wide, and too slow. This is a lot of negative toos. Moreover, it is unclear why this happens, because according to everything we know about the order of the truncation error, stability, and the CFL, there is no reason for this to occur. We saw the same thing when we used the upwind method to solve the advection equation in the last chapter. As before, the culprit is the specific form of the truncation error.

Based on this one example, the answer to the question posed in Figure 5.8 is that accuracy of the explicit method does depend on the relative size of h and k. Given a spatial resolution (h), then, the number of time points used to reach T should be chosen such that λ is close to one, but also satisfies $\lambda \leq 1$. A method that can be used to help support this conclusion involves plane waves, and this is the subject of the next section.

5.3 Numerical Plane Waves

The plane waves introduced earlier provide an effective tool in determining the wave properties of the solution of the wave equation. It is worth investigating what insights they can provide about the numerical solution. To start we assume a solution of the form

$$u_{ij} = e^{I(\overline{k}x_i - \overline{\omega}t_j)}. \tag{5.43}$$

Substituting this discrete plane wave into (5.31), one finds that

$$e^{-Ik\overline{\omega}} = \lambda^2 e^{Ih\overline{k}} + 2(1 - \lambda^2) + \lambda^2 e^{-Ih\overline{k}} - e^{Ik\overline{\omega}}. \tag{5.44}$$

Combining the exponentials, and using the identity $2\sin^2(\frac{\theta}{2}) = 1 - \cos(\theta)$, yields

$$\sin\left(\frac{\overline{\omega}k}{2}\right) = \pm\lambda\sin\left(\frac{\overline{k}h}{2}\right). \tag{5.45}$$

This is the numerical dispersion relation for the explicit method. It differs markedly from the actual dispersion relation for the wave equation, which is simply $\overline{\omega} = \pm c\overline{k}$. For one thing, given $\overline{k}$ there are an infinite number of solutions of (5.45) for $\overline{\omega}$. We are primarily interested in the case where h is small, and so it is assumed in what follows that $0 \leq \bar{k}h \leq \pi$. Also, we confine our attention to the case $-\pi \leq \bar{\omega}k \leq \pi$. With this, (5.45) can be written as

$$\bar{\omega} = \pm \frac{2}{k} \sin^{-1}\left(\lambda \sin\left(\frac{\bar{k}h}{2}\right)\right). \tag{5.46}$$

The question is, how well do the waves determined by (5.46) mimic the properties of the plane wave solutions? Numerical dispersion relations tend to be rather messy expressions, so to make things a bit more tractable we answer this by addressing a sequence of related, but simpler, questions.

1. Are there special values?
   Note that if $\lambda > 1$ in (5.46) then there are values of $\bar{k}$ for which $\lambda \sin(\bar{k}h/2) > 1$, and in such cases $\bar{\omega}$ is complex-valued. On the other hand, $\bar{\omega}$ is real-valued if $\lambda \leq 1$. Significantly, if $\lambda = 1$ then $\bar{\omega} = \pm c\bar{k}$, and this is in exact agreement with the dispersion for the wave equation.
2. Is the numerical method nondissipative?
   As pointed out above, $\bar{\omega}$ is real-valued for all $\bar{k}$ only if $\lambda \leq 1$. What this means is that in the stability region the numerical method has the same nondissipative property as the wave equation.
3. Is the numerical method nondispersive?
   To answer this question, we confine our attention to the stable region and assume $\lambda \leq 1$. In this case $\bar{\omega}$ is real-valued and therefore the numerical phase velocity is

$$v_{nph} = \frac{\bar{\omega}}{\bar{k}}$$
$$= \pm \frac{2}{k\bar{k}} \sin^{-1}\left(\lambda \sin\left(\frac{\bar{k}h}{2}\right)\right). \tag{5.47}$$

Assuming that h is small, and λ and $\bar{k}$ are fixed, then Taylor's theorem applied to (5.47) yields

$$v_{nph} \approx \pm c\left(1 - \frac{1}{12}(1 - \lambda^2)(\bar{k}h)^2\right). \tag{5.48}$$

Now, to be nondispersive, v_{nph} must be independent of $\bar{k}$, and the only time this happens in (5.47) and (5.48) is when $\lambda = 1$. Consequently, the numerical method is dispersive if $\lambda < 1$.

Is the fact that the explicit method is generally dispersive, yet the wave equation is nondispersive, significant? To answer this it is enough to consider only right-moving waves. From (5.48) we see that, except when $\lambda = 1$, the numerical phase velocity is less than the phase velocity for the wave equation. Therefore the numerical waves will lag behind the exact solution, and the shorter the wavelength, the more pronounced this will become. This observation is derived from the Taylor approximation in (5.48), but it applies to

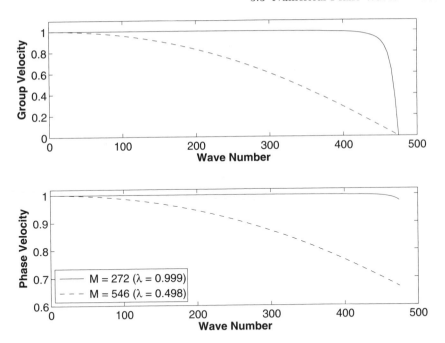

Figure 5.9. Numerical phase (5.45) and group (5.49) velocities for the explicit method when $0 \le \bar{k} \le \pi/h$. For comparison, for the wave equation $v_{ph} = v_g = 1$.

(5.47), as demonstrated in Figure 5.9. This explains the earlier calculations for the cosine bump in Figure 5.8. The waves needed to construct this bump all travel at different speeds. The fact that they all travel more slowly than they should, and when $\lambda = 0.498$ they are a lot slower, is why you see the bump coming apart at the later times and why the components are trailing the exact solution. This separation is not particularly evident in the computed solution curves for $\lambda = 0.999$, but this is not unexpected given the accuracy of the numerical phase velocity for λ values so close to one. However, if the calculation were allowed to continue, then eventually the differences would become apparent.

It is worth discussing the curves in Figure 5.9 a bit more. As already pointed out, the explicit method generally produces plane waves that travel more slowly than those for the wave equation. The curves in Figure 5.9 show that the numerical phase velocity for either M provides an accurate approximation only for smaller wave numbers. The implication of this, and the approximation in (5.48), is that the method is accurate only for longer wavelengths, and if we need to account accurately for short waves then it is necessary to use smaller values for h. This conclusion is not surprising because it is consistent with the idea that it takes more spatial points to describe sine or cosine curves with shorter wavelengths.

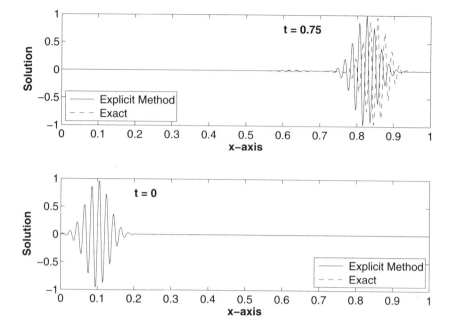

Figure 5.10. (**M**) Solution of the wave equation in the case where the initial profile is a wave packet as given in (5.51). Shown is the solution at the start and later at $t = 0.75$. In the calculation, $\gamma = 100\pi$, $\kappa = 500$, $x_0 = 0.1$, $N = 500$, and $M = 700$.

5.3.1 Numerical Group Velocity

Nothing has been said so far about the group velocity. To address this, note that the numerical group velocity can be determined using implicit differentiation on the numerical dispersion relation (5.45). The result is

$$v_{ng} = \frac{d\overline{\omega}}{d\overline{k}}$$

$$= \pm c \frac{\cos(\frac{\overline{k}h}{2})}{\sqrt{1 - \lambda^2 \sin^2(\frac{\overline{k}h}{2})}}. \qquad (5.49)$$

Assuming that h is small, and λ and $\overline{k}$ are fixed, then Taylor's theorem applied to (5.49) yields

$$v_{ng} \approx \pm c \left(1 - \frac{1}{8}(1 - \lambda^2)(\overline{k}h)^2 \right). \qquad (5.50)$$

As happened with v_{nph}, the group velocity for the numerical method depends on both the wave number and the grid. The exception is when $\lambda = 1$, in which case $v_{ng} = \pm c$. Moreover, from the Taylor approximations in (5.48) and (5.50) it is seen that for small h, the numerical group velocity is less than

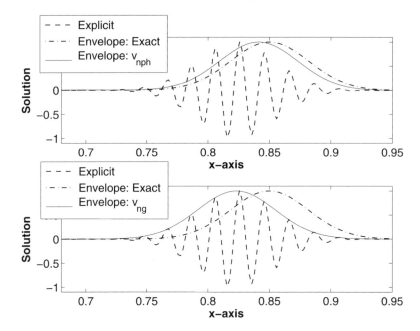

Figure 5.11. The wave packet from Figure 5.10 at $t = 0.75$ obtained using the explicit method. In the upper graph the exact envelope and the envelope predicted using the numerical phase velocity are shown. In the lower graph the exact envelope and the envelope predicted using the numerical group velocity are shown.

the numerical phase velocity and they are both smaller than the exact value. Again, the one exception to this is when $\lambda = 1$. As a final comment, note that the first wave number where $v_{ng} = 0$, which is $\bar{k}h = \pi$, is the same upper limit we introduced earlier when solving the dispersion relation to produce (5.46).

The numerical group velocity is shown in Figure 5.9 for $M = 273$ and $M = 546$ when the positive values are taken in (5.49). It is apparent that $M = 273$ does fairly well in producing the exact expression $v_g = 1$, while the $M = 547$ result is about as bad as it was for the phase velocity.

It is not particularly evident in the previous example why the group velocity is important. To address this suppose the initial condition is

$$f(x) = e^{-\kappa(x-x_0)^2} \sin(\gamma x). \tag{5.51}$$

Also, assume $g(x) = -cf'(x)$ and that the boundary conditions are $u(0,t) = f(-ct)$ and $u(1,t) = f(1-ct)$. In this case the solution is $u(x,t) = f(x-ct)$, which is nothing more than the initial profile moving to the right and centered at $x = x_0 + ct$. This solution is an example of a wave packet, where the exponential portion is the envelope, or pulse, and the sine term is the carrier wave. So, here is a question: presumably, the explicit method will produce a wave packet, and given what happened in the last example, it will probably be located a little bit behind the exact position. To demonstrate that this is indeed what happens, the numerical solution is shown in Figure 5.10, and as expected, the numerical packet lags the exact result when $t = 0.75$. Now, if we attempt to predict where the packet is located, should we use the numerical phase velocity or the numerical group velocity? The latter would predict it centered at $x = x_0 + v_{ng}t$ and the former would predict $x = x_0 + v_{nph}t$, where $\bar{k} = \gamma$. The results of these predictions are shown in Figure 5.11, and it is clear that the numerical group velocity is the winner. The reason it wins is that the group velocity is associated with the energy of the solution, and the envelope of the wave packet determines its energy.

5.4 Next Steps

The theoretical foundations for analyzing the numerical approximations for wave, or what are called hyperbolic, problems presents some interesting mathematical challenges. The conventional approach is to express the equation(s) in system form and then to use functional analytic methods to establish convergence. An introduction to this subject is given in ?. In terms of ideas for constructing an approximation to hyperbolic problems, one that is currently used by numerous codes involves finite volume methods. The derivation begins by expressing the problem in integral form, and then constructing approximations of the resulting integrals. The problems that arise in nature are usually nonlinear and multidimensional. As such they are very challenging and in many cases they are still active areas of research. A survey of finite volume methods can be found in LeVeque [2002].

Exercises

5.1. Find the dispersion relation, phase velocity, and group velocity for the plane wave solutions of the following equations. Also, state whether the equation is dispersive or dissipative.

(a) $u_t + gu_x + \mu u_{xxx} = 0$, where g, μ are positive constants. This equation describes the waves on a deep ocean.

(b) $c^2 u_{xx} = u_{tt} + \mu u_{xxxx}$, where c, μ are positive constants. This equation describes an elastic beam with an axial load. Also show that $v_{ph}^2 < v_g^2 < 4v_{ph}^2$.

(c) $I\hbar u_t = -\frac{\hbar^2}{2m}u_{xx} + Vu$, where $\hbar, m, V$ are positive constants. This is Schrödinger's equation, which arises in quantum mechanics.

5.2. Consider the problem of solving $\alpha u_{xx} + \beta u_{xt} + u_{tt} = 0$, where $u(x,0) = f(x)$ and $u_t(x,0) = g(x)$. Assume that α, β are constants, with α nonzero and $\beta^2 > 4\alpha$. Also, the spatial interval is $-\infty < x < \infty$.

(a) By factoring the equation in a manner similar to (5.7) solve the problem and show that the solution has the form

$$u(x,t) = \frac{b}{a+b}f(x+at) + \frac{a}{a+b}f(x-bt) + \frac{1}{a+b}\int_{x-bt}^{x+at} g(z)dz,$$

where a and b are constants. From this determine the domain of dependence for $(\overline{x},\overline{t})$.

(b) Find the dispersion relation, the phase velocity, and the group velocity for the equation. Is the equation dispersive or dissipative?

(c) What is the effect of a u_x term? In particular, is $\alpha u_{xx}+\beta u_{xt}+u_{tt}+\gamma u_x = 0$ dispersive or dissipative?

(d) What is the effect of a u_t term? In particular, is $\alpha u_{xx}+\beta u_{xt}+u_{tt}+\gamma u_t = 0$ dispersive or dissipative?

(e) What is the effect of a u term? In particular, is $\alpha u_{xx}+\beta u_{xt}+u_{tt}+\gamma u = 0$ dispersive or dissipative?

5.3. This problem investigates some of the consequences of using different approximations of the initial condition $u_t(x,0) = g(x)$.

(a) Use a one-sided $O(k^2)$ finite difference approximation of $u_t(x,0)$ to derive an approximation of the initial condition.

(b) The approximation from (a) involves $u_{i,1}$ and $u_{i,2}$. Explain how they are computed in conjunction with the finite difference equation for the wave equation.

(c) Does the approximation in (a) satisfy the CFL condition? Would an additional approximation, such as $y(x_i) = \frac{1}{2}\left(y(x_{i+1}) + y(x_{i-1})\right) + O(h^2)$, be of any help?

(d) Derive a $O(k^2)$ approximation of the initial condition by introducing a ghost point t_{-1}. Explain how this is used with the explicit method to compute the solution and also state whether the approximation is consistent with the CFL condition.

5.4. Consider the damped wave equation problem $u_{tt} + u_t = u_{xx}$, where the initial and boundary conditions are given in (5.2) and (5.3).

(a) Find an explicit finite difference approximation for this problem. Make sure to explain carefully how you approximate the initial conditions. All approximations must be $O(h^2)+O(k^2)$ and satisfy the CFL condition for this equation.

(b) What conditions, if any, are there on h and k so your scheme is stable?

(c) In the stability region, is your scheme dispersive or dissipative? Is the equation dispersive or dissipative?

5.5. The equation for the longitudinal motion in an elastic bar is

$$\frac{\partial}{\partial x}\left(E(x)\frac{\partial u}{\partial x}\right) = \frac{\partial^2 u}{\partial t^2},$$

where the initial and boundary conditions are given in (5.2) and (5.3). Assume that $E(x)$ is a smooth positive function. Find an explicit finite difference approximation for this problem that has truncation error $O(h^2) + O(k^2)$. Make sure to explain carefully how you approximate the initial conditions.

5.6. Consider the Klein–Gordon equation $c^2 u_{xx} = u_{tt} + bu$, where the initial and boundary conditions are given in (5.2) and (5.3) with $g(x) = 0$. Here c, b are positive constants. The exact solution is

$$u(x,t) = \sum_{n=1}^{\infty} a_n \sin(\lambda_n x) \cos\left(t\sqrt{b + \lambda_n^2 c^2}\right),$$

where $\lambda_n = n\pi$ and $a_n = 2\int_0^1 f(x)\sin(\lambda_n x)\,dx$.

(a) Find an explicit finite difference approximation for this problem. Make sure to explain carefully how you approximate the initial conditions. All approximations must be $O(h^2) + O(k^2)$ and satisfy the CFL condition for this equation.

(b) What conditions, if any, are there on h and k so your scheme is stable?

(c) In the stability region, is your scheme dispersive or dissipative?
    In the remainder of the problem assume $c = 1, b = 4$, and

$$f(x) = \begin{cases} \frac{1}{2}(1 - \cos(\frac{2\pi x}{a})) & \text{if } 0 \le x \le a, \\ 0 & \text{otherwise,} \end{cases}$$

where $a = 0.09$. Pick a value for the number of spatial points you feel is adequate for the spatial resolution of the solution (it must be greater than 150). This is to be used in (d) and (e).

(d) Pick a value for k that satisfies the stability condition (pick one close to the stability boundary). Calculate both the numerical and exact solutions up to $t = 1.8$, for $0 \le x \le 1$, and use this to plot profiles similar to those in Figure 5.8.

(e) Redo (d) but use twice the number of time points.

(f) On the same axes, plot v_{ph} for part (d) and for part (e).

(g) On the same axes, plot v_g for part (d) and for part (e).

(h) Explain the results in (d) and (e) using your results from (c), (f), (g) and anything else about the problem you think is relevant (e.g., is the numerical solution ahead of, or behind, the exact solution and why and how does this vary with k; what about the amplitude?).

5.7. This problem concerns the advection equation $u_t + au_x = 0$, where a is a positive constant.

(a) What are the dispersion relation and resulting phase and group velocities? Is the equation dissipative or dispersive?

(b) For the upwind method find the numerical dispersion relation and then solve this equation for $\bar{\omega}$.

(c) Assuming that h is small, and λ and $\overline{k}$ are fixed, find the first two terms in the Taylor approximation of $\overline{\omega}_r$ and the first term for $\overline{\omega}_i$. With this determine the corresponding approximations for the numerical phase and group velocities.

(d) Based on the Taylor approximations in (c), under what conditions will the upwind method be nondissipative? When will it be nondispersive? Will the numerical plane waves lead or lag their counterparts for the exact equation? Will long or short waves propagate further using this numerical method before they effectively dissipate?

(e) The square bump in Figure 4.2 is composed of waves covering a broad range of wavelengths. Use this, and your results from (c) and (d), to explain the properties of the computed solution when $M = 72$ and $M = 100$ in Figure 4.6.

(f) Suppose the upwind method is used to calculate the solution at $t = \frac{3}{4}$ in the case where the initial condition is

$$u(x,0) = e^{-\alpha(x-x_0)^2}\sin(\gamma x), \quad \text{for } 0 \le x \le 1,$$

where $\alpha = 500$, $x_0 = \frac{1}{10}$, and $\gamma = 100\pi$. Also, assume $c = 1$ and $h = \frac{1}{500}$. Use your results from (c) and (d) to estimate what k (or λ) must be so that the computed amplitude is within 90% of the exact value. What is the answer if γ is smaller, say $\gamma = 10\pi$?

5.8. Redo Exercise 5.7 for the Lax–Wendroff method, where Figure 4.11 is used for part (e).

5.9. A numerical approximation of the advection equation can produce an inaccurate result if the method is either dispersive or dissipative.

(a) Use the result of Exercise 5.8(c) to explain why, for small h, the dominate contribution to the error for the Lax–Wendroff method is dispersion (assume $\lambda < 1$).

(b) Use the result of Exercise 5.7(c) to explain why, for small h, dispersion and dissipation are generally of equal importance for the upwind method (assume $\lambda < 1$). However, show that there is a value of λ for which the dominate affect is dissipation.

(c) Suppose you could select between a method with no dispersion, or a method with no dissipation. Which one would you pick? Make sure to explain, in detail, why.

5.10. Suppose that to solve the advection equation one uses the finite difference equation

$$u_{i,j+1} = u_{i-1,j}.$$

(a) Is the method stable? Is it monotone?
(b) Show that the method is not dissipative but it is dispersive.
(c) What is the CFL condition?

(d) Is the method consistent? If so, what is the truncation error?

5.11. The domain of dependence (DoD) is connected with the speed at which information travels in the problem. Assume there are velocities v_m and v_M such that the DoD can be written as $\overline{x} - v_m \overline{t} \leq x \leq \overline{x} + v_M \overline{t}$. For example, for the wave equation, $v_m = v_M = c$. Using the Klein–Gordon equation as the test case, develop a hypothesis on how v_m and v_M depend on either the phase or group velocity.

5.12. This problem concerns the conservation of energy for the wave equation and its finite difference approximation.
(a) The energy is defined as

$$E(t) = \frac{1}{2} \int_0^\ell \left((u_t)^2 + (u_x)^2 \right) dx.$$

Show that $E(t)$ is constant.
(b) Using an approximation of the spatial derivative derive the following finite difference approximation of the wave equation

$$u_i'(t) = \frac{1}{h^2} \left(u_{i+1}(t) - 2u_i(t) + u_{i-1}(t) \right).$$

With this show that the discrete energy

$$E_h(t) = \frac{1}{4h} \left(u_1^2 + u_N^2 \right) + \frac{h}{2} \sum_{i=1}^N \left[(u_i')^2 + \frac{(u_{i+1} - u_i)^2 + (u_i - u_{i-1})^2}{2h^2} \right]$$

is constant.

6

Elliptic Problems

6.1 Introduction

One might title this chapter "The Challenges of Dimensionality," or perhaps "Why One-Dimensional Models Aren't So Bad After All." The reason is that we address how to solve boundary value problems with more than one spatial variable, and this will require us to consider some unique challenges. To introduce the ideas we will limit the development to two dimensions and consider how to find the function $u(x, y)$ that satisfies

$$\nabla \cdot (a\nabla u) + \mathbf{b} \cdot \nabla u + cu = f, \quad \text{for } (x, y) \in D, \qquad (6.1)$$

where

$$\nabla \equiv (\frac{\partial}{\partial x}, \frac{\partial}{\partial y}) \qquad (6.2)$$

is the gradient and D is a bounded domain in the xy-plane, as indicated in Figure 6.1. Also, the functions $a, \mathbf{b}, c, f$ are smooth with $a > 0$ and $c \leq 0$ on $\overline{D} = D \cup \partial D$, where ∂D is the boundary of D. We will use a Dirichlet boundary condition, which means that the solution is specified around the boundary, and the general form is

$$u = g(x, y), \quad \text{for } (x, y) \in \partial D, \qquad (6.3)$$

where g is given. The particular case of $a = 1, \mathbf{b} = \mathbf{0}, c = 0$ produces Poisson's equation. If, in addition, $f = 0$, one obtains Laplace's equation given as

$$\nabla^2 u = 0, \quad \text{for } (x, y) \in D, \qquad (6.4)$$

where

$$\nabla^2 \equiv \frac{\partial^2}{\partial x^2} + \frac{\partial^2}{\partial y^2} \qquad (6.5)$$

is the Laplacian.

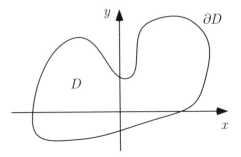

Figure 6.1. Schematic of the planar domain D and its boundary ∂D used for Laplace's equation.

The problem we are considering is so fundamental that it arises in most areas of science and engineering. For example, it plays a central role in electrostatics, where u represents the electrostatic potential and g is the surface charge. An interesting application of this arises in tomography, when electric current is used to determine location and shape information inside an object, such as in the human body as shown in Figure 6.2. It also arises in incompressible fluid dynamics, where u is the fluid potential. A novel use of this arises in robotics, when one tries to determine the motion, or path, of a robot through complex terrain using virtual potential theory (Figure 6.2). As a third example, the equation for multidimensional heat flow is $\nabla^2 u = \frac{\partial u}{\partial t}$ and when studying the steady-state problem (where $\frac{\partial u}{\partial t} = 0$) one obtains Laplace's equation.

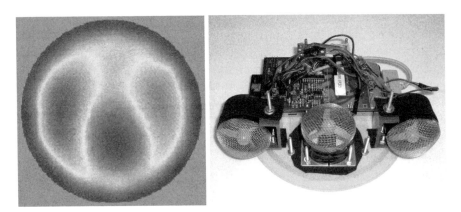

Figure 6.2. Examples where Laplace's equation arises. The picture on the left comes from electrical impedance tomography used to create images inside the human body (Isaacson et al. [2004]). Another application is path planning for robots, such as the robotic hovercraft pictured on the right (Connolly and Grupen [1993], Waydo and Murray [2003]).

The objective of this chapter is to develop some of the numerical methods necessary to deal with the additional spatial dimension. At the beginning the presentation follows the steps used for the one-dimensional boundary value problems in Chapter 2. We derive a finite difference approximation without much difficulty and then assemble the result into a matrix equation. The next step is the one that introduces new ideas. The additional dimension makes the algebraic problem rather large, and this will cause us to reconsider how best to find the solution. For this reason a large portion of this chapter is dedicated to deriving and then implementing effective methods for the large algebraic systems coming from the finite difference approximations of Laplace's equation.

6.1.1 Solutions

Given the diverse areas where Laplace's equation is found, and the centuries scientists have had to study it, one should not be too surprised to discover that there are numerous ways to find the solution. Our objective is to have a representation that can easily be used to compare with the numerical results, and for this, old-fashioned separation of variables will do nicely.

Suppose D is the rectangular domain shown in Figure 6.3 along with the indicated boundary conditions. Using separation of variables one finds that

$$u(x, y) = \sum_{n=1}^{\infty} a_n \sinh(\lambda_n y) \sin(\lambda_n x), \tag{6.6}$$

where $\lambda_n = n\pi/a$, and the Fourier coefficients are

$$a_n = \frac{2}{a \sinh(\lambda_n b)} \int_0^a g(x) \sin(\lambda_n x) dx. \tag{6.7}$$

To investigate the properties of the solution we consider three examples.

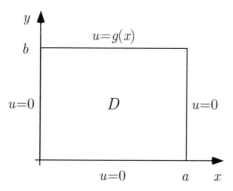

Figure 6.3. Rectangular domain and corresponding boundary conditions used to obtain (6.6).

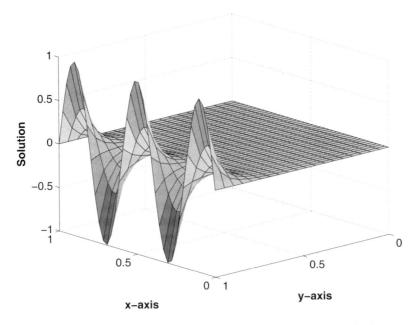

Figure 6.4. Solution of Laplace's equation as given in (6.8).

1. $g(x) = \sin(\lambda_5 x)$

   This corresponds to one of the modes in the series, and the solution in this case reduces to the single term

   $$u(x, y) = \frac{\sinh(\lambda_5 y)}{\sinh(\lambda_5 b)} \sin(\lambda_5 x). \tag{6.8}$$

   This function is shown in Figure 6.4 in the case $a = b = 1$. It is interesting how localized the solution appears in the sense that the oscillations in the boundary condition at $y = 1$ are almost nonexistent over much of the domain. Although the oscillations are present everywhere, the sinh term results in exponential decay away from $y = 1$, and this produces the localization seen in the figure. This effect is even more pronounced for the higher modes because of the larger value of λ_n in the sinh function. Another point to make is that to generate this plot 40 points were used along the x-axis and this number does reasonably well in rendering the sine function. For a higher mode it would be necessary to use a proportionally larger number of grid points. For example, for $g(x) = \sin(\lambda_{50} x)$, then 400 points would be used along the x-axis.

2. $g(x) = x(1-x)(\frac{4}{5} - x)e^{6x}$

   Assuming $a = b = 1$, the Fourier coefficient (6.7) is given as

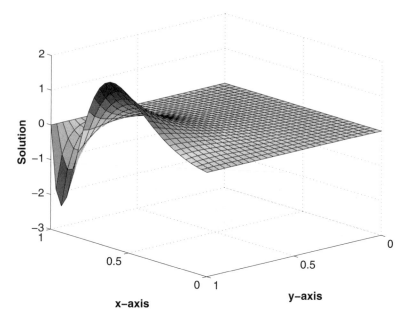

Figure 6.5. Solution of Laplace's equation calculated from the Fourier series (6.6) using the coefficients given in (6.9).

$$a_n = \frac{12\lambda_n \left(4\gamma_n(\lambda_n^2 - 36)(\lambda_n^2 + 6) - 672\,\lambda_n^2 - 5\,\lambda_n^4 - 26352\right)}{5\sinh(\lambda_n)\,(\lambda_n^2 + 36)^4}, \qquad (6.9)$$

where $\gamma_n = (-1)^n e^6$. The solution in this case is shown in Figure 6.5. This particular example makes for a good test case when evaluating how well a numerical method does in solving Laplace's equation. The reason is that it represents a typical problem with a smooth solution but it requires multiple modes in the series to accurately calculate the solution. As an example, to compute $g(x)$ with an error of less than 10^{-4} requires using up to about 400 terms in the series.

3. $g(x) = \begin{cases} 1 & \text{if } \frac{1}{4} \le x \le \frac{3}{4}, \\ 0 & \text{otherwise} \end{cases}$

Assuming $\frac{3}{4} < a$, the Fourier coefficient is given as

$$a_n = \frac{2}{a\lambda_n \sinh(\lambda_n b)}\left(\cos(\lambda_n/4) - \cos(3\lambda_n/4)\right). \qquad (6.10)$$

The solution is shown in Figure 6.6 in the case $a = b = 1$. It is interesting how similar this looks compared to the analogous problem for the heat equation, as given in Figure 3.3.

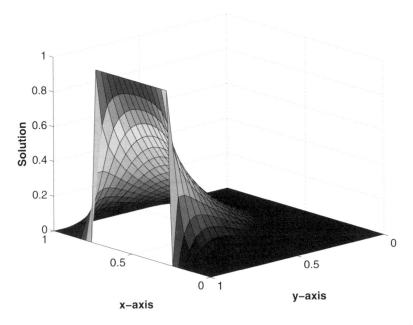

Figure 6.6. Solution of Laplace's equation calculated from the Fourier series (6.6) using the coefficients given in (6.10).

6.1.2 Properties of the Solution

The examples in Figures 6.4 to 6.6, along with the Fourier series solution (6.6), demonstrate several important properties of the solutions of Laplace's equation, when using a Dirichlet boundary condition, that are worth remembering when finding numerical solutions. Of particular importance are the following:

Smoothness: Even with a boundary condition containing jumps as in Figure 6.6, the solution away from the boundary is smooth.

Maximum and Minimum Principles: The maximum and minimum of the solution occur on the boundary. One consequence of this is that the problem is stable in the sense that if small changes are made in the boundary condition then the resulting changes in the solution over the entire domain are small. This follows because if u_1 and u_2 are solutions of Laplace's equation, then their difference $u_1 - u_2$ is also a solution. Therefore, the maximum and minimum values of $u_1 - u_2$ are determined by the differences between u_1 and u_2 on the boundary. If these two functions differ by only a small amount on the boundary then their difference $u_1 - u_2$ will be small throughout the entire domain.

Global Dependence: Changing the solution in a continuous manner anywhere in D or along the boundary will affect the solution throughout D. This can be seen in the Fourier series solution in (6.6). Because the coefficients in (6.7) depend on the value of $g(x)$ over the entire interval, changing this boundary condition over even a small portion of the boundary potentially affects every term in the series. This is important from a numerical point of view because if the method incorrectly calculates the solution at even one grid point, the effect of this will be felt everywhere. Fortunately, as seen in Figure 6.4, there is a certain amount of spatial localization if the imperfection involves shorter wavelengths. The latter occurs, for example, with round-off error (e.g., see Figure A.1).

The value of the above three observations is more qualitative than quantitative when one is computing the solution of Laplace's equation. For example, if the computed solution does not obey the maximum principle then we should seriously question whether the answer is correct. As we have seen earlier, such observational tests are easy and can prove invaluable when one is developing and testing a numerical method.

6.2 Finite Difference Approximation

We will derive a finite difference approximation for Laplace's equation when the domain is rectangular. The domain and boundary conditions are shown in Figure 6.7. As usual, we use a rectangular grid (x_i, y_j), where

$$x_i = ih, \quad \text{for } i = 0, 1, 2, \ldots, N+1, \tag{6.11}$$
$$y_j = jk, \quad \text{for } j = 0, 1, 2, \ldots, M+1. \tag{6.12}$$

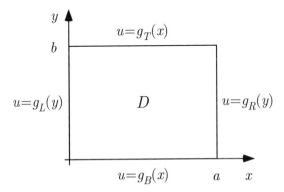

Figure 6.7. Rectangular domain and boundary conditions used to derive the finite difference approximation of Laplace's equation.

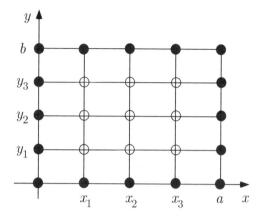

Figure 6.8. Example of the grid system used to find the numerical solution of Laplace's equation. As shown there are nine points, the hollow dots, where the solution is to be computed. The values of the solution at the grid points marked with solid dots are known from the given boundary conditions. As shown, $N = M = 3$.

The step sizes are $h = a/(N + 1)$ and $k = b/(M + 1)$. An illustration of the grid system we are using, in the particular case $N = M = 3$, is shown in Figure 6.8.

In preparation for introducing finite difference approximations, we evaluate the equation at (x_i, y_j) to obtain

$$u_{xx}(x_i, y_j) + u_{yy}(x_i, y_j) = 0. \tag{6.13}$$

We use centered differences to approximate the derivatives and this gives us

$$\frac{u(x_{i+1}, y_j) - 2u(x_i, y_j) + u(x_{i-1}, y_j)}{h^2}$$
$$+ \frac{u(x_i, y_{j+1}) - 2u(x_i, y_j) + u(x_i, y_{j-1})}{k^2} + \tau_{ij} = 0, \tag{6.14}$$

where $\tau_{ij} = O(h^2) + O(k^2)$ is the truncation error. The stencil for this approximation is shown in Figure 6.9. Rearranging things in (6.14), we have

$$-\lambda^2 u(x_{i+1}, y_j) + 2(1 + \lambda^2)u(x_i, y_j) - \lambda^2 u(x_{i-1}, y_j)$$
$$- u(x_i, y_{j+1}) - u(x_i, y_{j-1}) - k^2\tau_{ij} = 0, \tag{6.15}$$

where

$$\lambda = \frac{k}{h}. \tag{6.16}$$

Dropping the truncation error gives us the following finite difference approximation to the Laplace's equation

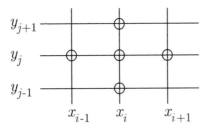

Figure 6.9. Stencil for the five-point finite difference approximation (6.17) of Laplace's equation.

$$-\lambda^2 u_{i+1,j} + 2(1+\lambda^2)u_{ij} - \lambda^2 u_{i-1,j} - u_{i,j+1} - u_{i,j-1} = 0, \quad \text{for} \quad \begin{array}{l} i = 1, \ldots, N, \\ j = 1, \ldots, M. \end{array} \quad (6.17)$$

The boundary conditions are

(i) $x = 0$: $u_{0,j} = g_L(y_j)$, for $j = 1, \ldots, M$,

(ii) $x = a$: $u_{N+1,j} = g_R(y_j)$, for $j = 1, \ldots, M$,

(iii) $y = 0$: $u_{i,0} = g_B(x_i)$, for $i = 1, \ldots, N$,

(iv) $y = b$: $u_{i,M+1} = g_T(x_i)$, for $i = 1, \ldots, N$.

The finite difference equation in (6.17) is known as the five-point scheme for Laplace's equation. Note that there are as many equations as there are points (x_i, y_j) within D. In our case, there are $n = N \times M$ equations. Also, it is interesting to observe that because of the shape of the stencil the values of the solution at the vertices of the domain do not contribute to the approximation.

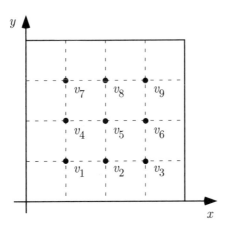

Figure 6.10. Grid and renumbering of the points to construct the vector **v** in the case $N = M = 3$.

$$
\begin{pmatrix}
\beta & -\lambda^2 & 0 & -1 & 0 & 0 & 0 & 0 & 0 \\
-\lambda^2 & \beta & -\lambda^2 & 0 & -1 & 0 & 0 & 0 & 0 \\
0 & -\lambda^2 & \beta & 0 & 0 & -1 & 0 & 0 & 0 \\
-1 & 0 & 0 & \beta & -\lambda^2 & 0 & -1 & 0 & 0 \\
0 & -1 & 0 & -\lambda^2 & \beta & -\lambda^2 & 0 & -1 & 0 \\
0 & 0 & -1 & 0 & -\lambda^2 & \beta & 0 & 0 & -1 \\
0 & 0 & 0 & -1 & 0 & 0 & \beta & -\lambda^2 & 0 \\
0 & 0 & 0 & 0 & -1 & 0 & -\lambda^2 & \beta & -\lambda^2 \\
0 & 0 & 0 & 0 & 0 & -1 & 0 & -\lambda^2 & \beta
\end{pmatrix}
\begin{pmatrix}
v_1 \\ v_2 \\ v_3 \\ v_4 \\ v_5 \\ v_6 \\ v_7 \\ v_8 \\ v_9
\end{pmatrix}
=
\begin{pmatrix}
g_B(x_1) + \lambda^2 g_L(y_1) \\
g_B(x_2) \\
g_B(x_3) + \lambda^2 g_R(y_1) \\
\lambda^2 g_L(y_2) \\
0 \\
\lambda^2 g_R(y_2) \\
g_T(x_1) + \lambda^2 g_L(y_3) \\
g_T(x_2) \\
g_T(x_3) + \lambda^2 g_R(y_3)
\end{pmatrix}
$$

Table 6.1. Matrix equation coming from the five-point scheme for Laplace's equation in the case $N = M = 3$. In the matrix, $\beta = 2(1 + \lambda^2)$.

6.2.1 Building the Matrix

The next step is to assemble all n equations into a matrix equation $\mathbf{Av} = \mathbf{b}$, where $\mathbf{A}$ is an $n \times n$ matrix. The vector $\mathbf{v}$ contains the unknowns, and we need to linearly order the u_{ij}'s to construct this vector. To explain how this will be done, the grid for the case $N = M = 3$ is shown in Figure 6.10. We will use what is known as a lexicographic ordering, which yields

$$
\begin{aligned}
v_1 &= u_{11}, & v_2 &= u_{21}, & v_3 &= u_{31}, \\
v_4 &= u_{12}, & v_5 &= u_{22}, & v_6 &= u_{32}, \\
v_7 &= u_{13}, & v_8 &= u_{23}, & v_9 &= u_{33}.
\end{aligned}
$$

In general, the formula connecting v_ℓ with u_{ij} is

$$v_\ell = u_{ij}, \quad \text{for } \ell = (j - 1)N + i. \tag{6.18}$$

It should be remembered that this formula applies to the grid points in D and not to those on the boundary.

With the given ordering of the grid points, the finite difference equation (6.17) now becomes

$$-\lambda^2 v_{\ell+1} + 2(1 + \lambda^2)v_\ell - \lambda^2 v_{\ell-1} - v_{\ell+N} - v_{\ell-N} = 0, \tag{6.19}$$

which can be written in matrix form as

$$\mathbf{Av} = \mathbf{b}. \tag{6.20}$$

The finite difference equation in (6.19) provides the ℓth row in (6.20) for those grid points that are not next to the boundary (e.g., the point for v_5 in Figure

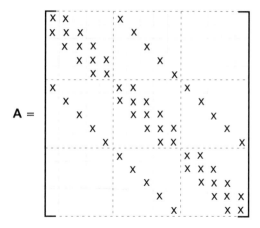

Figure 6.11. Schematic representation of the coefficient matrix obtained from (6.17) and (6.19).

6.10). At points next to a boundary one or more of the terms $v_{\ell\pm1}$ and $v_{\ell\pm N}$ must be modified to account for the boundary conditions. As an example, in the case $N = M = 3$ one obtains the grid in Figure 6.10 and the matrix equation given in Table 6.1. Both in this example and in general each row of the matrix $\mathbf{A}$ has $2(1 + \lambda^2)$ on the diagonal and -1 in the $\ell \pm N$ positions (when in range). On the sub- and superdiagonals there is $-\lambda^2$, except when the point is next to the boundary, in which case the entry is zero. For example, in Table 6.1 there is a zero in the $(4, 3)$ position of the matrix, and this gives rise to the $\lambda^2 g_L(y_2)$ term on the right-hand side. For the general case the matrix $\mathbf{A}$ has the structure shown in Figure 6.11. It is evident from this schematic representation that $\mathbf{A}$ has a block structure. Each block is an $N \times N$ symmetric matrix, and together they give $\mathbf{A}$ the following form

$$\mathbf{A} = \begin{pmatrix} \mathbf{T} & \mathbf{D} & & & \\ \mathbf{D} & \mathbf{T} & \mathbf{D} & & 0 \\ & \mathbf{D} & \mathbf{T} & \mathbf{D} & \\ & & \ddots & \ddots & \ddots \\ & 0 & & \mathbf{D} & \mathbf{T} & \mathbf{D} \\ & & & & \mathbf{D} & \mathbf{T} \end{pmatrix}. \tag{6.21}$$

In this expression $\mathbf{T}$ is an $N \times N$ symmetric tridiagonal matrix with $2(1+\lambda^2)$ on the diagonal and $-\lambda^2$ on the sub- and superdiagonals. The matrix $\mathbf{D}$ is an $N \times N$ diagonal matrix with -1 on the diagonal.

The matrix $\mathbf{A}$ plays such an important role in what follows that it is worthwhile to make a few observations about some of its properties that will have an impact on how we solve the matrix equation:

1. It is symmetric and positive definite (this will be shown later).

2. It is banded. This means that the entries of the matrix are zero outside a band along the diagonal. For $\mathbf{A}$, all entries more than N spaces away from the diagonal are zero. In the parlance of numerical linear algebra, it has bandwidth N.
3. It is sparse. This means that the number of nonzero entries is much less than the number of zero entries. For $\mathbf{A}$ there are fewer than $5n$ nonzero entries out of a total of n^2.
4. It has the potential to be quite large. For example, if we decide that 100 interior grid points should be used along each axis then $\mathbf{A}$ will be of size $10^4 \times 10^4$.

The last observation will force us to think very seriously about how to solve $\mathbf{Av} = \mathbf{b}$. In the earlier chapters, when faced with solving matrix equations we used a standard LU factorization method. However, this is not necessarily the best choice here, because the matrix can easily become so large that it is not possible even to store it in the computer's memory. One possibility is to make use of Observations (1) and (2) from above, and employ a banded version of the Cholesky factorization. This is a reasonable approach but one that will not be pursued here. Rather, we will investigate methods that take maximum advantage of the sparseness of $\mathbf{A}$. This will lead us into the world of iterative methods. There are numerous iterative solvers to choose from, having names such as Jacobi and Gauss–Seidel. For large spare systems, however, two stand out in terms of their computational effectiveness. These are the conjugate gradient method and the multigrid method. The former is based on a relatively simple idea that produces an extremely effective solver, and it is the one we will pursue here.

The remainder of the chapter concentrates on the matrix problem. The reason is that constructing a finite difference approximation to other elliptic equations, such as the one in (6.1), is straightforward and uses ideas that have been employed repeatedly in this text. So, the real question is not how to construct a finite difference approximation, but what to do with it once it is derived. Certainly complications arise, such as with variable boundaries, and these will be brought up later in the chapter. Before working on the matrix equation there are a few details to complete that are connected with proving that a matrix is positive definite.

6.2.2 Positive Definite Matrices

The methods to be used to solve the matrix equation will require $\mathbf{A}$ to be positive definite. It is therefore imperative that we have a way to determine whether the matrix has this property. In the discussion to follow it is assumed that $\mathbf{A}$ is symmetric. One consequence of this assumption is that $\mathbf{A}$ has only real eigenvalues.

There are various ways to define what it means to be positive definite. For example, the following statements are equivalent if $\mathbf{A}$ is a symmetric matrix:

PD1: $\mathbf{A}$ is positive definite,

PD2: $\mathbf{v}^T \mathbf{A} \mathbf{v} > 0,\ \forall\, \mathbf{v} \neq \mathbf{0}$,

PD3: $\mathbf{A}$ has only positive eigenvalues.

What this means is that we can use either PD2 or PD3 to establish whether a matrix is positive definite. The problem is that neither is particularly easy to use. Fortunately, there are special cases that will be sufficient for our problems and that can be used almost by inspection.

The first special case is actually a negative result in the sense that it can be used to determine when a matrix is not positive definite.

TEST 1: A matrix $\mathbf{A}$ is not positive definite if any diagonal entry is negative or zero.

The proof of this is outlined in Exercise 6.1. As an example of its usefulness, the matrix

$$\mathbf{A} = \begin{pmatrix} 4 & 1 \\ 1 & -1 \end{pmatrix} \tag{6.22}$$

is not positive definite because it has a negative diagonal entry. On the other hand, we can make no conclusion about the matrix in Table 6.1 or the one in (6.20). What we can say, however, is that if one does not like all the minus signs in this matrix and multiplies the equation by -1, the resulting matrix will not be positive definite. So, in terms of constructing matrix equations to solve elliptic problems, Test 1 can be thought of as a minus sign adjuster.

To produce a test that can determine whether a matrix is positive definite we will use the Gershgorin–Taussky theorem. The statement of this result introduces a requirement about the location of the nonzero entries in the matrix, particularly those off the diagonal. The specific condition is that the matrix must be irreducible. There is a relatively easy test for this that involves a paper-and-pencil like construction. To explain, suppose we are given an $n \times n$ matrix $\mathbf{A}$ with entries a_{ij}. The first step in the construction is to place n points $P_1, P_2, \ldots, P_n$ along a line as in Figure 6.12. For each nonzero entry a_{ij} in the matrix that is not on the diagonal, draw a directed curve from P_i to P_j (the curves can cross over each other but they are not considered to intersect when

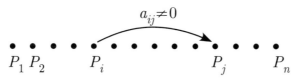

Figure 6.12. To test whether an $n \times n$ matrix is irreducible place n points $P_1, P_2, \ldots, P_n$ along a line and then connect them with directed line segments according to whether a_{ij} is nonzero.

they do). The set of curves obtained in this way is called the directed graph for the matrix.

Examples
Below are two matrices and their associated directed graphs.

$$\mathbf{A}_1 = \begin{pmatrix} 2 & 0 & 0 & 1 \\ 1 & 1 & 0 & 0 \\ 0 & 3 & 0 & 0 \\ 0 & 0 & 1 & 0 \end{pmatrix}$$

$$\mathbf{A}_2 = \begin{pmatrix} 2 & 0 & 1 & 0 \\ 1 & 1 & 0 & 0 \\ 0 & 3 & 0 & 0 \\ 0 & 0 & 1 & 0 \end{pmatrix}$$

With the directed graph, $\mathbf{A}$ is said to be irreducible if it is possible to start at any point P_i and then follow the directed curves and arrive at any other point P_j. Based on this, in the above example $\mathbf{A}_1$ is irreducible but $\mathbf{A}_2$ is not because it is not possible to go from P_3 to P_4. It might seem strange that this graphical procedure can be of any help for what we are trying to do, but the connectivity of the path is actually saying something important about the location of the nonzero entries in the matrix. Those who are interested in the theory underlying this method should consult Varga [2000]. One last comment is that this definition does not require, in contrast to the other results in this section, that the matrix be symmetric.

To state the theorem we will also need to introduce the row sums

$$r_i = \sum_{\substack{j=1 \\ j \neq i}}^{n} |a_{ij}|, \tag{6.23}$$

and the intervals $I_i = [a_{ii} - r_i, a_{ii} + r_i]$. So, I_i is the closed interval centered at the diagonal value of the ith row, with radius equal to the sum of the remaining row entries (in absolute value). Gershgorin–Taussky states the following:

GT1: Every eigenvalue of $\mathbf{A}$ is in at least one of the intervals I_i.
GT2: If an eigenvalue of $\mathbf{A}$ is on the boundary of one of the intervals, and if $\mathbf{A}$ is irreducible, then the eigenvalue is on the boundary of all the intervals.

The proof of this result can be found in Varga [2000]. As an example of its usefulness consider the following two symmetric 3×3 matrices, and their corresponding Gershgorin intervals.

$$\mathbf{A}_3 = \begin{pmatrix} 2 & 1 & -2 \\ 1 & 7 & 2 \\ -2 & 2 & 12 \end{pmatrix} \quad \Rightarrow \quad \begin{array}{l} I_1 = [-1, 5], \\ I_2 = [4, 10], \\ I_3 = [8, 16], \end{array}$$

$$\mathbf{A}_4 = \begin{pmatrix} 2 & 1 & 0 \\ 1 & 7 & 2 \\ 0 & 2 & 12 \end{pmatrix} \quad \Rightarrow \quad \begin{array}{l} I_1 = [1, 3], \\ I_2 = [4, 10], \\ I_3 = [10, 14]. \end{array}$$

The three intervals for $\mathbf{A}_4$ are positive, but this is not the case for $\mathbf{A}_3$. There-fore, using GT1, of the two matrices only $\mathbf{A}_4$ is guaranteed to have positive eigenvalues. This observation can be used to check on positive definiteness using PD3. In particular, if all the intervals are positive then the eigenvalues are positive, and this gives rise to the following result.

TEST 2: A matrix $\mathbf{A}$ is positive definite if $r_i < a_{ii} \ \forall \ i$.

Matrices that satisfy the inequality in this test are strict diagonally dominant. This was introduced earlier, in Chapter 2, when we were studying tridiagonal matrices. Unfortunately, this test does not work on the matrix in (6.20) for larger values of N and M, because there are one or more rows where $r_i = a_{ii}$. For example, this happens for the fifth row in the matrix in Table 6.1. Consequently, we need a slightly stronger test, and this comes from combining GT1 and GT2 to produce the following result.

TEST 3: A matrix $\mathbf{A}$ is positive definite if the following hold: (i) $r_i \leq a_{ii} \ \forall \ i$, (ii) $0 < a_{ii} \ \forall \ i$, (iii) there is at least one row where $r_i < a_{ii}$, and (iv) $\mathbf{A}$ is irreducible.

In this test, conditions (i) and (ii) guarantee that the Gershgorin intervals are at least nonnegative. This still leaves open the possibility that zero is an eigenvalue, but conditions (iii) and (iv) prevent this from happening. It is not hard to show that the matrix in (6.20) satisfies the conditions of this test, and it is therefore positive definite.

There is an additional piece of useful information concerning the eigenval-ues of $\mathbf{A}$ that will come in handy later in the chapter. If the domain in Figure 6.7 is the unit square, and if $h = k$, then the eigenvectors of $\mathbf{A}$ can be deter-mined from the eigenfunctions of the Laplacian. To explain how, recall that the eigenvalue problem for the Laplacian is $\nabla^2 w = \bar{\lambda} w$, where $w = 0$ around the boundary. Using separation of variables one finds that the eigenfunctions are $w(x, y) = \sin(p\pi x)\sin(q\pi y)$, where $p, q = 1, 2, 3, \ldots$. For any given p, q satisfying $p, q \leq N$ one can show that the vector $\mathbf{w}$, formed from evaluating w at the grid points, is an eigenvector for $\mathbf{A}$. The corresponding eigenvalues for $\mathbf{A}$ are

$$\bar{\lambda}_{pq} = 4\left[\sin^2\left(\frac{p\pi h}{2}\right) + \sin^2\left(\frac{q\pi h}{2}\right)\right], \quad \text{for } \begin{array}{l} p = 1, 2, \ldots, N, \\ q = 1, 2, \ldots, N. \end{array} \quad (6.24)$$

From this we get that the smallest eigenvalue of $\mathbf{A}$ is $\lambda_m = 8\sin^2(\pi h/2)$ and the largest is $\lambda_M = 8\sin^2(N\pi h/2)$. This is potentially useful information, because using the Euclidean norm, the condition number $\mathrm{cond}_2(\mathbf{A})$ for $\mathbf{A}$ is the ratio of these two eigenvalues. What we have found is that for large systems,

$$\mathrm{cond}_2(\mathbf{A}) = \frac{\lambda_M}{\lambda_m} \approx \frac{4N^2}{\pi^2}. \tag{6.25}$$

Consequently, as the number of grid points increases the matrix becomes more ill-conditioned.

6.3 Descent Methods

The objective of this section is to develop an effective method for solving matrix equations involving large sparse matrices. The first step in this effort is to rewrite the equation as a minimization problem. To explore how this might be done suppose we start with the $n = 1$ case, where the equation is simply $av = b$. We want to find a function $F(v)$ that has a minimum exactly when $av = b$. One possibility is to use a least-squares formulation and take $F(v) = (av - b)^2$. Another approach is to recall that at a minimum of $F(v)$ one solves $F'(v) = 0$. If $av - b = 0$ corresponds to the equation $F'(v) = 0$ then we need $F'(v) = av - b$. Integrating gives us $F(v) = \frac{1}{2}av^2 - bv$. For this to correspond to a minimum problem the quadratic must open upward, which means we have the positivity requirement $a > 0$. To see what happens when there are more variables, let $n = 2$, so the equations have the form

$$a_{11}v_1 + a_{12}v_2 = b_1,$$
$$a_{21}v_1 + a_{22}v_2 = b_2.$$

We want to find a smooth function $F(\mathbf{v})$ that has a minimum exactly when $\mathbf{Av} = \mathbf{b}$. To achieve this we require that F satisfy

$$\frac{\partial F}{\partial v_1} = a_{11}v_1 + a_{12}v_2 - b_1, \tag{6.26}$$

$$\frac{\partial F}{\partial v_2} = a_{21}v_1 + a_{22}v_2 - b_2. \tag{6.27}$$

For F to be smooth we need

$$\frac{\partial^2 F}{\partial v_1 \partial v_2} = \frac{\partial^2 F}{\partial v_2 \partial v_1},$$

and so from (6.26), (6.27) we conclude $a_{12} = a_{21}$. In other words, our construction requires that the matrix $\mathbf{A}$ be symmetric. Assuming this holds, then integrating (6.26) and (6.27) it follows that

$$F(\mathbf{v}) = \frac{1}{2}\mathbf{v}^T\mathbf{A}\mathbf{v} - \mathbf{b}\cdot\mathbf{v}. \tag{6.28}$$

To guarantee that this has a minimum there is a positivity requirement on $\mathbf{A}$, namely, it must be positive definite (as well as symmetric). Although this result has been derived in the particular case of $n = 2$, the quadratic form in (6.28) is what is obtained for general n and is the function used in what follows.

The quadratic form in (6.28) will be used to solve the matrix equation coming from the finite difference approximation of Laplace's equation. By doing this we have transformed the equation into a minimization problem. Is this an improvement? Is this going to help us deal with the large sparse system we obtained from (6.17)? The answer to both questions is yes. Rewriting the problem in this way opens up new avenues for finding the solution and we are going to explore some of them.

The objective now is to find the vector $\mathbf{v} = \mathbf{v}_m$ that minimizes the function $F(\mathbf{v})$ given in (6.28). This will be done using an iteration method in which we pick a starting vector $\mathbf{v}_1$ and then construct a sequence $\mathbf{v}_2, \mathbf{v}_3, \mathbf{v}_4, \ldots$ that converges to $\mathbf{v}_m$. In doing this we will require the value of F to decrease with each step. The idea underlying how we will determine $\mathbf{v}_{k+1}$ from $\mathbf{v}_k$ is fairly simple. As an analogy, if you were standing on a hillside and wanted to go to the bottom of the valley you would pick a direction of descent and then start walking downhill. This is basically what we will do to solve the minimization problem. Starting at $\mathbf{v}_k$ we will identify a direction of descent $\mathbf{d}_k$ and then "walk" a distance α_k in that direction to reach a minimum value. If this point is not the global minimum $\mathbf{v}_m$ then it will be necessary to select a direction of descent at this new point and repeat the procedure. In mathematical terms we have the following formula

$$\mathbf{v}_{k+1} = \mathbf{v}_k + \alpha_k\mathbf{d}_k, \quad \text{for } k = 1, 2, 3, \ldots. \tag{6.29}$$

In this expression, α_k is the value of α that minimizes $q(\alpha) = F(\mathbf{v}_k + \alpha\mathbf{d}_k)$. This can be found in the usual way by solving the equation $q'(\alpha) = 0$. Carrying out the calculations, one finds that

$$\alpha_k = \frac{\mathbf{d}_k\cdot\mathbf{r}_k}{\mathbf{d}_k\cdot\mathbf{q}_k}, \tag{6.30}$$

where $\mathbf{q}_k = \mathbf{A}\mathbf{d}_k$ and

$$\mathbf{r}_k = \mathbf{b} - \mathbf{A}\mathbf{v}_k. \tag{6.31}$$

To avoid the matrix multiplication in (6.31) one can use (6.29) to rewrite (6.31) as

$$\mathbf{r}_k = \mathbf{r}_{k-1} - \alpha_{k-1}\mathbf{q}_{k-1}, \quad \text{for } k = 2, 3, \ldots. \tag{6.32}$$

With (6.29), (6.30), and (6.32) we have the basic template for our descent algorithm. It remains to specify the descent directions, and we will consider various possibilities below.

In this construction the vector $\mathbf{r}_k$ is the residual at $\mathbf{v}_k$ and it provides a measure of the error. If we are fortunate enough to have $\mathbf{r}_k = \mathbf{0}$, then the problem is solved and $\mathbf{v}_m = \mathbf{v}_k$. Since $\|\mathbf{r}_k\| = 0$ only when $\mathbf{v}_k$ is the solution, having $\|\mathbf{r}_k\|$ small can be used as a stopping condition for the iteration. This works as long as $\mathbf{A}$ is not ill-conditioned, and we will return to this topic later.

6.3.1 Steepest Descent Method

Our descent algorithm is coming along nicely. The only thing left to do is decide on what to use for a descent direction. With the objective of getting the algorithm to converge as quickly as possible, the most obvious choice is to pick the direction of steepest descent. In other words, we pick $\mathbf{d}_k$ to be the negative of the gradient of F at $\mathbf{v} = \mathbf{v}_k$. From (6.28) one finds that $\nabla F = \mathbf{A}\mathbf{v} - \mathbf{b}$, and therefore the direction of steepest descent at $\mathbf{v}_k$ is determined by the residual (6.31). The resulting iteration formula is

$$\mathbf{v}_{k+1} = \mathbf{v}_k + \alpha_k \mathbf{r}_k, \quad \text{for } k = 1, 2, 3, \ldots, \tag{6.33}$$

where $\mathbf{r}_k$ is given in (6.31) and α_k is given in (6.30). This is known as the method of steepest descents (SDM).

Because the only significant operations involved with SDM are vector and matrix multiplication it is very easy to code the method. Moreover, it is capable of taking maximum advantage of the sparseness of $\mathbf{A}$. For example, to calculate $\mathbf{q}_k = \mathbf{A}\mathbf{d}_k$ we need to store, and use, only the nonzero entries in $\mathbf{A}$. Also, there is only one matrix multiplication per iteration, and this multiplication involves approximately $9n$ operations versus the usual $2n^2$ operations for full matrix multiplication.

How effective is SDM in solving matrix equations? Let's find out by trying a couple of examples.

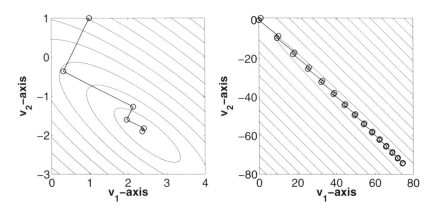

Figure 6.13. Contours of $F(v_1, v_2)$ and the first few steps taken by the method of steepest descent in the two example problems.

Examples
In the two examples below, the starting vector is $\mathbf{v}_1 = (1,\ 1)^T$ and the stopping condition is $\|\mathbf{v}_{k+1} - \mathbf{v}_k\| < 10^{-4}$. Once the SDM has stopped, then the error $\|\mathbf{v}_m - \bar{\mathbf{v}}_k\|$, which is the difference between the exact and computed solutions, is computed as a check to see how well the method has worked.

1. $\begin{pmatrix} 2 & 2 \\ 2 & 3 \end{pmatrix} \begin{pmatrix} v_1 \\ v_2 \end{pmatrix} = \begin{pmatrix} 1 \\ -1 \end{pmatrix}$

   The first few steps produced with SDM are shown in the contour plot on the left in Figure 6.13. It is found that the method takes 16 iterations and the resulting error is 1.3×10^{-4}. For later reference, the eigenvalues of the matrix are, approximately, $\lambda_m = 0.4$ and $\lambda_M = 4.6$.

2. $\begin{pmatrix} 5 & 4.99 \\ 4.99 & 5 \end{pmatrix} \begin{pmatrix} v_1 \\ v_2 \end{pmatrix} = \begin{pmatrix} 1 \\ -1 \end{pmatrix}$

   The first few steps produced with SDM are shown in the contour plot on the right in Figure 6.13. The contours are so elongated in this example that the elliptical curves around the minimum are not evident (as they are in the contour plot to the left). It is found that the method takes a whopping 200 iteration steps and the resulting error is a mediocre 9×10^{-3}. For the record, the eigenvalues of the matrix are, approximately, $\lambda_m = 0.01$ and $\lambda_M = 10$.

In reviewing the results from the above examples one might wonder why anyone would ever consider using SDM because the method can be painfully slow to converge. As seen in Figure 6.13, the more elongated the contours, the slower the convergence. To state this in more mathematical terms, from the Principal Axis Theorem it is known that the ratio of the semimajor to semiminor axes of the elliptical contours in the above two examples equals $\sqrt{\lambda_M/\lambda_m}$, where λ_m, λ_M designate the smallest and largest eigenvalues, respectively. Therefore, the larger the ratio λ_M/λ_m, the more elongated the contours. As an illustration of this, in the first example $\lambda_M/\lambda_m \approx 10$, and in the second example $\lambda_M/\lambda_m \approx 1000$. Our results indicate that the larger the ratio λ_M/λ_m, the slower SDM converges. This observation is important because the condition number of a symmetric positive definite matrix, using the Euclidean norm, is λ_M/λ_m. Therefore, if the above examples are any indication, it would appear that SGM can converge painfully slowly for matrices usually considered well conditioned and is probably worthless for matrices with moderately large condition numbers.

6.3.2 Conjugate Gradient Method

To improve the convergence of SDM we need to modify the direction of descent. In looking at Figure 6.13 it appears that the search path is almost in the right direction, but the method continually makes small adjustments (every other step) to keep heading toward the minimum. To improve on this, when picking $\mathbf{d}_k$ we will still use the steepest descent direction $\mathbf{r}_k$ but modify it slightly to reduce this readjustment. Our choice will still be to start off with $\mathbf{d}_1 = \mathbf{r}_1$ but from then on take

$$\mathbf{d}_k = \mathbf{r}_k + \beta_{k-1}\mathbf{d}_{k-1}, \quad \text{for } k = 2, 3, \ldots. \tag{6.34}$$

The β_{k-1} in this formula will be used to prevent the problem seen in the last example. The choice is based on the observation that in the SDM the direction of steepest descent at $\mathbf{v}_2$ is orthogonal to the steepest descent direction at $\mathbf{v}_1$; that is, $\mathbf{r}_1 \cdot \mathbf{r}_2 = 0$. Similarly, one finds that $\mathbf{r}_2 \cdot \mathbf{r}_3 = 0$, but that it is not necessarily true that $\mathbf{r}_1 \cdot \mathbf{r}_3 = 0$. In fact, in the earlier example it turns out that $\mathbf{r}_3$ is parallel to $\mathbf{r}_1$. This can be seen in Figure 6.13, where the corners in the path form right angles but the path from $\mathbf{v}_1$ to $\mathbf{v}_2$ is parallel to the path from $\mathbf{v}_3$ to $\mathbf{v}_4$. Consequently, in this example, SDM ends up using only two distinct directions of descent and produces a sequence of alternating parallel lines forming the search path. It is this parallelism in the residual vectors that will be avoided using β_{k-1}.

The way we will prevent $\mathbf{r}_3$ from being parallel to $\mathbf{r}_1$ is to select β_1 so that $\mathbf{r}_3$ is actually orthogonal to $\mathbf{r}_1$. To determine how this can be accomplished, note that no matter how $\mathbf{d}_k$ is selected, the formula for the residual in (6.32) gives us $\mathbf{r}_2 = \mathbf{r}_1 - \alpha_1\mathbf{q}_1$, and $\mathbf{r}_3 = \mathbf{r}_2 - \alpha_2\mathbf{q}_2$. We know that $\mathbf{r}_1 \cdot \mathbf{r}_2 = 0$, and as for the other possible inner products, a short calculation shows that $\mathbf{r}_1 \cdot \mathbf{r}_3 = -\alpha_2\,\mathbf{r}_1 \cdot \mathbf{q}_2$ and $\mathbf{r}_2 \cdot \mathbf{r}_3 = (const)\,\mathbf{r}_1 \cdot \mathbf{q}_2$. Therefore, if we take β_1 such that $\mathbf{r}_1 \cdot \mathbf{q}_2 = 0$, we will end up with the residuals being mutually orthogonal. Working out the details, one finds that $\beta_1 = \mathbf{r}_2 \cdot \mathbf{r}_2 / \mathbf{r}_1 \cdot \mathbf{r}_1$. Continuing in this way yields the following formula

$$\beta_k = \frac{\mathbf{r}_{k+1} \cdot \mathbf{r}_{k+1}}{\mathbf{r}_k \cdot \mathbf{r}_k}, \quad \text{for } k = 1, 2, 3, \ldots. \tag{6.35}$$

It is assumed here that $\mathbf{r}_k \cdot \mathbf{r}_k \neq 0$, but if this were not the case we would have solved the equation exactly at the previous step and there would be no need to calculate β_k. This choice of β_k, along with the descent direction in (6.34), produces what is known as the conjugate gradient method (CGM), and the steps involved are summarized in Table 6.2. In terms of operations per iteration step it is not much more than SDM. To calculate β_k and then construct $\mathbf{d}_k$ adds about $4n$ flops per step.

The CGM has a remarkable property that has profound consequences for how well it works. By construction $\mathbf{r}_k$ is orthogonal to all of the previous residuals. In other words, $\mathbf{r}_1, \mathbf{r}_2, \mathbf{r}_3, \ldots$ are mutually orthogonal, which means that $\mathbf{r}_i \cdot \mathbf{r}_j = 0, \forall i \neq j$. In n dimensions, the only way for $n + 1$ vectors to be mutually orthogonal is that one of them is the zero vector. If a residual is zero we have found the exact solution and therefore the conjugate gradient method produces the exact solution in no more than n iteration steps. For this reason CGM is not an iteration method in the traditional sense. To demonstrate the

(1) picking $\mathbf{x}_1$, set $\mathbf{r}_1 = \mathbf{b} - \mathbf{A}\mathbf{x}_1$ and $\mathbf{d}_1 = \mathbf{r}_1$

(2) for $k = 1, 2, 3, \ldots$

$$\mathbf{x}_{k+1} = \mathbf{x}_k + \alpha_k \mathbf{d}_k$$
$$\mathbf{r}_{k+1} = \mathbf{r}_k - \alpha_k \mathbf{q}_k$$
$$\mathbf{d}_{k+1} = \mathbf{r}_{k+1} + \beta_k \mathbf{d}_k$$

where

$$\mathbf{q}_k = \mathbf{A}\mathbf{d}_k$$
$$\alpha_k = \frac{\mathbf{r}_k \cdot \mathbf{r}_k}{\mathbf{d}_k \cdot \mathbf{q}_k}$$
$$\beta_k = \frac{\mathbf{r}_{k+1} \cdot \mathbf{r}_{k+1}}{\mathbf{r}_k \cdot \mathbf{r}_k}$$

Table 6.2. Outline of the conjugate gradient method (CGM) used to solve the linear system $\mathbf{A}\mathbf{x} = \mathbf{b}$, where $\mathbf{A}$ is an $n \times n$ symmetric positive definite matrix.

significance of this finite termination property we return to the examples used for SDM.

Examples
In the two examples below, the starting vector is $\mathbf{v}_1 = (1, 1)^T$ and the stopping condition for the CGM iteration is $\|\mathbf{v}_{k+1} - \mathbf{v}_k\| < 10^{-4}$.

1. $\begin{pmatrix} 2 & 2 \\ 2 & 3 \end{pmatrix} \begin{pmatrix} v_1 \\ v_2 \end{pmatrix} = \begin{pmatrix} 1 \\ -1 \end{pmatrix}$

All of the steps produced with CGM are shown in the contour plot on the left in Figure 6.14. It is found that the method takes two iteration steps and the resulting error is 4×10^{-16}, which is as accurate as can be expected using double precision.

2. $\begin{pmatrix} 5 & 4.99 \\ 4.99 & 5 \end{pmatrix} \begin{pmatrix} v_1 \\ v_2 \end{pmatrix} = \begin{pmatrix} 1 \\ -1 \end{pmatrix}$

All of the steps produced with CGM are shown in the contour plot on the right in Figure 6.14. It is found that the method takes two iteration steps and the resulting error is 7×10^{-12}.

In comparing the search paths in Figures 6.13 and 6.14, the contrast between SDM and CGM is striking. As advertised, the CGM took two steps to solve the 2×2 matrix equations, and in doing so it showed itself to be far superior to SGM. However, the reality is that round-off generally prevents

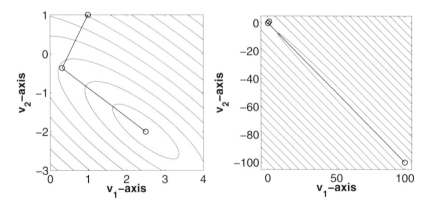

Figure 6.14. Contours of $F(v_1, v_2)$ and the paths taken by the conjugate gradient method in the two example problems.

CGM from producing the exact result. This is not unexpected, nor a criticism of the method, because as with all floating-point computations we can only get as close as round-off permits. The issue with CGM is that it is built on the assumption of exact orthogonality of the residuals at each step. With round-off, however, this will not happen. For well-conditioned matrices this is not an issue, and the above examples demonstrate this fact. However, if the condition number of the matrix gets too large then it is an issue and CGM will lose its finite termination property. We will discuss later how this situation can be corrected, but it is not a pressing problem for the matrices coming from Laplace's equation.

Convergence of CGM

To investigate the effectiveness of the CGM for larger systems we will solve $\mathbf{A}\mathbf{v} = \mathbf{b}$, where $\mathbf{A}$ is the same matrix as in (6.20) but in the special case that we know the solution $\mathbf{v}_m$. This will be done by specifying $\mathbf{v}_m$ and then setting $\mathbf{b} = \mathbf{A}\mathbf{v}_m$. This way we can observe the accuracy, and rate of convergence, of the CGM as it progresses. With the intention of picking a representative solution for Laplace's equation, the vector $\mathbf{v}_m$ is taken from the series solution in (6.6) when the Fourier coefficients are as given in (6.9). Now, there are various ways to measure error, and for this example we will consider the following

$$\text{Error:} \quad E_k = \|\mathbf{v}_k - \mathbf{v}_m\|, \tag{6.36}$$

$$\text{Iteration Error:} \quad I_k = \|\mathbf{v}_k - \mathbf{v}_{k-1}\|, \tag{6.37}$$

$$\text{Residual Error:} \quad R_k = \|\mathbf{r}_k\|. \tag{6.38}$$

The reason for considering I_k and R_k is that we will usually not know $\mathbf{v}_m$ and consequently we would like to have some idea of whether I_k or R_k can be used

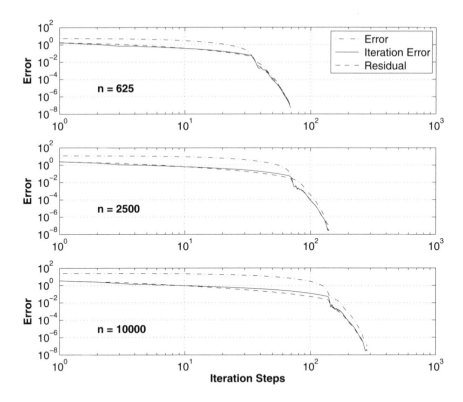

Figure 6.15. Error obtained at each iteration step of CGM to solve a matrix equation involving (6.21), for various sizes n of the matrix. The errors are defined in (6.36)–(6.38).

as a reasonable substitute for E_k. In any case, taking $h = k$, the results when $N = M = 25$ $(n = 625)$, $N = M = 50$ $(n = 2500)$, $N = M = 100$ $(n = 10000)$ are shown in Figure 6.15. From these plots one sees that for this example the residual and iteration errors are effectively equivalent in their estimation of E_k. Both consistently underestimate the error and neither decreases monotonically (the latter is not easy to see in the plot, but if you look closely at the curves near the 10^{-2} level there are multiple local maximum and minimum points on the two curves). There are also two stages in the convergence. In the first the error reduction is modest. For example, with $n = 10000$ there is not much improvement in the solution for the first 100 steps. Shortly after that things change, and CGM converges relatively quickly. This acceleration in the convergence rate as the iteration proceeds is called superlinear convergence. Accurate predictions of where in the calculation this begins to become evident require information about the distribution of the eigenvalues of $\mathbf{A}$ and are therefore not easy to obtain. However, a useful and well-known upper bound on how fast the method converges is the following

$$\|\mathbf{v}_k - \mathbf{v}_m\| \leq 2\|\mathbf{v}_1 - \mathbf{v}_m\|\sqrt{\kappa}\left(\frac{\sqrt{\kappa}-1}{\sqrt{\kappa}+1}\right)^{k-1}, \tag{6.39}$$

where $\kappa = \lambda_M/\lambda_m$ (a proof of this can be found in Luenberger [2002]). This result shows that the better the conditioning of the matrix (i.e., the closer κ gets to one), the faster the convergence. However, as shown in Exercise 6.22, this inequality does not give a very accurate estimate of the rate of convergence, so its use as a predictive tool is a bit limited.

Another important point to make about Figure 6.15 concerns the number of iterations. The CGM has produced a reasonably accurate solution of the matrix equation in significantly fewer than n iterations. It is interesting to note that the point in the iteration where the error reaches, say, 10^{-6} scales approximately with N (and M) and not with n. For example, when going from $N = 25$ (so $n = 625$) to $N = 50$ (so $n = 2500$) the number of iterations increases from 63 to 125. In other words, if N and M are doubled the number of iterations needed to reach 10^{-6} increases by about a factor of two. This observation is consistent with (6.39). This inequality indicates that the rate of convergence depends on $\sqrt{\kappa}$, and from (6.25), for larger matrices, $\sqrt{\kappa} = O(N)$.

To conclude the development of CGM we consider the question of how it compares with a direct solver when n is large. In terms of memory, CGM requires storage of only the nonzero entries of $\mathbf{A}$ along with five n-vectors. Direct solvers cannot do better than this. For example, the usual banded Cholesky algorithm ends up storing an $N \times n$ array along with a handful of n-vectors. In terms of computational effort, calculating $\mathbf{v}_{k+1}$ takes approximately $19n$ flops. The stopping condition on the iteration can raise this total, although using the residual adds an insignificant amount of effort. For direct solvers that make use of the banded structure of the matrix, the flop total is approximately $2N^2n$. In round numbers, for the case $M = N$, this means that if CGM takes approximately $n/10$ iterations then the computational effort is close to what it would take a direct solver. Based on this estimate the advantage of using CGM is clearly evident in Figure 6.15 as n increases. This is the point where one is supposed to say "your mileage might vary." In other words, a lot of assumptions have gone into this estimate, and everything from efficient coding to chip design can significantly affect the actual result. Nevertheless, it is useful to have some quantitative measure of computational effort when testing the CGM and the flop count will be used in this role.

6.4 Numerical Solution of Laplace's Equation

It is time to reap the rewards of all the hard work we have invested in deriving the CGM and now actually solve Laplace's equation. The domain and boundary conditions of the first example are shown in Figure 6.3, where $a = b = 1$ and $g(x) = x(1-x)(\frac{4}{5}-x)e^{6x}$. The solution as obtained using a Fourier series solution is shown in Figure 6.5. We already know what the solution looks

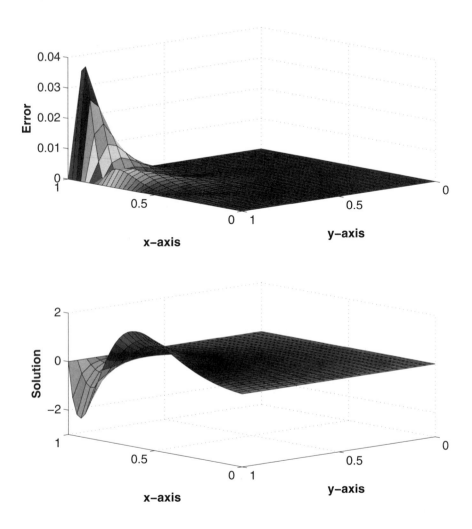

Figure 6.16. Numerical solution of Laplace's equation and the resulting spatial distribution of the pointwise error. The five-point scheme (6.17) and the CGM were used in the calculation.

like, and so the suspense here centers on how accurately the finite difference approximation and the CGM solve the problem. In other words, we are interested in the error. Because the solution varies more in the x direction, in the calculations to follow we take $M = N/2$. The stopping condition used with the CGM is $\|\mathbf{v}_{k+1} - \mathbf{v}_k\| \leq 10^{-6}$. In Figure 6.16 the numerical solution and the pointwise error $E(x_i, y_j)$ are shown in the case $N = 40$. The error in this case is simply the difference between the numerical and Fourier series solutions, in absolute value. Specifically, $E(x_i, y_j) = |u(x_i, y_j) - \bar{u}_{ij}|$, where $\bar{u}_{ij}$ is the computed value of the solution at (x_i, y_j). As can be seen, the spatial distribution of the error corresponds to the variation in the solution.

Very little has been said so far about the rate of convergence of the numerical approximation. As shown in Exercise 6.14, the error in the solution is determined by the truncation error. To verify this numerically, the maximum error $E_{\max} = \max_{i,j}|u(x_i, y_j) - \bar{u}_{ij}|$ is plotted as a function of N in Figure 6.17. For a little variety, the errors when $M = N$ and when $M = N/2$ are shown. The fact that this is a second-order method is clearly seen in both curves.

Although it is nice that CGM has this finite termination property, it is our hope not to have to rely on this property to solve our matrix equation. We saw that n can easily be 10^4 or larger, and we certainly do not want to carry out this many iterations to solve the problem. It is more important for us that the method converge quickly. For example, if $n = 10^4$ then it is our hope that a reasonably accurate solution is obtained in, say, 10^2 iterations. The question is, what is meant by accurate? We need to solve the problem only to an accuracy that is consistent with how well we are approximating Laplace's equation. This is another advantage of using an iterative method. It makes little sense to push the CGM to generate an answer correct to 12 digits if the step size used in the finite difference approximation is capable of getting us to within only 4 digits of the solution of Laplace's equation. Based on this observation, it would seem we have overcomputed the solution in Figure 6.16. The reason is that the plot shows that the maximum error in the finite difference approximation is about

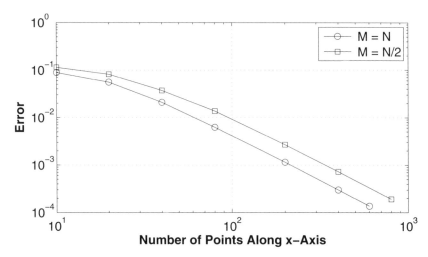

Figure 6.17. Maximum pointwise error as a function of the number of points N used along the x-axis.

10^{-2}. Consequently, the stated stopping condition that $\|\mathbf{v}_{k+1} - \mathbf{v}_k\| \leq 10^{-6}$ is overly restrictive. Based on the information in Figure 6.15 it would have been sufficient to use $\|\mathbf{v}_{k+1} - \mathbf{v}_k\| \leq 10^{-4}$.

6.5 Preconditioned Conjugate Gradient Method

With CGM we have a fairly robust method for solving the matrix equation coming from Laplace's equation. With an eye on improving things a bit, the results in Figure 6.15 indicate that as n increases, the number of iterations increases. This is due, in part, to the matrix becoming less well conditioned with increasing n. Said another way, the method slows down as the ratio λ_M/λ_m of the largest to smallest eigenvalues of $\mathbf{A}$ increases. It is possible to quantify this observation in the particular case where D is the unit square and $h = k$. The eigenvalue ratio in this case is given in (6.25), where for large N,

$$\frac{\lambda_M}{\lambda_m} \approx \frac{4N^2}{\pi^2}. \tag{6.40}$$

The N^2 term in this result indicates that larger matrices will almost invariably lead to problems for CGM.

The question considered here is, can we do anything to the matrix equation to improve the convergence of CGM? As an example, for the symmetric and positive definite matrix

$$\mathbf{C} = \begin{pmatrix} 1 & 3-s \\ 3-s & 9 \end{pmatrix}, \tag{6.41}$$

where $s = 10^{-10}$, one finds that $\lambda_M/\lambda_m \approx 2 \times 10^{11}$. Given the rather large value of this ratio, the matrix is ill-conditioned. However, setting

$$\mathbf{B} = \begin{pmatrix} 9 & s-3 \\ s-3 & 1 \end{pmatrix},$$

then the product $\overline{\mathbf{C}} = \mathbf{BC}$ has $\lambda_M/\lambda_m = 1$ and this certainly qualifies as a well-conditioned matrix. Therefore, the original matrix has been made more tractable by simply multiplying it by a well-chosen matrix. So the question arises as to whether it might be possible to improve the convergence of CGM by multiplying the equation by a well-chosen matrix $\mathbf{B}$. What $\mathbf{B}$ is supposed to do is keep the resulting condition number λ_M/λ_m from getting too large. To see whether this can work we simply multiply $\mathbf{Av} = \mathbf{b}$ by $\mathbf{B}$ and obtain $\overline{\mathbf{A}}\mathbf{v} = \overline{\mathbf{b}}$, where $\overline{\mathbf{A}} = \mathbf{BA}$ and $\overline{\mathbf{b}} = \mathbf{Bb}$. Unfortunately, $\overline{\mathbf{A}}$ is not necessarily symmetric unless we severely restrict $\mathbf{B}$. The needed remedy is to note that

$$\mathbf{BAv} = (\mathbf{BA})(\mathbf{B}^T\mathbf{B}^{-T})\mathbf{v}$$
$$= (\mathbf{BAB}^T)\mathbf{B}^{-T}\mathbf{v},$$

where $\mathbf{B}^{-T} = \left(\mathbf{B}^T\right)^{-1}$. Therefore, we can transform $\mathbf{A}\mathbf{v} = \mathbf{b}$ into $\overline{\mathbf{A}}\overline{\mathbf{v}} = \overline{\mathbf{b}}$, where $\overline{\mathbf{A}} = \mathbf{B}\mathbf{A}\mathbf{B}^T$, $\overline{\mathbf{v}} = \mathbf{B}^{-T}\mathbf{v}$, and $\overline{\mathbf{b}} = \mathbf{B}\mathbf{b}$. In this case $\overline{\mathbf{A}}$ is symmetric and it is also positive definite if $\mathbf{B}$ is invertible.

At the moment we do not have a clue what to select for $\mathbf{B}$, but assuming that good choices are available, we need to determine what happens to the steps making up the CGM. The algorithm is still given in Table 6.2, but the equation to be solved is $\overline{\mathbf{A}}\overline{\mathbf{v}} = \overline{\mathbf{b}}$. If one writes out the CGM algorithm for $\overline{\mathbf{A}}\overline{\mathbf{v}} = \overline{\mathbf{b}}$ and then expresses the variables in terms of the original $(\mathbf{A}, \mathbf{v}, \mathbf{b})$, one obtains the algorithm in Table 6.3, where $\mathbf{M} = \mathbf{B}^{-1}\mathbf{B}^{-T}$. This is known as the preconditioned conjugate gradient method (PCGM) and $\mathbf{M}$ is the preconditioner. In terms of flops, the only significant difference between CGM and PCGM is the need to solve $\mathbf{M}\mathbf{z} = \mathbf{r}$ at each iteration step. It is interesting that $\mathbf{B}$ appears in this procedure only through the matrix $\mathbf{M}$. Therefore to use this algorithm we need only the preconditioner $\mathbf{M}$ and not the matrix $\mathbf{B}$.

We now turn our attention to specifying the preconditioner. There are a few general rules to keep in mind while doing so, and they are as follows:

1. $\mathbf{M}$ must be symmetric and positive definite.
2. Solving $\mathbf{M}\mathbf{z} = \mathbf{r}$ must not add significant computational time or storage to the iteration step.
3. PCGM takes less computing time than CGM.

These requirements are listed in their order of difficulty, and the first two can be determined more or less immediately. The third requirement is paramount and is the very reason for introducing PCGM in the first place. The reality is, however, that it is usually difficult to determine whether a particular $\mathbf{M}$ reduces the total computational time until the algorithm is coded and run on a few example problems. This is exactly the approach we will take here as well.

The matrix $\mathbf{M}$ is derived from $\mathbf{A}$, but exactly how is a topic that needs some discussion. One can think of $\mathbf{M}\mathbf{z} = \mathbf{r}$ as a reduced version of $\mathbf{A}\mathbf{v} = \mathbf{b}$, and the better $\mathbf{M}$ approximates $\mathbf{A}$, the fewer the iteration steps that will be needed. For example, if $\mathbf{M} = \mathbf{A}$ then PGCM produces the exact solution in one step. This is useful information, because it indicates that if we carefully select entries from $\mathbf{A}$ when constructing $\mathbf{M}$, we might be able to significantly reduce the computational difficulty of the problem. As a first attempt, $\mathbf{M}$ could be taken to be diagonal with its diagonal entries equal to those from $\mathbf{A}$, something known as the Jacobi preconditioner. This is not such a great choice for our problem, and the reasons are investigated in Exercise 6.20. An improvement on this idea is to use the tridiagonal entries from $\mathbf{A}$, and some of the consequences of this choice this will be examined in the example below.

There is another approach for selecting $\mathbf{M}$ that is based on the observation that if $\mathbf{M} = \mathbf{A}$, then, because $\mathbf{M} = \mathbf{B}^{-1}\mathbf{B}^{-T}$, we have $\mathbf{A} = \mathbf{C}\mathbf{C}^T$, where $\mathbf{C} = \mathbf{B}^{-1}$. This formula for $\mathbf{A}$ is the same sort of expression obtained when solving $\mathbf{A}\mathbf{x} = \mathbf{b}$ using a Cholesky factorization. It is possible to take advantage of this observation by noting that if we can find an approximate factorization

(0) pick a symmetric positive definite matrix $\mathbf{M}$

(1) picking $\mathbf{x}_1$, set $\mathbf{r}_1 = \mathbf{b} - \mathbf{A}\mathbf{x}_1$, solve

$\mathbf{M}\mathbf{z}_1 = \mathbf{r}_1$ and then let $\mathbf{d}_1 = \mathbf{z}_1$

(2) for $k = 1, 2, 3, \ldots$

$$\begin{aligned}
\mathbf{x}_{k+1} &= \mathbf{x}_k + \alpha_k \mathbf{d}_k \\
\mathbf{r}_{k+1} &= \mathbf{r}_k - \alpha_k \mathbf{q}_k \\
\mathbf{M}\mathbf{z}_{k+1} &= \mathbf{r}_{k+1} \\
\mathbf{d}_{k+1} &= \mathbf{z}_{k+1} + \beta_k \mathbf{d}_k
\end{aligned}$$

where

$$\begin{aligned}
\mathbf{q}_k &= \mathbf{A}\mathbf{d}_k \\
\alpha_k &= \frac{\mathbf{z}_k \cdot \mathbf{r}_k}{\mathbf{d}_k \cdot \mathbf{q}_k} \\
\beta_k &= \frac{\mathbf{r}_{k+1} \cdot \mathbf{z}_{k+1}}{\mathbf{r}_k \cdot \mathbf{z}_k}
\end{aligned}$$

Table 6.3. Outline of the preconditioned conjugate gradient method (PCGM) used to solve the linear system $\mathbf{A}\mathbf{x} = \mathbf{b}$, where $\mathbf{A}$ is an $n \times n$ symmetric positive definite matrix.

$\mathbf{A} \approx \mathbf{C}\mathbf{C}^T$ then $\mathbf{M} = \mathbf{C}\mathbf{C}^T$ is a candidate for a preconditioner. The reason this is worth pursuing is that we do not need to find an exact factorization as must be done using Cholesky but only something that approximates the exact result and that is easy to calculate. In the jargon of the subject, this type of preconditioner is said to be an incomplete Cholesky factorization of the matrix. This idea will be demonstrated in the examples below.

Examples
For the matrix in (6.21) the following are possibilities for a preconditioner:

1. $\mathbf{M} = \mathbf{M}_1$ where $\mathbf{M}_1$ is tridiagonal

   The tridiagonal entries for $\mathbf{M}_1$ are taken directly from $\mathbf{A}$. This particular choice is a variant of what is called a block Jacobi preconditioner. It satisfies Condition (1), and with the tridiagonal solver in Table 2.1 this choice increases the operational count by about $8n$ per iteration step. This would seem to satisfy Condition (2). The only way to determine whether Condition (3) holds is to actually compute the solution. However, to obtain at least an approximate idea of how well this choice works we will consider the flops. Since the flop count for CGM is $19n$, we want this version of PCGM to reduce the number of iteration steps by more than a factor of

19/27. Simplifying this, we are looking for a reduction of the number of iteration steps of at least 5/7 of the original.

2. $\mathbf{M} = \mathbf{M}_2$, where $\mathbf{M}_2 = (\mathbf{D} + \mathbf{L})\mathbf{D}^{-1}(\mathbf{D} + \mathbf{L}^T)$

The matrices in this expression come from writing $\mathbf{A} = \mathbf{D} + \mathbf{L} + \mathbf{L}^T$, where $\mathbf{D}$ is diagonal and $\mathbf{L}$ is strictly lower triangular (so it has zeros on the diagonal). Since $\mathbf{M}_2 = \mathbf{A} + \mathbf{L}\mathbf{L}^T$, this choice for the preconditioner assumes that $\mathbf{L}\mathbf{L}^T$ can be dropped and a reasonable approximation of $\mathbf{A}$ remains. It is possible to write this preconditioner as $\mathbf{M}_2 = \mathbf{C}\mathbf{C}^T$, where $\mathbf{C}$ is lower triangular, which means that it is an example of an incomplete Cholesky factorization. However, for computational purposes it is better to leave it as is or, even better, write it as $\mathbf{M}_2 = (\mathbf{D} + \mathbf{L})(\mathbf{I} + \mathbf{D}^{-1}\mathbf{L}^T)$, where $\mathbf{I}$ is the identity matrix. The flops needed to solve $\mathbf{M}\mathbf{z} = \mathbf{r}$ are approximately $10n$, which brings the per iteration total to $29n$. For this version of PCGM we look for a reduction of the number of iteration steps to be at least 2/3 of CGM. For the record, this particular choice for $\mathbf{M}$ is called the symmetric successive overrelaxation (SSOR) preconditioner.

The first set of computational results using these two preconditioners is given in Figure 6.18. In these calculations, to be able to evaluate the effectiveness of the methods, the solution $\mathbf{v}_m$ is specified and $\mathbf{b} = \mathbf{A}\mathbf{v}_m$. As was done for Figure 6.15, with the intention of picking a representative solution for Laplace's equation, the vector $\mathbf{v}_m$ is taken from the series solution in (6.6) when the Fourier coefficients are given in (6.10). In looking at the curves in Figures 6.18 it is apparent that both preconditioners succeed in reducing the error. However, it is unclear from this graph whether there is any real improvement

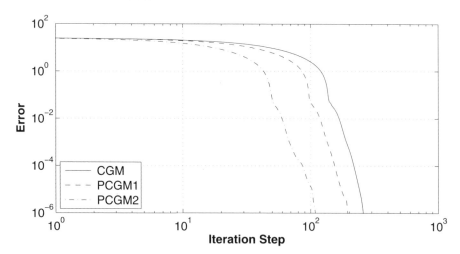

Figure 6.18. Error $E_k = \|\mathbf{v}_m - \mathbf{v}_k\|$ in the computed solution during each step of the iteration for CGM and PCGM. The matrix is given in (6.21) with $n = 10^4$ ($N = M = 100$). PCGM1 uses preconditioner $\mathbf{M}_1$ and PCGM2 uses $\mathbf{M}_2$.

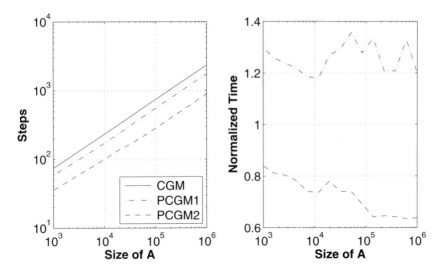

Figure 6.19. Number of iteration steps required to achieve an error of 10^{-6} as the size of the matrix increases using CGM and PCGM. Also shown is the time it takes PCGM to achieve this result in comparison to CGM. The matrix is given in (6.21); PCGM1 uses preconditioner $\mathbf{M}_1$ and PCGM2 uses $\mathbf{M}_2$.

in the amount of time it takes to compute the solution. For example, a reasonable question to ask is, just how long does it take each method to achieve an error of, say, 10^{-6}? The answer is given in Figure 6.19, which shows the number of iteration steps required to achieve an error of 10^{-6} as the size n of the matrix increases. Also shown is the normalized time it takes using the preconditioner, which is determined by taking the computing time for PCGM and dividing it by the time taken by CGM. It is evident from these graphs that $\mathbf{M} = \mathbf{M}_1$ results in PCGM taking fewer iteration steps but it requires more computing time. It is interesting to note that the number of iteration steps using $\mathbf{M}_1$ comes close to what was hoped for using a simple flop count. For example, if $n = 10^4$, the number of iteration steps for CGM is 234, while with PCGM1 it is 172. Based on the flop count, we were looking for the number of steps to be less than 167. The fact that the PCGM1 number is so close to the break-even point means that other factors, such as the time necessary to simply run through the loops in the preconditioner algorithm when such large vectors are involved, will affect the outcome. This also clearly demonstrates the pitfalls of using flops as a timing benchmark, particularly when the predicted improvement is as modest as it is with this version of the PCGM. In contrast to this situation, when the preconditioner $\mathbf{M} = \mathbf{M}_2$ is used the total number of iterations for $n = 10^4$ is reduced to 100, and this is significantly less than 156, which is the number required using a flop count ($156 = \frac{2}{3} \times 234$). In conjunction with this, the overall computing time for this version of PCGM is reasonably faster than CGM and improves as the matrix gets larger. Therefore, it is easy to recommend this as a choice for a preconditioner for this problem.

With PCGM we have an effective method for solving Laplace's equation. It is straightforward to extend what we did to the more general problem given in (6.1), and this is explored in the exercises.

6.6 Next Steps

When studying multidimensional problems it is common to have complicated domains, something similar to what is shown in Figure 6.1. One way to handle such situations is to find a change of variables that transforms the region into one that is rectangular. Generally this produces equations with nonconstant coefficients, but from a numerical point of view dealing with this is easy. An example of this is explored in Exercise 6.15, and a more sophisticated example that involves determining the flow of air over an airplane wing can be found in Cole and Cook [1986]. The limitation is that it can be difficult, if not impossible, to find such a change of variables. An alternative is to just solve the problem using the original domain. The issue in this case is how to satisfy the boundary condition when the grid points do not lie on the boundary. The situation is illustrated in Figure 6.20. To construct an approximation for u_{xx} at, say, point C one can use points L, C, and R'. The formula that applies in the case is given in Exercise 1.30. Another option is to extend the boundary condition to R and then use a standard centered difference formula for u_{xx}. Both methods are only $O(h)$ accurate, but there are ways to construct approximate boundary conditions that produce a $O(h^2)$ truncation error. The theory underlying this requires assumptions related to the connectivity of the grid and is beyond the scope of this text. An introduction to the material can be found in Morton and Mayers [2005].

The long history of solving elliptic equations has resulted in several innovative methods for constructing a numerical approximation besides the finite difference approach examined in this chapter. As an example, it is possible

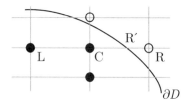

Figure 6.20. Grid and boundary when solving a problem with a nonrectangular domain. In this figure, the grid points in D have solid dots and those outside D have hollow dots. Also, R' is the point where the boundary curve intersects the grid between C and R.

to use the Divergence Theorem and write the solution of Laplace's equation, with the boundary condition in (6.3), as

$$u(\mathbf{x}) = \int_{\partial D} v(\mathbf{y})\, \Phi(\mathbf{x} - \mathbf{y}) ds_y - \int_{\partial D} g(\mathbf{y})\, \frac{\partial \Phi}{\partial n}(\mathbf{x} - \mathbf{y}) ds_y, \quad \text{for } \mathbf{x} \in D, \ (6.42)$$

where n is the unit outward normal, $v = \frac{\partial u}{\partial n}$, and Φ is a fundamental solution of Laplace's equation and is known. The function v, however, is not known. If it can be determined then the problem is solved and the solution is given in (6.42). With the goal of finding v, we evaluate (6.42) on the boundary and obtain

$$\frac{1}{2} g(\mathbf{x}) = \int_{\partial D} v(\mathbf{y})\, \Phi(\mathbf{x} - \mathbf{y}) ds_y - \int_{\partial D} g(\mathbf{y})\, \frac{\partial \Phi}{\partial n}(\mathbf{x} - \mathbf{y}) ds_y, \quad \text{for } \mathbf{x} \in \partial D.$$
$$(6.43)$$

The factor $\frac{1}{2}$ in this expression is not a mistake; it is due to a singularity in the function Φ. The derivation requires a certain smoothness in the boundary and an extended discussion of this is given in Colton [2004]. With (6.43), we have a Fredholm integral equation of the first kind for the function v. It is interesting as it replaces the differential equation over a two-dimensional domain with a one-dimensional integral equation. In other words, all of the issues discussed in the previous paragraph that arise when approximating a complicated two-dimensional domain have disappeared. The resulting problem involves integrals, and so constructing an approximation appears to be relatively straightforward. However, as is usually the case, there is a conservation of difficulty. The complication is that Φ is singular on the boundary, so constructing an approximation requires some care. Another issue is that integral equations of the first kind are susceptible to ill-conditioning. Nevertheless, the idea is workable and is the basis for what are called boundary integral methods. An introduction to this subject and its applications can be found in Bonnet [1999], Sloan [1992], and McLean [2000].

Another interesting alternative for solving elliptic equations is multigrid, which can be used in conjunction with the finite difference approximation. The idea underlying this method starts with the observation that when using an iterative matrix solver such as Gauss–Seidel, the initial iterations reduce the error a lot but the later iterations show a much slower reduction. Also, the error depends strongly on what is used as the starting value for the iterative solver. Multigrid takes advantage of these two observations by using a hierarchy of grids, from coarse to fine. The procedure begins by running the solver for a few iterations on the coarse grid and then using the computed solution to build a starting value for the successively finer grids. During the process one cycles back and forth between coarse and fine grids in such a way that the error is reduced very quickly. A nice discussion of the method and its applications can be found in Wesseling [2004].

One of the more frequent methods used for solving elliptic equations is finite elements. This generally uses an integral formulation of the problem in conjunction with an approximation similar to the one used for residual methods in Chapter 2. One of the strengths of finite elements is the ability of building accurate approximations for very complicated geometries. An introduction to this subject can be found in Ciarlet [2002] and Brenner and Scott [2002]. Finite difference approximations also work on complicated geometries and an introduction to this is given in Petersson [1999] and Henshaw [2005].

Exercises

6.1. This problem examines some of the properties of symmetric positive definite matrices.

(a) Using the coordinate vectors e_i, where $e_{ij} = 0$ if $i \neq j$ and $e_{ii} = 1$, and PD2 show that if A is positive definite then $a_{ii} > 0 \; \forall \; i$. From this establish Test 1.

(b) Show that Test 2 cannot be reversed by giving an example of a positive definite matrix that is not strict diagonally dominant.

(c) Suppose A, B are symmetric and positive definite. Show that if they have the same dimension and if α, β are positive constants then $\alpha A + \beta B$ is symmetric positive definite.

(d) Show that if A, B are symmetric positive definite then the block matrix

$$\begin{pmatrix} A & 0 \\ 0 & B \end{pmatrix}$$

is symmetric positive definite.

6.2. This problem considers some of the properties of the matrix equation (6.20).

(a) What contributions, if any, do the boundary conditions have to the matrix being positive definite?

(b) What contributions, if any, do the step sizes h and k have to the matrix being positive definite?

(c) What contributions, if any, do the size and location of the domain have to the matrix being positive definite?

6.3. This problem concerns the equation

$$u_{xx} + u_{yy} + \alpha u_x + \beta u_y + \gamma u = 0,$$

where α, β, γ are constant. The domain and boundary conditions are given in Figure 6.3.

(a) Derive a finite difference approximation of this problem that has truncation error $O(h^2) + O(k^2)$ and express the result in matrix form. The matrix must reduce to the one in (6.20) when $\alpha = \beta = \gamma = 0$.

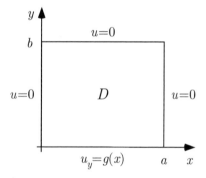

Figure 6.21. Domain for Exercise 6.4.

(b) Suppose $\alpha = \beta = 0$. Using the tests listed in Section 6.2.2, what condition(s), if any, must be imposed on γ so the CGM can be used to solve the matrix equation in part (a)?

(c) Suppose $\gamma = 0$. Using the tests listed in Section 6.2.2, what condition(s), if any, must be imposed on α, β so the CGM can be used to solve the matrix equation in part (a)?

(d) Combine the results from (b) and (c) to determine condition(s), if any, on α, β, γ so the CGM can be used to solve the matrix equation in part (a).

6.4. This problem considers Laplace's equation using the domain and boundary conditions shown in Figure 6.21. The issue is how best to treat the derivative boundary condition.

(a) One option is to use a $O(k^2)$ centered difference to approximate the condition at $y = 0$. In this case (6.17) is extended to $j = 0$ with an appropriate substitution for $u_{i,-1}$. Write the result in matrix form. Is the matrix symmetric and positive definite?

(b) A second option is to use a $O(k^2)$ one-sided difference to approximate the condition at $y = 0$. In this case (6.17) applies with an appropriate substitution for $u_{i,0}$. Explain why the resulting matrix is not symmetric.

6.5. This problem concerns the equation

$$\frac{\partial}{\partial x}\left(P\frac{\partial u}{\partial x}\right) + \frac{\partial}{\partial y}\left(Q\frac{\partial u}{\partial y}\right) = 0,$$

where $P(x, y)$ and $Q(x, y)$ are given smooth positive functions. Use the rectangular domain and boundary conditions in Figure 6.3.

(a) After expanding the derivatives using the product rule, derive a finite difference approximation of this problem that has truncation error $O(h^2) + O(k^2)$.

(b) Write the result from part (a) in matrix form $\mathbf{Av} = \mathbf{b}$, where the diagonals of the matrix are positive.

(c) Use the integration method to derive a finite difference approximation by integrating the differential equation over the rectangular region $x_i - h/2 \leq x \leq x_i + h/2$, $y_j - k/2 \leq y \leq y_j + k/2$. Each term in the equation should reduce to an expression involving a single derivative and a single integration. Using the midpoint rule for the integrals and a centered difference for the derivatives derive a finite difference approximation for the equation. Does this have the same truncation error as in part (a)?

(d) Write the result from part (c) in matrix form $\mathbf{Av} = \mathbf{b}$, where the diagonals of the matrix are positive.

(e) The differential equation, and boundary conditions, form a self-adjoint (symmetric) and positive definite problem. Does either of the matrix equations in (b) or (d) have these properties?

6.6. This problem considers Laplace's equation using the domain and boundary conditions shown in Figure 6.22.

(a) For the grid shown in the figure, determine the entries in the matrix equation (6.20).

(b) Suppose someone solves the equation in part (a) using the CGM and they claim it took 204 iteration steps. Why would you question this assertion?

6.7. This problem examines special cases for the descent methods.

(a) If $\mathbf{A}$ is a diagonal matrix and the diagonals are positive, show that $\mathbf{A}$ is positive definite.

(b) For a matrix as in part (a), explain why SDM produces the exact solution in one iteration step.

(c) Explain why if the SDM converges in one step then the CGM must converge in one step (for any symmetric positive definite matrix).

6.8. This problem considers the example matrix given in Table 6.1 that is obtained from the finite difference approximation of Laplace's equation.

(a) Construct the directed graph for this matrix and then explain why the matrix is irreducible.

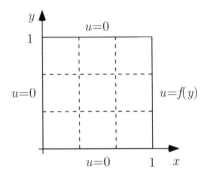

Figure 6.22. Domain for Exercise 6.6.

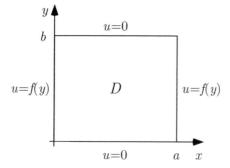

Figure 6.23. Domain for Exercise 6.10.

(b) Use the Gershgorin–Taussky theorem to find positive numbers λ_L and λ_R such that given any eigenvalue λ of the matrix you have the bounds $\lambda_L \leq \lambda \leq \lambda_R$.

6.9. This problem considers some of the properties of irreducible matrices. The matrices are not necessarily symmetric.

(a) What is the minimum number of nonzero entries in an $n \times n$ irreducible matrix?

(b) If $\mathbf{A}$ is irreducible explain why $\mathbf{A}^T$ and $\alpha\mathbf{A}$ are irreducible (assuming $\alpha \neq 0$).

(c) Show that a tridiagonal matrix is irreducible if the super- and subdiagonal entries are all nonzero.

(d) Explain why if $\mathbf{A}$ and $\mathbf{B}$ are irreducible, it is not necessarily true that $\mathbf{A} + \mathbf{B}$ is irreducible.

(e) Show that if a matrix has either a row or column in which all the nondiagonal entries are zero, then the matrix cannot be irreducible.

6.10. This problem considers Laplace's equation using the domain and boundary conditions shown in Figure 6.23.

(a) Explain how to rewrite the problem to effectively reduce the computational cost to approximately half of what it takes to solve the original problem.

(b) What assumptions on a, b, and $f(y)$ are needed to be able to rewrite the problem to effectively reduce the computational cost by approximately one-quarter of the original problem?

(c) Suppose the boundary conditions at $y = 0$ and $y = b$ are not zero but functions of x. What is required of these two functions so the reduction in (a) still works?

6.11. Some care is needed when manipulating a finite difference approximation, because it is possible that the matrix $\mathbf{A}$ ends up losing some of its more important properties. This problem demonstrates this for the following system

$$4v_1 - v_2 = b_1,$$
$$-v_1 + 5v_2 = b_2.$$

(a) Show that the coefficient matrix is symmetric and positive definite.
(b) Show that by multiplying the first equation by a nonzero constant, the resulting matrix can be nonsymmetric and not positive definite. What if both equations are multiplied by the same nonzero constant? Is it guaranteed that the coefficient matrix is symmetric and positive definite?
(c) Show that if the equations are interchanged, the resulting coefficient matrix is nonsymmetric and not positive definite.
(d) What consequences do (b) and (c) have for solving Laplace's equation? Specifically, in regard to the symmetry and positive definiteness of $\mathbf{A}$, does it make any difference (i) how the nodes are numbered to form $\mathbf{v}$, (ii) what order the v_ℓ equations in (6.19) are put into the matrix $\mathbf{A}$, or (iii) by what constant (6.19) is multiplied?

6.12. Suppose Laplace's equation is to be solved in the region shown in Figure 6.24, using the boundary conditions shown in the figure.
(a) In the case $a = b = 1$ explain why it is a good idea to take $N = M$. Your reason should also explain why $N = M + 1$ and $N = M - 1$ are not such great ideas.
(b) Write down the resulting matrix equation in part (a) when $N = M = 3$.
(c) If $a \neq b$ does your reason for taking $N = M$ in part (a) still apply?

6.13. This problem examines the error $e_{ij} = u(x_i, y_j) - u_{ij}$ and its connection with the truncation error τ_{ij}.
(a) Write the errors in vector form by taking $e_\ell = e_{ij}$ and $\tau_\ell = \tau_{ij}$ and then showing that $\mathbf{Ae} = k^2\tau$.
(b) Using the Euclidean norm show that $\|\mathbf{e}\| \leq k^2 \|\mathbf{A}^{-1}\| \cdot \|\tau\|$ and from this conclude that $\|\mathbf{e}\| \leq k^2 \|\tau\| \frac{1}{\lambda_m}$.

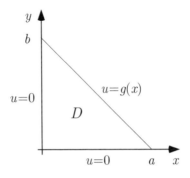

Figure 6.24. Domain D and boundary conditions for Exercise 6.12.

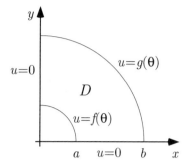

Figure 6.25. Domain for Exercise 6.15.

(c) Use the inequality $\frac{2}{\pi}\theta \le \sin(\theta)$, when $0 \le \theta \le \frac{\pi}{2}$, to show that $\|\mathbf{e}\| \le \frac{1}{8}\|\tau\|$ when D is the unit square with $h = k$. From this conclude that the truncation error determines the convergence rate of the finite difference approximation (assuming round-off is not significant).

6.14. This exercise addresses the "Are we there yet?" question, because it examines the error and how it is estimated in the CGM. In this problem $\mathbf{v}_m$ is the exact solution and $\mathbf{e}_k = \mathbf{v}_m - \mathbf{v}_k$ is the error vector for the kth iteration step.
(a) Show that $\mathbf{e}_k = \mathbf{A}^{-1}\mathbf{r}_k$ and from this conclude that $\|\mathbf{e}_k\| \le \|\mathbf{A}^{-1}\| \cdot \|\mathbf{r}_k\|$.
(b) Explain why $\|\mathbf{r}_k\|$ small does not necessarily mean that $\|\mathbf{e}_k\|$ is small.
(c) What must be assumed about $\mathbf{A}$ or $\mathbf{A}^{-1}$ for it to be possible to conclude that $\|\mathbf{e}_k\|$ is small if $\|\mathbf{r}_k\|$ is small?
(d) When implementing the algorithm for the CGM it is necessary to specify a stopping tolerance *tol*. If n is large, is it better to use $\frac{\mathbf{r}_k \cdot \mathbf{r}_k}{n} \le tol$ rather than $\mathbf{r}_k \cdot \mathbf{r}_k \le tol$?
(e) The underlying assumption of any descent method is that $F(\mathbf{v}_{k+1}) < F(\mathbf{v}_k)$. Can this fact be used in some way to develop a stopping condition on the method?

6.15. This problem examines solving Laplace's equation in the region shown in Figure 6.25, using the boundary conditions shown in the figure.
(a) Rewrite the problem in polar coordinates and sketch the rectangular domain in the $r\theta$-plane where the problem is to be solved.
(b) Derive a finite difference approximation of the polar coordinate problem in part (a) that is second-order accurate.
(c) Sketch D and the grid system used in part (b) in the xy-plane.
(d) Write the result from (b) in matrix form $\mathbf{A}\mathbf{v} = \mathbf{b}$, where the matrix is positive definite (you need to explain why it has this property).

6.16. Show that the CGM algorithm can be used, without change, to solve $\mathbf{A}\mathbf{v} = \mathbf{b}$ when $\mathbf{A}$ is symmetric and negative definite ($\mathbf{A}$ is negative definite if $-\mathbf{A}$ is positive definite).

6.17. The Fourier series solution associated with Figures 6.16 and 6.17 is given in (6.6) with the coefficients given in (6.9). To compute $g(x)$ with an error of 10^{-6} requires using 300 terms in the series, whereas an error of 10^{-2} requires using 30 terms. It is reasonable to assume that it takes at least $7n$ spatial points to resolve $\sin(\lambda_n x)$. In this case one would predict having to use about 200 spatial points to achieve an error of 10^{-2} and 2100 points for an error of 10^{-6}. How do these estimates line up with the results in Figure 6.17? Is there any connection between the number of modes needed in the series solution and the number of points needed in the numerical approximation?

6.18. This problem concerns Poisson's equation

$$\nabla^2 u = f(x, y), \quad \text{for } (x, y) \in D,$$

where D is the unit square $0 < x < 1, 0 < y < 1$. The boundary condition is that $u = 0$ around the boundary of the square.

(a) Write down a finite difference approximation of this problem that has a truncation error of $O(h^2) + O(k^2)$. Your final answer should be in the form of a matrix equation, where the matrix is positive definite (you need to explain why it has this property).

(b) If the exact solution is $u(x, y) = \sin(3\pi x)\sin(2\pi y)$ then what is $f(x, y)$? Assume that this is the function $f(x, y)$ in the remainder of the problem.

(c) Taking $N = M = 40$, plot both the numerical solution obtained using the CGM and the resulting pointwise error as a function of x and y (as is done in Figure 6.16). The error in this case is defined as $E(x_i, y_j) = |u(x_i, y_j) - \bar{u}_{ij}|$, where $\bar{u}_{ij}$ is the computed solution. Is there anything distinctive, or unusual, about where the maximum error occurs in D? Is there anything distinctive, or unusual, about how the error is distributed over D?

(d) Letting $E_{\max} = \max_{i,j} E(x_i, y_j)$ be the maximum error, use your code to complete the entries in Table 6.4. Does the error improve as expected? What about the change in the number of iteration steps for the CGM? Does the number change the way you would expect as the size of the problem increases? What about the amount of cpu time? Does it increase as expected?

(e) Using the PCGM, redo (d). Make sure to state what preconditioner you use (it cannot be a diagonal matrix).

6.19. This problem concerns Helmholtz's equation $\nabla^2 u + \omega u = 0$, where ω is a constant. Assuming that the domain and boundary conditions are as in Figure 6.3, with $a = b = 1$ and $\omega < \pi$, then the exact solution is

$$u(x, y) = \sum_{n=1}^{\infty} a_n \sinh(\gamma_n y) \sin(\lambda_n x),$$

where $\lambda_n = n\pi$, $\gamma_n = \sqrt{\lambda_n^2 - \omega}$, and

N	E_M	Total Iterations	Total Time
20			
40			
80			
160			
320			

Table 6.4. Table used for Exercises 6.18 and 6.19.

$$a_n = \frac{2}{\sinh(\gamma_n)} \int_0^1 g(x) \sin(\lambda_n x) dx.$$

(a) Find a finite difference approximation of this problem that has a truncation error of $O(h^2) + O(k^2)$. Your final answer should be in the form of a matrix equation, where the matrix is positive definite (you need to explain why it has this property).
In the rest of the problem assume $w = 2$, $g(x) = \sin(3\pi x)$, and $M = N$.
(b) Taking $N = 40$, plot both the numerical solution obtained using the CGM and the resulting pointwise error as a function of x and y (as is done in Figure 6.16). The error in this case is defined as $E(x_i, y_j) = |u(x_i, y_j) - \bar{u}_{ij}|$, where $\bar{u}_{ij}$ is the computed solution. Is there anything distinctive, or unusual, about where the maximum error occurs in D? Is there anything distinctive, or unusual, about how the error is distributed over D?
(c) Letting $E_{\max} = \max_{i,j} E(x_i, y_j)$ be the maximum error, use your code to complete the entries in Table 6.4. Does the error improve as expected? What about the change in the number of iteration steps for the CGM? Does the number change the way you would expect as the size of the problem increases? What about the amount of cpu time? Does it increase as expected? For example, do the error, number of iteration steps, and cpu time appear to be determined by some property of the problem (N, $n = N \times M$, flops/iteration, etc.) that would enable you to approximately predict their values for $N = 640$?
(d) Using the PCGM, redo (c). Make sure to state what preconditioner you use (it cannot be a diagonal matrix).

6.20. Suppose for a preconditioner for Laplace's equation one takes $\mathbf{M}$ to be diagonal and the diagonal entries to be the same as those in the matrix $\mathbf{A}$. If the diagonals are equal, as they are for this matrix, this preconditioner does absolutely nothing to improve the convergence rate of CGM. Explain why.

6.21. This problem investigates variations of the preconditioner $\mathbf{M}_2$.
(a) Suppose $\mathbf{M}_3 = (\mathbf{D} + \mathbf{L})(\mathbf{D} + \mathbf{L}^T)$. (i) Is this matrix symmetric and positive definite? (ii) What is the approximate flop count to solve $\mathbf{Mz} = \mathbf{r}$?

(iii) Plot the error as a function of the iteration step for the example in Figure 6.18. How does this preconditioner compare with $\mathbf{M}_2$?

(b) Suppose $\mathbf{M}_4 = (\mathbf{I}+\mathbf{L}\mathbf{D}^{-1})\mathbf{D}(\mathbf{I}+\mathbf{D}^{-1}\mathbf{L}^T)$, where $\mathbf{I}$ is the identity. (i) Is this matrix symmetric and positive definite? (ii) What is the approximate flop count to solve $\mathbf{M}\mathbf{z} = \mathbf{r}$? (iii) Plot the error as a function of the iteration step for the example in Figure 6.18. How does this preconditioner compare with $\mathbf{M}_2$?

6.22. Given the upper bound in (6.39) it is worth considering how accurately it predicts the convergence rate. Using (6.24) to determine λ_M and λ_m, answer the following questions.

(a) In Figure 6.15, $E_1 \approx 6$ when $n = 625$. What is the upper bound in (6.39) when $k = 10$ and when $k = 100$? How do these values compare to the values in Figure 6.15?

(b) In Figure 6.15, $E_1 \approx 13$ when $n = 2500$ and $E_1 \approx 26$ when $n = 10000$. For each, what is the upper bound in (6.39) when $k = 10$ and when $k = 100$? How do these values compare to the values in Figure 6.15?

(c) Using your results from (a) and (b) discuss the limitations of using (6.24) to predict the convergence rate of CGM.

6.23. According to the tests given in Section 6.2.2, for what values of a, b, c can the following tridiagonal matrix be used as a preconditioner?

$$\begin{pmatrix} 4 & a & & & & & \\ b & 4 & 1 & & 0 & & \\ & 1 & 4 & 1 & & & \\ & & \ddots & \ddots & \ddots & & \\ & 0 & & 1 & 4 & 1 \\ & & & & 1 & c \end{pmatrix}$$

A

Appendix

A.1 Order Symbols

One of the central questions arising when one is deriving a numerical method concerns the accuracy of the approximation. It is almost inevitable that the answer to this question involves order symbols in one form or another. As an example, a standard difference formula for the first derivative is

$$f'(x_i) = \frac{f(x_{i+1}) - f(x_i)}{h} + \tau_i, \tag{A.1}$$

where $\tau_i = -hf''(\eta_i)/2$ is the truncation error for this approximation. To emphasize how the error depends on the step size h, we write $\tau_i = O(h)$. In words, one says that τ_i is big-O of h. The precise definition of what this means is given below.

Definition A.1. *The statement that $\tau = O(h^n)$ as $h \downarrow 0$ means that there are constants A, h_0 such that $|\tau| \leq Ah^n$ for $0 < h < h_0$.*

Usually this notation will be used in conjunction with Taylor's theorem, in which case the order is determined by simply looking for the smallest exponent in the expression.

Examples

1. If $\tau = -3h^5 + 2h^8$ then $\tau = O(h^5)$. It is also true that $\tau = O(h)$ and $\tau = O(h^3)$.
2. Letting $\tau_1 = O(h)$ and $\tau_2 = O(h^3)$ set $\tau = \alpha\tau_1 + \beta\tau_2$, where α, β are independent of h. As long as α is nonzero, the conclusion is $\tau = O(h)$. For example, $-50\tau_1 + \tau_2 = O(h)$ and $\tau_1 + 314\tau_2 = O(h)$.
3. Letting $\tau_1 = h - 3h^2$ and $\tau_2 = -h^3$, set $\tau = \tau_1\tau_2$. In this case $\tau = O(\tau_1)O(\tau_2) = O(h^4)$.

4. In the text there are numerous occasions when τ depends on a step size h in addition to depending on a second step size k. An example of this situation, reminiscent of what is found for the heat equation, is $\tau = k - 5h^2$. In this case $\tau = O(k) + O(h^2)$.
5. If $\tau = k + k^3 - h^2 + kh^2$ then $\tau = O(k) + O(h^2)$. However, if we link the step sizes and take $k = h^2$, then $\tau = O(h^4)$, or equivalently, $\tau = O(k^2)$.

A.2 Taylor's Theorem

The single most important result needed to develop finite difference approximations is Taylor's theorem. Given its role, it is odd, or at least the author thinks it is odd, that web sites listing the top 100 theorems in mathematics almost inevitably have Taylor's theorem toward the middle of the list. This is a serious misjudgment, because it should easily make anyone's top 10! Anyway, the statement of the theorem is below.

Theorem A.1. *Given a function $f(x)$ assume that its $(n + 1)$st derivative $f^{(n+1)}(x)$ is continuous for $x_L < x < x_R$. In this case, if x and $x + h$ are points in the interval (x_L, x_R) then*

$$f(x + h) = f(x) + hf'(x) + \frac{1}{2}h^2 f''(x) + \cdots + \frac{1}{n!}h^n f^{(n)}(x) + R_{n+1}, \quad \text{(A.2)}$$

where the remainder is

$$R_{n+1} = \frac{1}{(n+1)!}h^{n+1} f^{(n+1)}(\eta), \quad \text{(A.3)}$$

and η is a point between x and $x + h$.

The two-variable version of the expansion in (A.2) is

$$
\begin{aligned}
f(x + h, t + k) = {} & f(x, t) + hf_x(x, t) + kf_t(x, t) \\
& + \frac{1}{2}h^2 f_{xx}(x, t) + hk f_{xt}(x, t) + \frac{1}{2}k^2 f_{tt}(x, t) + \cdots .
\end{aligned}
\quad \text{(A.4)}
$$

As an example of how this theorem is used, note that we get

$$f(x + 2h) = f(x) + 2hf'(x) + 2h^2 f''(x) + \frac{4}{3}h^3 f'''(\eta_2)$$

and

$$f(x + h) = f(x) + hf'(x) + \frac{1}{2}h^2 f''(x) + \frac{1}{6}h^3 f'''(\eta_1).$$

Multiplying the last equation by -4 and adding the result to the first equation, we obtain

$$
\begin{aligned}
f'(x) &= \frac{-f(x + 2h) + 4f(x + h) - 3f(x)}{2h} - \frac{1}{3}h^2 f'''(\eta_1) + \frac{2}{3}h^2 f'''(\eta_2) \\
&= \frac{-f(x + 2h) + 4f(x + h) - 3f(x)}{2h} + O(h^2).
\end{aligned}
$$

Other differentiation formulas can be derived in a similar manner.

A.3 Round-Off Error

Round-off error arises in scientific computation because most computer systems use a finite number of digits when representing real numbers. The potential consequences of this are demonstrated below.

A.3.1 Function Evaluation

Even simple-looking expressions can cause numerical problems, and an example is the power function

$$f(x) = (x - 1)^8. \tag{A.5}$$

This can be multiplied out to yield the expanded form

$$g(x) = x^8 - 8x^7 + 28x^6 - 56x^5 + 70x^4 - 56x^3 + 28x^2 - 8x + 1. \tag{A.6}$$

One might ask why $g(x)$ is used to designate the expanded version of the function when it is nothing more than $f(x)$. As it turns out, $g(x)$ is not the same as $f(x)$ when one attempts to evaluate the function using a computer. To demonstrate this, both functions are shown in the upper plot in Figure A.1, and as expected, there are no apparent differences between the two curves. If the interval is reduced, as in the middle plot in Figure A.1, one starts to notice small differences between the curves. If the interval is reduced further, as in the lower plot, then significant differences exist between the two functions. This brings up several observations and questions related to function evaluation on a computer:

- The curves for $f(x)$ behave as expected, in particular, they are smooth, have a minimum at $x = 1$, are never negative, and are symmetric about $x = 1$.
- The curves for $g(x)$, at least in the lower two plots, are not smooth, have multiple minimum points, are sometimes negative, and are not symmetric about $x = 1$.

Mathematically the functions $f(x)$ and $g(x)$ are equivalent, but according to the computer they are not. What is so bad about the formula for $g(x)$ that these problems arise, and is it possible to tell this before attempting to evaluate it? As once stated by Yogi Berra, "In theory, theory and practice are the same. In practice they aren't." This bit of wisdom, apparently, applies to scientific computation. A hint on what might be the source of the problem is found by noticing the small function values in the lower two plots. The values are close to the numerical resolution of the computer when double precision is used, which generally is accurate to 15 or 16 digits (this is discussed later). Consequently, the repeated adds/multiplies/exponentiations in the formula for $g(x)$ combine to cause the computer difficulty in accurately calculating the function in this range of values.

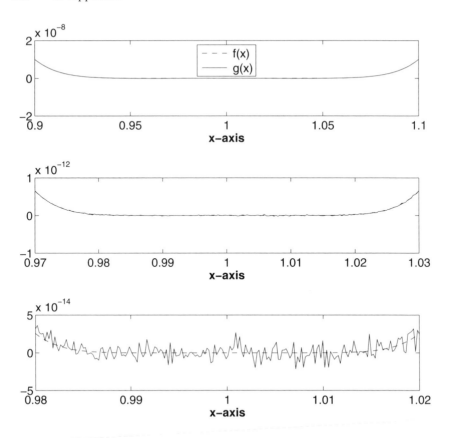

Figure A.1. Evaluation of (A.5) and (A.6) over progressively smaller intervals on the x-axis. Even though $f(x)$ and $g(x)$ are equal, due to the finite number of digits used in double precision the computer produces different values for the two functions.

A.3.2 Numerical Differentiation

Another interesting example of not getting what you might expect from a computer arises when a finite difference approximation is used to calculate a derivative. Two used extensively in this book are

$$\text{Forward Difference: } f'(x) = \frac{f(x+h) - f(x)}{h} + O(h), \tag{A.7}$$

$$\text{Centered Difference: } f'(x) = \frac{f(x+h) - f(x-h)}{2h} + O(h^2). \tag{A.8}$$

If we take $f(x) = \sqrt{x}$ and $x = 1$ then the forward difference approximation of the derivative is $(\sqrt{1+h} - 1)/h$ and the centered difference approximation is $(\sqrt{1+h} - \sqrt{1-h})/(2h)$. The differences between these expressions and the exact value $f'(1) = \frac{1}{2}$ are shown in Figure A.2(a). What is seen is that starting at $h = 10^{-1}$ the error decreases as expected as h decreases based on the truncation terms in the above difference formulas. However, for small values of h the error actually gets worse. The reason is that the values of h drop below the resolution of the computer's number system and it rounds $\sqrt{1+h}-1$ to zero. This fact is seen in Figure A.2(b), which gives the values of $\sqrt{1+h}-1$ over a portion of the h interval used earlier. The staircasing in this plot is due to the finite set of numbers the computer has to work with, and it rounds the function to the closest value it has available. For $h < 10^{-16}$ this value is zero, and in this case the computer calculates the forward difference formula to be zero (instead of a value very close to $\frac{1}{2}$). It is for this reason that the error in Figure A.2(a) is constant for $h < 10^{-16}$, because the computer is calculating the error to be $\frac{1}{2}$.

A.4 Floating-Point Numbers

Nonzero floating-point numbers have the normalized form

$$x_f = \pm m \times 2^E, \text{ for } E_m \leq E \leq E_M, \tag{A.9}$$

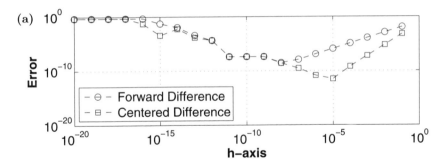

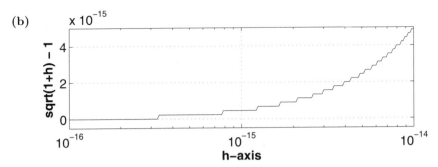

Figure A.2. In (a) the difference between the values of the difference approximations of the derivative as calculated by the computer and the exact value are given. The function is $f(x) = \sqrt{x}$ and $x = 1$. In (b) the values of the numerator for the forward difference formula over a portion of the h interval are shown.

where
$$m = 1 + \frac{b_1}{2} + \frac{b_2}{2^2} + \cdots + \frac{b_{N-1}}{2^{N-1}}.$$

In the above expressions, E is the exponent, m is the significand, or mantissa, and the b_i's make up the fractional part of m. Note that E is an integer and the upper and lower limits on the exponent are given as $E_M = 2^{M-1} - 1$ and $E_m = -E_M - 1$. The b_i's are either zero or one, which means that the significand is a real number satisfying $1 \le m < 2$. In addition to the normalized numbers given above, the floating-point system includes $0, \pm\infty$, and NaN (not a number). Including something called NaN might seem odd, but it is very useful, because the computer uses this in response to expressions that are undefined. For example, a NaN is produced for $0 \times \infty, 0/0, \infty - \infty$, etc. A few other mileposts for the number system are discussed below. A more in-depth presentation of the properties and limitations of IEEE floating-point arithmetic can be found in Overton [2001].

Machine Epsilon
A particularly important number in scientific computation is machine ϵ. This is defined as the distance between $x = 1$ and the next-largest machine number. Using the representation in (A.9), it follows that $\epsilon = 1/2^{N-1}$. The reason this is important is that it is used to determine the numerical resolution that is possible with the particular computer system being used. For example, in MATLAB machine ϵ is assigned to the variable *eps*. Since MATLAB uses double precision, one finds that $\epsilon = 2.22 \cdots \times 10^{-16}$.

Largest Floating-Point Number and Overflow
The largest positive floating-point number is $x_M = (1 - 2^{-N}) \times 2^{E_M+1}$. Generally, when a computer encounters a number larger than x_M it assigns it the value ∞. As with NaNs, this indicates that there is a problem with the calculation. With MATLAB, because it uses double precision, $x_M = 1.796 \cdots \times 10^{308}$. If one enters $y = 1.790 \times 10^{308}$ then $y + 100$ is computed with no problem but the computer calculates $2y$ to be ∞.

Smallest Floating-Point Number and Underflow
Based on the representation in (A.9), the smallest positive floating-point number is $x_m = 2^{E_m}$. The expectation therefore is that the computer will round $\frac{1}{10}x_m$ to be either zero or x_m. However, most computer systems include what are known as subnormal numbers that are located between zero and x_m. These are designed to allow for what is called gradual underflow, but they do not provide the accuracy of regular floating-point numbers. For example, using double precision, $x_m = 2.2 \cdots \times 10^{-308}$. One finds that $10^{-n}x_m$ is calculated correctly for $n = 0$ but the answer becomes progressively worse as n is increased. To illustrate, when $n = 15$ the answer is $1.976 \cdots \times 10^{-323}$, whereas if $n = 16$ then it calculates the result to be simply zero.

Precision	N	M	E_m	E_M	Smallest Positive	Largest Positive	Machine Epsilon	Decimal Digits	$C,$ C^{++}	FORTRAN	MATLAB
Single	24	8	-126	127	1.2×10^{-38}	3.4×10^{38}	1.2×10^{-7}	6	Float	Real*4	
Double	53	11	-1022	1023	2.2×10^{-308}	1.8×10^{308}	2×10^{-16}	15	Double	Real*8	Default
Quadruple	113	15	-16382	16383	3.4×10^{-4932}	1.2×10^{4932}	10^{-34}	33	Long double	Real*16	

Table A.1. Values for various floating-point systems specified by IEEE-754 and its extensions. The values for the smallest positive, largest positive, and machine epsilon $\epsilon = 1/2^{N-1}$ are given to only one or two significant digits. Similarly, the number of decimal digits is also approximate.

References

U. M. Ascher, R. M. M. Mattheij, and R. D. Russell. *Numerical Solution of Boundary Value Problems for Ordinary Differential Equations*. Society for Industrial and Applied Mathematics, Philadelphia, 1995.

K. Atkinson and W. Han. *Theoretical Numerical Analysis: A Functional Analysis Framework*. Springer, New York, 2005.

M. Bonnet. *Boundary Integral Equation Methods for Solids and Fluids*. Wiley, New York, 1999.

W. E. Boyce and R. C. DiPrima. *Elementary Differential Equations and Boundary Value Problems*. Wiley, New York, 2004.

F. Brauer and C. Castillo-Chavez. *Mathematical Models in Population Biology and Epidemiology*. Springer-Verlag, New York, 2001.

S. C. Brenner and L. R. Scott. *The Mathematical Theory of Finite Element Methods*. Springer, New York, 2002.

J. C. Butcher. *The Numerical Analysis of Ordinary Differential Equations: Runge–Kutta and General Linear Methods*. Wiley, Chichester, UK, 1987.

P. G. Ciarlet. *The Finite Element Method for Elliptic Problems*. Society for Industrial and Applied Mathematics, Philadelphia, 2002.

B. Cipra. You can't always hear the shape of a drum. In *What's Happening in the Mathematical Sciences*, volume 1, Providence, RI, 1993. American Mathematical Society.

J. D. Cole and L. P. Cook. *Transonic Aerodynamics*. North-Holland, Amsterdam, 1986.

T. Colonius. Modeling artificial boundary conditions for compressible flow. In *Annual Review of Fluid Mechanics*, pages 315–345. Annual Reviews, CA, 2004.

D. Colton. *Partial Differential Equations: An Introduction*. Dover, New York, 2004.

C. Connolly and R. Grupen. The application of harmonic potential functions to robotics. *J. Robotic Systems*, 10(7):931–946, 1993.

R. Courant and D. Hilbert. *Methods of Mathematical Physics, Vol 2*. Wiley-Interscience, New York, 1989.

G. Dahlquist. A special stability problem for linear multistep methods. *BIT*, 3:27–43, 1963.

S. A. David, S. S. Babu, and J. M. Vitek. Welding: Solidification and microstructure. *JOM*, 55(6):14–20, 2003.

J. W. Demmel. *Applied Numerical Linear Algebra*. Society for Industrial and Applied Mathematics, Philadelphia, PA, 1997.

F. M. Denaro, F. S. Marra, and G. Continillo. High-order numerical schemes with positivity-preserving properties for the solution of the transport equations in reactive flows. In *1995 Meeting of Italian Section of the Combustion Institute*, page Internet Poster Session, 1995.

P. Deuflhard, F. Bornemann, and W. C. Rheinboldt (translator). *Scientific Computing with Ordinary Differential Equations*. Springer, New York, 2002.

D. R. Durran. *Numerical Methods for Wave Equations in Geophysical Fluid Dynamics*. Springer, New York, 1998.

L. C. Evans. *Partial Differential Equations*. American Mathematical Society, Providence, RI, 2002.

W. J. F. Govaerts. *Numerical Methods for Bifurcations of Dynamical Equilibria*. Society for Industrial and Applied Mathematics, Philadelphia, 2000.

R. Haberman. *Applied Partial Differential Equations*. Prentice Hall, New York, 2003.

E. Hairer and G. Wanner. *Solving Ordinary Differential Equations II : Stiff and Differential–Algebraic Problems*. Springer-Verlag, New York, 2004.

E. Hairer, S. P. Norsett, and G. Wanner. *Solving Ordinary Differential Equations I : Nonstiff Problems*. Springer-Verlag, New York, 2002.

E. Hairer, C. Lubich, and G. Wanner. Geometric numerical integration illustrated by the Stormer–Verlet method. In *Acta Numerica 12*, pages 399–450. Cambridge University Press, Cambridge, UK, 2003.

J. Han, A. Globus, R. Jaffe, and G. Deardorff. Molecular dynamics simulation of carbon nanotube based gears. *Nanotechnology*, 8(3):95–102, 1997.

M. Hauth and O. Etzmuß. A high performance solver for the animation of deformable objects using advanced numerical methods. In A. Chalmers and T.-M. Rhyne, editors, *Proc. Eurographics 2001*, volume 20(3) of *Computer Graphics Forum*, pages 319–328, 2001.

W. Henshaw. On multigrid for overlapping grids. *SIAM J. of Scientific Computing*, 26:1528–1538, 2005.

W. Henshaw and D. Schwendeman. Moving overlapping grids with adaptive mesh refinement for high-speed reactive and non-reactive flow. *J. of Comp. Physics*, 2007.

M. H. Holmes. *Introduction to Perturbation Methods*. Springer-Verlag, New York, 1995.

T. J. R. Hughes. *The Finite Element Method : Linear Static and Dynamic Finite Element Analysis*. Dover, Mineola, New York, 2000.

D. Isaacson, J.L. Mueller, J. C. Newell, and S. Siltanen. Reconstructions of chest phantoms by the d-bar method for electrical impedance tomography. *IEEE Trans. Med. Imaging*, 23(7):821–828, 2004.

E. Isaacson and H. B. Keller. *Analysis of Numerical Methods*. Wiley, New York, 1966. Reprinted by Dover, New York, 1994.

A. Iserles. *A First Course in the Numerical Analysis of Differential Equations*. Cambridge University Press, Cambridge, UK, 1996.

M. Kac. Can one hear the shape of a drum? *American Mathematical Monthly*, 73:1–23, 1966.

H. B. Keller. *Numerical Methods for Two-Point Boundary-Value Problems*. Dover, New York, 1992.

H. Kudela. Viscous flow simulation of a two-dimensional channel flow with complex geometry using the grid-particle vortex method. In *ESAIM: Proceedings, Third International Workshop on Vortex Flows and Related Numerical Methods*, pages 215–224. EDP Sciences, France, 1999.

J. D. Lambert. *Numerical Methods for Ordinary Differential Systems : The Initial Value Problem*. Wiley, Chichester, UK, 1991.

R. J. LeVeque. *Finite Volume Methods for Hyperbolic Problems*. Springer, New York, 2003.

R. J. LeVeque. *Finite Volume Methods for Hyperbolic Problems*. Cambridge University Press, Cambridge, UK, 2002.

D. G. Luenberger. *Linear and Nonlinear Programming*. Cambridge University Press, Cambridge, UK, 2002.

W. McLean. *Strongly Elliptic Systems and Boundary Integral Equations*. Cambridge University Press, Cambridge, 2000.

J. J. H. Miller, E. O'Riordan, and G. I. Shishkin. *Numerical Methods For Singular Perturbation Problems: Error Estimates in the Maximum Norm for Linear Problems in One and Two Dimensions*. World Scientific, Singapore, 1996.

K. W. Morton and D. F. Mayers. *Numerical Solution of Partial Differential Equations*. Cambridge University Press, Cambridge, 2005.

M. L. Overton. *Numerical Computing with IEEE Floating Point Arithmetic*. Society for Industrial and Applied Mathematics, Philadelphia, PA, 2001.

N. A. Petersson. An algorithm for assembling overlapping grid systems. *SIAM J. of Scientific Computing*, 20:1995–2022, 1999.

Pixar Animation Studios. *for the Birds*. Emeryville, CA, 2000.

P. M. Prenter. *Splines and Variational Methods*. Wiley, New York, 1989.

J. D. Pryce. *Numerical Solution of Sturm-Liouville Problems*. Oxford University Press, New York, 1994.

R. D. Richtmyer and K. W. Morton. *Difference Methods for Initial-Value Problems*. Krieger, Florida, 1994.

S. M. Ross. *An Elementary Introduction to Mathematical Finance: Options and Other Topics*. Cambridge University Press, Cambridge, UK, 2002.

L. F. Shampine. *Numerical Solution of Ordinary Differential Equations*. Chapman and Hall, New York, 1994.

I. H. Sloan. Error analysis of boundary integral methods. *Acta Numerica*, 1: 287–339, 1992.

J. Stoer and R. Bulirsch. *Introduction to Numerical Analysis*. Springer-Verlag, New York, 2002.

A. Stuart and A. R. Humphries. *Dynamical Systems and Numerical Analysis*. Cambridge University Press, Cambridge, UK, 1998.

E. Süli and D. F. Mayers. *An Introduction to Numerical Analysis*. Cambridge University Press, Cambridge, UK, 2003.

E. F. Toro. *Riemann Solvers and Numerical Methods for Fluid Dynamics: A Practical Introduction*. Springer-Verlag, New York, 1999.

G. Turk. Generating textures on arbitrary surfaces using reaction-diffusion. *Computer Graphics*, 25(4):289–298, 1991.

R. S. Varga. *Matrix Iterative Analysis*. Springer-Verlag, New York, 2000.

C. R. Vogel. *Computational Methods for Inverse Problems*. Society for Industrial and Applied Mathematics, Philadelphia, 2002.

S. Waydo and R. M. Murray. Vehicle motion planning using stream functions. In *2003 IEEE International Conference on Robotics and Automation*, pages 2484–2491. IEEE, 2003.

P. Wesseling. *An Introduction to Multigrid Methods*. R.T. Edwards, New York, 2004.

J. Wisdom and M. Holman. Symplectic maps for the n-body problem. *Astron. J.*, 102:1528–1538, 1991.

Index

Texts in Applied Mathematics

Needs a little skill 100

Up for a challenge 148

Welcome bakers!

When you're learning to bake, there's no better place to start than with a cake, and there are 40 absolute classics in this book.

As well as being great bakes, the recipes have been carefully chosen to introduce you to all the key techniques, such as creaming, whisking, rubbing in and folding that not only set you up to bake better cakes, but which you will find invaluable for all baking.

Start with the 'Easy does it' section and master the basics with recipes like Quick Berry Muffins, Sticky Gingerbread or Apple and Maple Syrup Traybake. As you grow in confidence you will feel ready to move to the recipes that 'Need a little skill' – a Bitter Chocolate Roulade perhaps, a perfect Dundee Cake or a sumptuous Devil's Food Cake. The more you bake, the sooner you will be 'Up for a challenge', testing your decorating skills with the Mile-high Chocolate Cake or whipping up a light-as-air sponge for the classic Fraisier.

The colour strip on the right-hand side of the page tells you at a glance the level of the recipe (from one spoon for easy to three spoons for a challenge), and gives you a helpful checklist of the skills and special equipment you will use.

Before you begin, have a look at the Baker's Guide at the beginning of the book. That will tell you what equipment you need to get started (just a bowl, a spoon and a cake tin will do!), introduce you to the most important ingredients, and explain some terms and techniques in more detail.

We have chosen Classic Cakes as the subject of our first book in the Bake It Better series because even the simplest of cake recipes gives impressive results. It's amazing what you can do with just flour, butter, sugar and eggs when you know how. So dive in, and get baking!

HOW TO USE THIS BOOK

SECTION 1: BAKER'S GUIDE
Read this section before you start baking.

The Baker's Guide contains key information on ingredients (pages 10–15), equipment (pages 16–21) and techniques (pages 22–35) relevant to the recipes in the book.

Refer back the Baker's Guide when you're baking if you want a refresher on a particular skill. In the recipes the first mention of each skill is highlighted in bold.

SECTION 2: RECIPES
Colour strips on the right-hand side and 1, 2 or 3 spoons show the level of the recipe.

Within the colour strips you'll find helpful information to help you decide what to bake: Hands-on time; Baking time; Makes; Special equipment; Sponge used and Storage.

Refer back to the Baker's Guide when a skill is highlighted in bold in the recipe if you need a reminder.

Try Something Different options are given where the recipe lends itself to experimenting with ingredients or decorations.

BAKE IT BETTER

Baker's Guide

Ingredients

The most important piece of advice we can give when you're checking the ingredients list of a recipe is don't be tempted to substitute. This is particularly true if you're new to baking, or it's a new recipe for you.

Below are the most frequently used cake ingredients, with advice on buying, storing and using them – the more you understand your ingredients, the easier it will be to avoid problems and get consistently good bakes.

BAKING POWDER, BICARBONATE OF SODA AND CREAM OF TARTAR

Some bakes need the help of a chemical raising agent to increase their lightness. The two most common are **bicarbonate of soda** (an alkali) and **cream of tartar** (an acid). **Baking powder** is a mixture of both.

Raising agents work by reacting together with moisture and heat to release small bubbles of carbon dioxide, which lighten the crumb of your bake. Some recipes use a slightly acidic ingredient, such as buttermilk or yoghurt, in combination with bicarbonate of soda to produce the bubbles of gas. Make sure you use the exact amount stated, and discard out-of-date or damp raising agents as they won't give you the best bake. To check if baking powder is still active mix a teaspoonful into a glass of warm water – if it bubbles up nicely it's fine to use, otherwise, throw it out.

If you want to you can make your own by combining 1 teaspoon bicarbonate of soda with 2 teaspoons cream of tartar, but baking powder is easy to come by so it's not really necessary. More useful to know is that you can make your own self-raising flour by adding 4 teaspoons baking powder to every 225g plain flour.

BUTTER

Most bakers use **unsalted butter**, which has a lovely rich flavour and gives a more evenly coloured bake because it contains less whey than salted butters. Some **salted butters** also have a strong taste that can be overpowering in a sweet bake, but if the urge to bake takes you and you only have salted butter, it's really not a big problem; just remember not to add any additional salt to sweet recipes.

Wrap butter well and store it in the fridge away from strong flavours, or freeze it for up to a month. Remove from the fridge in plenty of time so that it's the right consistency for your recipe. Creamed and all-in-one sponges (see page 24) use butter at room temperature so that it's easier to smoothly incorporate into dry ingredients; chilled and diced butter is needed for rubbed-in sponges (see page 22). In an emergency you can soften diced butter cubes for a few seconds in the microwave, but it takes much longer to firm up over-softened butter.

CHOCOLATE

The flavour of chocolate cakes and ganaches, toppings, fillings and frostings really depends on the chocolate you use. Good-quality chocolate is widely available in supermarkets these days and you can get chips in larger bags from online suppliers. Store bars of chocolate well wrapped in a cool, dry, dark cupboard, and away from strong-flavoured ingredients.

Take care when melting chocolate (see page 34) as it easily scorches if it gets too warm and then becomes unusable. Chocolate chunks and chocolate chips are useful for adding to sponge mixes and for toppings/decorations, but you can also use bars chopped into similar-sized pieces.

Dark chocolate is most widely used in this book. One with around 70 per cent cocoa solids will give the best flavour. Anything over 75 per cent can be too dry and bitter for general baking.

Milk chocolate has a much lower cocoa content and a milder, sweeter taste that can be slightly greasy. Good-quality milk chocolate has a higher percentage of cocoa solids, giving better flavour and a slightly firmer set when used as a coating, or decoration.

White chocolate doesn't contain any cocoa solids, so cocoa butter content is what you're looking for here: 30 per cent or more (children's bars usually have hardly any actual cocoa butter). Because there is more fat in white chocolate it sets less firmly than dark or milk chocolates.

COCOA POWDER
A dark, unsweetened powder made from pure cocoa with nearly all the cocoa butter removed – it is very bitter and powerfully flavoured, and adds an excellent chocolate taste to lightly textured sponges and butter icings. Don't use drinking chocolate, which has had sugar and dried milk powder added to it, as a substitute.

CREAM
Always use the cream recommended in the recipe – the fat content varies significantly and can have a huge effect on your bake.

Single cream has at least 18 per cent butterfat and is good for pouring and adding richness to rubbed-in mixtures (see page 22), but it is not suitable for whipping.

Double cream has at least 48 per cent butterfat and whips well (see page 35) when thoroughly chilled. Don't use the extra-thick double cream labelled 'for spooning' as you won't be able to whip it; nor is it suitable for making ganache.

Whipping cream does exactly what it says on the tub. It has at least 35 per cent butterfat. For best results, chill it thoroughly before whipping (see page 35).

Buttermilk, often found in supermarkets as 'cultured buttermilk', is low-fat or non-fat milk plus another lactic acid-producing culture to give it an acidic 'tang'. It is often used, along with bicarbonate of soda, to add lightness and flavour to cakes.

Clotted cream has at least 55 per cent butterfat and is best for serving on the side with a slice of cake, not for whipping or ganache-making.

Sour cream has only around 18 per cent butterfat and is made by introducing a bacterial culture to give it a naturally 'soured' tang.

Crème fraîche is cream that has been soured in a similar way to sour cream, but with a richer, milder flavour.

DRIED FRUIT
Vine fruits, such as raisins, sultanas and currants, are preserved but still soft. They add sweetness and moisture, as well as a fruity flavour, to cakes and, soaked overnight in alcohol, tea or fruit juice, plump up deliciously. Soft-dried apricots, prunes, figs, cranberries, blueberries, sour cherries and dates are also useful additions and can

replace vine fruits in many recipes. Candied peel is widely available ready chopped, but can also be found in boxes as whole pieces of orange, lemon or citrus peel, allowing you to cut the pieces into a size and mix that you prefer. Store opened packets in a screw-top jar to stop the fruit getting hard.

EGGS

All the recipes in this book use medium-sized eggs (about 62–65g each). Eggs help your cake rise. Their size is important as they work in ratio with other ingredients (fats, sugar and flour). Using a different-sized egg might affect results – you might need more liquid, or it may not bind together well, rise properly or cook all the way through.

Store eggs in the fridge, pointed-side down, to protect the yolk from drying out and spoiling. Keep them in the box they came in and in the cooler body of the fridge, not the door, and use by the best before date. Spare egg whites freeze well for up to a month – mark the quantity and date on the container and defrost thoroughly before use.

Eggs should be used at room temperature as they give a greater volume when beaten, so always try to bring them out of the fridge 30–60 minutes before using. If you've left it to the last minute you can gently warm up eggs by placing them in a bowl of lukewarm water for 10 minutes.

EXTRACTS AND FLAVOURINGS

Try to avoid synthetic flavourings as they can give your bake a rather unpleasant 'fake' taste.

Vanilla extract and **almond extract** are concentrated liquids, so use them in tiny quantities. You'll find 'essences' and 'flavours' in the shops, but these won't have the same effect so try not to use them.

Vanilla paste is made from the seeds of vanilla pods and provides an even more concentrated flavour than vanilla extract.

Coffee flavour can come from bottled coffee essence, although some people find it a rather strange flavour so you can instead use instant espresso powder or granules dissolved in boiling water.

Ground spices should be measured carefully and kept in screw-topped jars rather than open packs. Try to use them when they are still fresh, preferably within a few months of opening; it's probably best to buy them in small quantities.

Alcohol in the form of liqueurs or brandies is used in cake-making. Génoise sponges are often brushed with sugar syrup mixed with a dash of liqueur or brandy. When making a rich fruitcake, dried fruit is usually soaked in alcohol to balance the sweetness and help the cake keep for longer.

FLOUR

Use flour when it's still fresh and store it correctly – keep opened packs either in storage jars, plastic food boxes or plastic food bags to stop the flour getting damp. Don't add new flour to old in storage jars, and aim to use it within a month of opening or by its best before date.

Wheat flours are the most frequently used flours in cake-making. **Plain flour** is just that: flour with nothing added. **Self-raising flour** is made from plain flour with baking powder added so that the mixture expands and rises in the oven. Make your own self-raising flour by adding 4 teaspoons baking powder to every 225g plain flour.

The flours used for cake making are known as 'soft' flours, as they have a relatively low proportion of protein to starch – usually 8–10 per cent protein (protein is the gluten-forming part of the flour which is necessary for bread-making, but is too much for making cakes). Finely milled **wholemeal** and **spelt flours** labelled 'plain' or 'self-raising' are great for more robust cakes as they add a slightly richer and nutty flavour, but they are too heavy for finer bakes like creamed sandwich cakes (see page 24).

Gluten-free flours are now readily available and inexpensive. They are usually made from a combination of rice, potato, tapioca, maize, chickpea, broad bean, white sorghum or buckwheat flours, and vary in taste and texture from brand to brand, so it's worth trying a few out. A few are specially made for cake-making and can be substituted fairly easily, although some suggest adding xanthum gum (which comes in powder form) to help the structure of the bake and allow it to rise. Others recommend adding more liquid to your recipe, so make sure you follow the advice on the packet. (It's also worth checking whether your baking powder is gluten-free.)

HONEY
Look out for honey that comes from the nectar of a single variety of flower or plant (such as orange blossom), as the flavour should be more distinct, but be careful that it doesn't overpower the flavours in your bake. As a general rule, the paler the honey, the milder the flavour. Soft-set honey (but not honeycomb) is easiest to blend in, but solid honey can be used if it is softened first (by gentle warming in the microwave or in a small dish set in a bowl of hot water). Honey also adds moisture to cake mixtures, but too much can make it dense, so make sure you follow the recipe.

ICING SUGAR
Icing sugar comes as either a very finely powdered white sugar or a lightly golden unrefined sugar, which dissolves readily and can be used for piped or spreadable icings, frostings and toppings. Refined white will give you pure white icing, while unrefined will give a hint of gold. Sift icing sugar well to remove any lumps before use (see page 30). **Fondant icing sugar** is a combination of icing sugar and dried glucose syrup that you mix with water or juice to make a glossy, satiny icing that's slightly more substantial, or to make a dough-like modelling paste for edible decorations. **Royal icing sugar** is a mixture of icing sugar and dried egg white that can be mixed with water to make a stiff white icing. **Ready-made ready-to-roll** and **rolled icing** can be used to cover and decorate cakes and cupcakes.

MARGARINE AND SPREADS
Margarines are based on vegetable oils, with added salt and flavourings. Some are made specifically for baking and can be used straight from the fridge; they give good results but won't taste quite the same as bakes made with butter. Spreads designed for use on breads and crackers are not meant for baking and won't give a good bake as they contain too much water and not enough fat. Check the pack as most say whether they're suitable for baking or not.

MARZIPAN
Made from a sweet paste of ground almonds, this can be rolled out to cover and decorate, or used as a filling ingredient.

NUTS
With distinct and varied flavours and textures, the type of nut you use is very important to your bake. Some nuts are oilier than others, which will alter the crumb and can make it heavier than expected. Their consistency, whether whole, chopped or ground, is also crucial to baking success, so check the recipe to make sure you are using what is needed. Toasting nuts increases flavour, but they burn easily, so watch them carefully. Nuts can quickly turn rancid and bitter, so store them in a screw-topped jar or airtight container in a cool, dark spot and use before they hit their best before date.

OILS
American-style cake recipes, like Apple and Maple Syrup Traybake (page 66), often use **vegetable oil** instead of butter. The advantage is that it speeds up the method because you don't have to cream butter; plus it is a handy store-cupboard ingredient. **Sunflower oil** gives the best results as it has very little flavour, but a light, mild **olive oil** can also work very well. Don't use vegetable frying oil, which will give a distinctive, unpleasant 'savoury' flavour to your baking.

SUGAR
It's important to use the type of sugar specified in the recipe. They all combine with other ingredients in slightly different ways and this affects the end result.

Caster sugar is best for most sponges as the grains are fine and quickly break down when beaten with butter. **Golden caster sugar** is less refined than caster sugar, which gives it a pale golden colour and very slightly richer flavour. **Granulated sugar** takes much longer to dissolve and often gives a speckled finish to the tops of cakes.

Brown muscovado sugars are available in both light and dark; they add a toffee-ish flavour and colour to cake mixes, but they also tend to make them slightly more moist and heavy. They will form into lumps during storage, so you'll need to sift/press out the lumps before use.

SYRUP AND TREACLE
As with sugar, it's important to use the syrup or treacle specified, as they have quite different flavours. **Golden syrup** is a sticky, pale gold syrup made from sugar cane sap. It's sweeter than sugar and gives cakes a moist, dense texture. **Black treacle** is much darker brown, thicker and stickier than golden syrup and has a very strong, almost bitter, flavour. It gives the characteristic flavour and dark colour to gingerbreads (see page 62). **Maple syrup** is the boiled-down sap of sugar maple trees and has a wonderful red-gold colour and unsurpassed flavour. It is expensive, and do beware of cheaper 'maple-flavour' syrups that won't produce the same results.

Syrup and treacle can be awkward and messy to measure and weigh, but if you sit the whole tin in a bowl of just-boiled water, or warm the measuring spoon in a mug of boiled water beforehand, you will find it much easier and less messy by far.

Equipment

The beauty of baking is that you don't need huge amounts of kit to get started; in fact you can probably produce your first bake with items you already have – a bowl, a cake tin or two, scales for weighing your ingredients accurately and a wooden spoon should just about cover it. The other bits of kit you get will depend as much on the size of your kitchen as your enthusiasm for cake-making. Bear in mind that for more complex recipes you may need more than one tin or baking sheet, although you may be able to re-use what you already have. Read the equipment list alongside each recipe carefully before you start to make sure you have everything you need.

BAKING PAPER AND LINERS

Lining papers make it easier to remove your bake from the tin once it's cooked (see page 29). Some recipes require a double layer of lining paper to give extra protection from the heat of the oven. **Non-stick baking paper** or **baking parchment** is suitable for most purposes and both are good for delicate mixtures. **Parchment-lined foil** is heavier and can be folded to make a sturdy cake case – useful for making Battenberg Cake (see page 100).

Ready-made cake tin liners and discs for lining the bases of tins will save you time and effort if you bake dozens of cakes, and are available from most supermarkets. **Re-usable silicone liners** are excellent for lining baking sheets and can be cut to fit other tins you use regularly.

Greaseproof paper is best kept for wrapping cooked food as it is water resistant, but its waxy coating doesn't stand up well to heating and your cakes will just stick to it.

BOWLS

You'll probably end up with a variety of bowls. It's useful to have more than one and some different sizes, and there are pros and cons to the different types. **Heatproof glass** bowls are probably the best all-purpose choice for mixing, whisking and melting chocolate over hot water (see page 34). **Stainless steel** bowls are unbreakable and dishwasher-proof, but they won't go in the microwave. **Ceramic** bowls are pretty, but can break quite easily and can be heavy. **Plastic** are all-purpose and cheap, and ones with rubber bases are non-slip. (You can solve the non-slip issue by placing a damp cloth underneath any bowl.) **Anodised aluminium** bowls are very durable and will last a lifetime but, again, they're no good for the microwave.

CAKE BOARD

A cake board isn't essential when you're starting out – you can use a plate or the base of a cake box in most cases – but it is useful for more advanced cake recipes as it will protect the delicate sponge while you decorate. You can get disposable and re-usable cake boards.

CAKE TINS

It is very important to use the tin that's specified in the recipe. The quantities and baking time have been calculated to work with that particular cake tin and your bake won't turn out the way it's supposed to if you use a different one.

A really solid, heavy-duty tin will last forever, withstand repeated baking without scorching or warping, and stay rust-free if you take care of it, which means washing and drying it thoroughly after use.

Heavy-duty metal cake tins are probably

the most reliable and durable, but other options are non-stick metal tins, heavy aluminium, glass, ceramic or silicone. Avoid flimsy non-stick tins as they will quickly lose their coating. Never use a tin that has started to shed its coating as bits of it could end up in your bake. If you opt for silicone moulds, a baking sheet underneath will help stabilise the mould before you pour in the mixture, and when you're transferring it to the oven.

When you're starting out you'll get the most use out of a couple of 20.5cm straight-sided deep sandwich tins, a 450g loaf tin and maybe a square 20–20.5cm traybake tin. With these you'll be able to make most of the Easy Does It recipes. Unless you buy a tin with the size stamped on the base it's a good idea to mark the size underneath with an indelible marker to save getting out a tape measure each time you bake.

Deep round/square cake tins are good for richer cakes and fruitcakes, such as the Cherry Cake or the Porter Cake (see pages 112 or 152) and are available with fixed or loose bases. You can get 'push-up' loose-based tins with a watertight silicone seal for easy release.

Loaf tins can be used for cakes like Lemon Drizzle Loaf Cake or Double Marble Cake (see page 44 or 86), as well as for bread-making. The 450g (about 19 × 12.5 × 7.5cm) and 900g (about 26 × 12.5 × 7.5cm) tins are the most commonly used sizes.

Sandwich tins are the ones to start with. A pair of round, deep, straight-sided sandwich tins 20.5cm across, with sides 4–5cm high, will allow you to make most classic cakes. A third tin makes for quicker baking when

making American-style layer cakes, such as the Three-layer Banana Cake (page 98), but you can always just re-use your other tin while the first layer is cooling.

Specialist tins are useful if you want to bake recipes such as the Lemon-scented Madeleines (page 64), for which you'll need a madeleine mould tray. Recipes such as the Battenberg Cake (page 100) and Bitter Chocolate Roulade (page 128) also use specialist tins, although you can use standard tins instead (see the recipe methods for instructions on adapting regular tins).

Springclip tins have a spring release, a base that clamps in place when the clip is fastened and a deep metal ring which lifts off when unfastened. These are really helpful when making cheesecakes and cakes that are not inverted, or are particularly fragile, such as the Apple and Ginger Crumble Cake (page 54). They come in many sizes, but 20.5cm and 22–23cm tins are the most useful.

Swiss roll tins are rectangular with shallow sides about 2cm high. The most useful sizes are 20 × 30cm and 23 × 33cm.

Traybake tins are square or rectangular with sides that are about 4cm high. They are endlessly versatile and are used in this book for making Chocolate Brownies (page 60) as well as basic sheet cakes like the Vanilla Traybake (page 42). The most useful ones are loose-based, or slide apart for the easy removal of your bake. A 20–20.5cm square tin and a 20.5 × 25.5cm rectangular tin will come in handy most often for the bakes in this book.

6- or 12-hole deep muffin tins/cupcake tins are what you need for small bakes, muffins and cupcakes. You can get non-stick or silicone versions (choose wire-framed for stability), which may not need lining with paper cases.

COOLING RACKS

A large wire cooling rack with legs allows air to circulate underneath your cooling cakes, preventing condensation (and the dreaded 'soggy bottom'). If needs be, you can improvise with a clean grill-pan rack, but the finer wires on a collog rack are more effective.

FOOD-PROCESSOR

A food-processor makes light work of chopping and blending. Some will also make cake mixtures with a plastic blade or paddle, but results vary, so go on personal recommendation if you wish to use the machine for cake-making. Generally, they'll work better for some of the wetter mixtures than for recipes where you need to incorporate a lot of air.

ICING TURNTABLE

This is a non-essential but useful bit of equipment that enables rotation of the cake while you're decorating. Pick one with a non-slip base.

KNIVES

A **large sharp knife** for chopping nuts and neat slicing is vital (as is a **knife sharpener**). A **long-bladed serrated bread knife** is good for slicing cakes into layers (see page 34). A **palette knife** is good for spreading, and an **off-set palette knife** (one with a kink in the blade) is essential for more advanced cake-decorating.

Knives are made from different materials. The main ones to consider are stainless steel, which is cheaper but needs to be sharpened regularly; carbon steel, which is more expensive, harder and easier to keep sharp; and ceramic, which is far harder than carbon steel ones, much lighter and don't require sharpening – but can chip easily.

LARGE METAL SPOON

A large metal spoon with a long handle is useful for folding in (see page 33) and for transferring your cake mixtures.

MEASURING JUG

Pick a heat-resistant and microwave-safe jug that has both metric and imperial measures, starting from 50ml if you can find one, otherwise 100ml, and going up to 2 litres. A small jug or cup that measures from 1 teaspoon (5ml) up to 4 tablespoons (60ml) is a very useful extra. You can also get measuring jugs with cup measurements – useful if you bake a lot of North American recipes.

MEASURING SPOONS

Everyday teaspoons, dessertspoons and tablespoons vary enormously in size so you shouldn't use them to measure ingredients. Baking is such an exact science that it's worth investing in proper measuring spoons for small amounts of liquids and dry ingredients (such as baking powder, spices, salt and sugar), ranging from 1/8 teaspoon to 1½ tablespoons. Go for spoons with narrow ends that will fit into fiddly spice jars. Unless the recipe says otherwise, all spoon measures in these recipes are level – skim off the excess with a finger or the back of a knife.

OVEN THERMOMETER

Baking requires accuracy and ovens vary; their internal thermostats can be notoriously unreliable. If you really get the baking bug you might want to invest in an oven thermometer to double-check it's the correct temperature and to work out where in your oven the hotter and cooler spots are located.

PASTRY BRUSH

Available in a variety of widths and bristles, pick a brush in a medium width for brushing on melted butter, beaten egg, glaze or sugar syrup. Make sure whatever type you buy is heat resistant and dishwasher-proof.

PIPING BAGS

Disposable plastic piping bags in various sizes are available from most supermarkets. Generally the ones with the non-slip exteriors are easiest to use. You can also find reusable **nylon piping bags** from specialist shops and cake-decorating suppliers. They have a little more weight and strength to them and don't have seams for the mixtures to leak through. Most can be rinsed and then washed inside out in very hot water. Always make sure they are completely dry before putting them away. (See page 35.)

PIPING NOZZLES

These conical tubes fit into the end of piping bags and are available in scores of shapes and sizes from the finest writing tip to large, sharp-toothed star nozzles for piping buttercream. The best value are the sets that provide a reusable bag plus a set of stainless steel nozzles, either just the small size for decorating or an all-round set that includes large (1.5cm/2cm), plain and star nozzles.

ROLLING PIN

In cake making a rolling pin is used for rolling out marzipan and icing. Choose a long, fairly heavy one that is about 6–7cm in diameter – ones without handles are generally easier to use.

SCALES

Baking is really a science, so it pays to be accurate if you want perfect results every time. As you'll be dealing with some quite small quantities, **digital** scales are preferable to **spring** or **balance** scales as they are much more precise and can weigh ingredients that are as little as 1 gram. You can see the weight easily at a glance and add multiple items to one bowl simply by resetting the balance to zero after adding each ingredient. Always keep a spare battery on standby.

SIEVE

Essential for removing lumps from icing sugar and other ingredients, and to bring air into mixtures. A stainless steel wire sieve with a large bowl is the most versatile and should last longer than plastic. A smaller tea-strainer sized sieve is also a good piece of kit to have, for dusting cakes and bakes with icing sugar and cocoa powder.

SPATULA

Flexibility is key here. You want a good-sized rubber or plastic spatula that's heat resistant for mixing ingredients together, cleaning out bowls and spreading mixtures. A smaller one is perfect for fiddly amounts. A metal spatula is good for spreading.

STORAGE CONTAINERS

It's not essential to go to the expense of buying a special container to store your bakes, but if you find yourself trying to transport your cakes a lot it's probably worth investing in a cake carrier, which makes transportation a breeze. Choose one that has secure locking clips and can go in the dishwasher. Otherwise, use a good stainless steel tin with a tight-fitting lid, or a heavy-duty plastic container with a secure lid. Store your bake well away from any heat sources (radiators, sunlight, kitchen light fittings, your fridge or cooker) and mark on a sticker the day of baking so you know when your bake is still good to eat.

TIMER

A digital kitchen timer, with seconds as well as minutes, is another good purchase for achieving that all-important accuracy. Get one with a loud ring and set it for 1 minute less than the suggested time in the recipe, especially if you are unsure of your oven temperature – you can always increase the cooking time if needed.

WHISKS AND MIXERS

These range from the most basic, which means the baker has to do the energetic whisking, to free-standing food mixers that do all the hard work for you.

Wire whisks can be balloon-shaped or flat. A sturdy hand-held wire whisk with an easy-grip handle that fits your hand is ideal for whisking mixtures on and off the heat.

Hand-held rotary whisks have two beaters in a metal frame, which are turned by hand. They're perfect for whisking egg whites (see page 31), whisking mixtures over heat (no trailing leads) and whisking out lumps in batters.

Hand-held electric whisks are more expensive, but much more powerful and can also be used for creamed cake mixtures (see page 24) and more general mixing. Look for models with a set of attachments and a retractable cord for easy storage.

Free-standing mixers really do save time and energy if you do a lot of baking. A large free-standing model with attachments for beating, whisking and making dough is a great investment. If possible, buy an extra bowl too, as it helps when making cakes with multiple elements; a bowl with a snap-on lid is very useful too. Large mixers do all the beating and whisking for you, so you will get more volume into your meringues and whisked egg mixtures in a much shorter time, and are great for batch-baking. They do take up space though, so they're not ideal for a small kitchen – an electric hand whisk is probably the best choice here.

WOODEN SPOONS

You can never have enough wooden spoons – they're heat-resistant, won't scratch non-stick pans and are ideal for beating mixtures (see page 31). It's a good idea to keep ones for baking separate from those that are used for savoury cooking, as wooden spoons are porous and will absorb strong flavours.

ZESTER

The zest of lemons, limes and oranges is full of flavour and often used in baking recipes. A zester is the best way to remove the zest so you don't end up with lots of bitter pith as well. A long zester with a sturdy, easy-grip handle is easiest to use and clean. If you can, go for one of the new types with a ceramic-coated stainless steel surface and slide-on zest catcher.

Skills

Once your ingredients are lined up and equipment sorted, you're ready to get baking. This section covers everything from lining your tin to mixing and folding, and it's worth reading at least once. Follow this advice and you can enjoy your baking safe in the knowledge that you'll be happy with the end results.

All the recipes in the book tell you exactly what you need to do step-by-step, but you'll notice that some of the baking terms are highlighted in bold, which means you can refer back to this section if you want a bit more detail, or to refresh your memory.

THE 5 CAKE-MAKING METHODS

The key to any good cake is the texture of the sponge. Whether it's the delicate crumb of a Victoria sponge, a denser fruitcake or light-as-air Swiss roll, it's all down to the way the ingredients are combined. Once you've got to grips with these techniques the world of cakes will be there for the baking.

Below each method are some examples of recipes in the book that use them. Try these and in no time you'll be an expert in each method.

RUBBED-IN METHOD
This refers to the action of combining fat and flour to add air to the mix, making the final bake lighter. It's used for fairly robust, lower-fat cakes like rock cakes and simple fruitcakes, and is quick to do, either with your hands and a bowl, or using a mixer or food-processer. Rubbed-in mixtures use cold butter, straight from the fridge; try to keep your palms clean while you combine the ingredients as your fingertips are cooler and won't melt the butter.

Because these mixtures use about half fat to flour they lose their freshness quickly, so cakes should be eaten on the same day. *Learn with: Spicy Scottish Rock Cakes (page 40), Simple and Pretty Fruitcake (page 46)*

How to rub in by hand
1. **Sift** the flour (and any spices or raising agents) into a mixing bowl.
2. Make sure the butter is cool and firm and cut it into small cubes. It helps to have floury hands to do this to stop it melting and sticking as soon as you touch it.
3. Add the butter pieces to the flour and quickly toss them around in the flour with your fingertips.

4. Once the butter pieces are coated, pinch them into the flour using just your fingertips and thumbs. Lift your hands to the top of the bowl and gently rub more of the mixture between your fingers and thumbs so that it combines as it falls back into the bowl.

5. Keep doing this until the mixture has a finer, crumb-like consistency (*see photo, right*). To check if you've missed any clumps, give the bowl a shake – you'll see them come to the surface.

6. Once you're happy with the consistency, quickly stir, or mix in, any other flavourings with a spoon or fork, plus eggs and/or milk, to make a fairly stiff cake mixture, but don't over-beat it as you don't want to develop the gluten in the flour and toughen the crumb.

How to rub in using a food-processor
1. Tip the flour into your food-processor bowl, add the chilled, diced butter and pulse in short bursts until there are no visible butter pieces remaining.
2. Add the sugar and pulse again until combined, then slowly pour the liquid through the feeder tube, pulsing until the dough starts to clump together.
3. Tip into a bowl and use your hands to very gently knead and gather the mixture into a ball.

How to rub in using a free-standing mixer
1. Fit the machine with the creamer/ paddle attachment and slowly mix your dry ingredients with the chilled, diced butter until the butter is no longer visible.
2. Add the sugar and mix again. You can then add the liquid and gently mix until starting to clump together.
3. Gather the mixture into a smooth ball.

ALL-IN-ONE METHOD

The recipes that use this method are the heroes of the cake world, offering speed, simplicity and reliability. All the ingredients are combined in one go by beating with a wooden spoon, electric beater or whisk. Make sure the ingredients are at room temperature so they can be smoothly combined and air can be easily incorporated into the mix. Take the butter and eggs from the fridge a good hour before you need them.

1. Sift the flour and any spices, cocoa powder or raising agents (a little baking powder is usually included even with self-raising flour to help the mixture rise well) into a large bowl, then add the sugar, butter (or oil), beaten eggs and any liquid (milk, coffee, fruit juice) and start **beating** everything together fairly slowly.

2. Once all the dry ingredients have been incorporated, scrape down the sides of the bowl and beat for a couple of minutes at a fast speed until it is smooth and creamy (*see photo, left*).

Learn with: Apple and Ginger Crumble Cake (page 54), Fudgy Chocolate Birthday Cake (page 82)

CREAMED METHOD

This is the method used for most of our best-loved cakes. As with the all-in-one mixtures, make sure the ingredients are at room temperature so they quickly combine and take up as much air as possible. You should always use the type of sugar specified in the recipe, but with creamed mixtures this is particularly important.

1. Put the softened butter into a large bowl and **beat** with a wooden spoon, whisk or electric mixer until the butter is very creamy – almost like mayonnaise (*see photo, right*).

2. Gradually beat in the sugar, scraping

down the sides of the bowl every minute or so, until fully combined, then add any small amounts of flavouring, if using. (Save larger amounts of flavouring, such as coffee or alcohol, until later). Keep on beating and scraping down the sides of the bowl, until the mixture is very pale in colour and the texture is fluffy and lump-free.

3. Beat the eggs with a fork until well broken up, then beat them into the mixture with a wooden spoon, a tablespoon at a time. Take your time to beat in thoroughly after each addition as you're incorporating air into the mixture, which will make your finished cake much lighter. Add 1 tablespoon of the weighed flour from your recipe with each of the last two portions of egg to prevent the mixture from curdling. If it does start to separate a little it's not a disaster – the cake will taste just the same, but it might not rise quite so well (see page 36).

4. Sift the remaining flour and salt into the bowl (plus any spices) and gently **fold** in using a large metal spoon, until you can no longer see any specks or streaks of flour.

5. At this point you can add any additional flavours (such as coffee, juice or alcohol), edible food colouring, fruit, nuts, milk or melted chocolate, as per the recipe. Again, for best results, make sure all additions are at room temperature.

6. Make sure all your ingredients are combined and you have a soft mixture; a good test for a creamed sponge mixture is that it should be just soft enough to drop off a wooden spoon when lightly shaken.

7. Transfer to the prepared tin and the oven immediately, as the raising agents will be getting to work already.

Learn with: Victoria Sandwich Cake (page 74), Whisky Dundee Cake (page 104) and Porter Fruitcake (page 152)

MELTED METHOD

Sponges that use the melted method are usually pretty straightforward to make and are good for hard-to-combine ingredients with a high moisture content, like treacle or fruit juice. They are usually runnier mixtures than those on previous pages and make cakes with a soft, moist crumb.

You will need a medium to large-sized pan (large enough to hold all the ingredients) and a wooden spoon.

1. Dice the fat (usually butter) and put it into the pan with the sugar(s) and golden syrup or black treacle, or a liquid like tea or fruit juice. Stir over a low heat until the butter has melted and the mixture is very smooth and lump-free. Some mixtures then need to be boiled or simmered for a few minutes.

2. To avoid scrambling the eggs, leave the mixture to cool until barely warm before adding the beaten eggs, flour and any other ingredients, then pouring the mixture into your tin (*see photo, left*).

Learn with: Malted Tea-loaf Cake (page 56), Sticky Gingerbread (page 62) and Chocolate Brownies (page 60)

WHISKED METHOD

Whisked sponges rely on whisked eggs, rather than the addition of raising agents, to make them light and fluffy. It's a very versatile method used to produce cakes of varying levels of complexity, from the fairly simple Classic Swiss Roll (page 94) to the challenging Gâteau de l'Opéra (page 168).

Cake mixtures where whole eggs and sugar are whisked together are also known as 'génoise' sponges and may, or may not, have a small quantity (around 15g per egg) of melted, cooled (or very creamy and pourable) butter folded into

the mixture right at the end, after the flour. This makes the sponge softer and richer.

'Biscuit' sponges like the Fraisier (page 162) are also whisked, but the yolks and whites are whisked separately with sugar, and combined later. These sponges are slightly drier and are often used for multi-layered creations because they are a bit more robust.

The key to success with whisked sponges is good technique (see page 31). Also make sure your eggs are at room temperature to aid expansion. Ideally, you should use an electric hand whisk or a large free-standing mixer fitted with a whisk attachment for the task. If you only have a rotary whisk, it's best to stand the mixing bowl over a pan of steaming hot water (but don't let the base of the bowl touch the hot water), as the gentle warmth increases the expansion of the eggs (be very careful!).

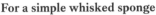

For a simple whisked sponge
1. Whisk the eggs and sugar at high speed for at least 5 minutes, until the colour changes from bright yellow to a very pale, creamy colour and the volume increases about five-fold. You should have a thick, mousse-like foam consistency. You can tell that the mixture is ready when it passes the 'ribbon test'. To do this, lift the whisk out of the mixture – if a very distinct ribbon-like trail of mixture falls back into the bowl, you can stop whisking (*see photo, right*).
2. At this point the carefully **sifted** flour is added to the bowl and gently **folded** into the foamy mixture. This very light sponge is raised entirely by the air bubbles, so the folding in of the flour must be done with care and a light hand.
Learn with: Classic Swiss Roll (page 94), Lemon Curd Layer Cake (page 144)

For a 'biscuit' sponge

1. Carefully separate the egg whites from the yolks and melt the butter.

2. Whisk the whites to soft peaks (*see photo, left*), then gradually whisk in half the sugar, a couple of teaspoons at a time, to make a light meringue.

3. Add the remaining caster sugar to the yolks, along with any other ingredients, and whisk until you have a mousse-like mixture that falls in thick ribbons form the whisk.

4. Fold the meringue into the mousse mixture in batches.

5. Sift the flour on top and gently **fold** it into the mixture.

6. Drizzle the melted butter over the top and fold that in, too.

Use for: Fraisier (page 162), Gâteau de l'Opéra (page 168)

EXPERT ADVICE FROM START TO FINISH

This section takes you through every stage of the cake-making process, explaining the how, what and why behind the key techniques.

HOW TO LINE CAKE TINS

Most recipes ask you to prepare your cake tin, which usually involves greasing and lining it in some shape or form. This stops the cake sticking to the tin. If you double-line your tin, the paper will also act as protection from the oven's heat. Prepare your tin before you start so that your mixture doesn't have to sit and wait later.

When you're getting ready to bake, always make sure your tin is clean and dry.

Shallow tins like a sandwich tin or a traybake tin should be greased and base-lined. Lightly brush melted butter inside the base, sides and rim (the butter makes a better barrier than oil and tastes nicer than lard or vegetable fat). Set the tin on a sheet of baking paper and draw around it (*see photo, above right*), then cut out the shape and press it onto the base, taking care that there are no creases (you can also buy ready-cut discs – see page 17).

Loaf tins should first be lightly brushed inside with melted butter. Then cut a long strip of baking paper that is the same width as the tin base and twice its length (the extra paper each end will help you lift the baked loaf from the tin to save damaging it). Press the paper onto the base and up the short sides to line it (*see photo, right*) – the long sides will not be covered by paper (they are well-greased and once you loosen the cake after baking it will lift out easily).

Circular deep tins should be greased and lined on the base and sides. First brush with melted butter then cut out two rounds of baking paper very slightly smaller than the base of the tin (see instructions above). Next cut a strip of baking paper that's long enough to go all round the sides, and high enough to stand about 5cm above the rim.

Make a fold 3cm deep along one long edge of the strip and open it back out so that you have a crease. Snip all the way along the folded edge up to the crease, at 1cm intervals, to create a fringe. Press one of the paper rounds into the base of the tin, then press the long strip around the sides up to the crease, so that the paper fringes fold inside the tin (*see photo, left*), slightly overlapping; be careful to avoid creases.

Brush a little melted butter onto the first base round, then put the second round onto it, covering the fringe. Lightly brush a little more melted butter on the sides and base lining to hold them in place (you can also buy cake tin liners that need no extra preparation, rather like huge cupcake cases – see page 17).

HOW TO SIFT DRY INGREDIENTS

It might often seem unnecessary, but sifting dry ingredients like flour, raising agents, cocoa powder, icing sugar and spices, not only removes any large lumps, but also introduces air and disperses the ingredients throughout the mixture, which can make a big difference to the evenness of your bake.

Tip your ingredients into a large dry sieve over a large bowl and gently tap the sieve with your hand so the ingredients settle into the bowl.

You can also use a food-processor to combine dry ingredients effectively – this is great for mixtures using muscovado sugar, which tends to form hard lumps. The machine will add air automatically, so you won't need to sift ingredients into it.

HOW TO BEAT A MIXTURE

Recipes often tell you to beat an ingredient or mixture and what this usually means is to add air by vigorous mixing. You can beat individual ingredients like butter or eggs, or whole cake mixtures as you do in the all-in-one method (see page 24).

There are various kit options available for beating, ranging from the most basic wooden spoon, to a food-processor. An electric whisk or mixer will add air into the mixture for you, but to do this by hand, angle the bowl slightly away from you and beat in small circles, rotating the spoon or whisk away from you, lifting it slightly out of the mixture in the upward movement to trap the air that will then be incorporated into the mixture in the downward movement (*see photo, right*).

HOW TO WHISK EGGS

Whether you're whisking whole eggs or just their whites, eggs should always be at room temperature to help them expand to their maximum volume.

Whisking egg whites needs a large, spotlessly clean and grease-free bowl (any trace of fat or yolk stuck to the bowl or whisk will prevent the whites from being beaten successfully – you can cut a lemon in half and run the cut side around the inside of the bowl and over the whisk to be really sure).

Put the egg whites in the bowl and whisk on a low speed (or slowly by hand) for about 30 seconds so they become frothy and the structure starts to develop.

If you add a pinch of cream of tartar or a drop of lemon juice at this point the slight acidity will help the structure to stiffen, which helps you achieve the maximum volume. Increase the speed and continue whisking until the mixture is a mass of tiny bubbles with a very smooth and fine texture. To tell if the whites have reached **soft peak stage**, lift the whisk out of the mixture – you should get a peak of egg whites that slightly droops down (*see photo, left*).

The next stage, after a little more whisking, is **stiff peak**, when the peak should stand upright, with no droop (*see photo, below left*). At this stage you should be able to turn the bowl upside down and hold it over your head (or the worktop, if you're not feeling quite so adventurous) without the whites falling out.

Whisking egg whites for meringues involves adding sugar, which you do when the whites reach soft peak stage. If you add the sugar too early it will dissolve quickly and make the mixture soft and damp. But don't wait until stiff peak stage because the structure won't be elastic enough to hold the sugar, resulting in a lumpy, bobbly meringue instead of a shiny, glossy, smooth one.

Whisking to the 'ribbon stage' is used for delicate sponges where eggs and sugar are whisked at high speed to build up a thick, mousse-like mix. Use a large bowl to allow for increased volume and whisk on a high speed for 4–5 minutes until the mixture becomes so thick that when the whisk is lifted out of the bowl the mixture falls back into it to leave a thick, ribbon-like trail on the surface. (See page 26 for more on this and for photograph.)

HOW TO FOLD IN

This is the way to delicately combine two or more ingredients – for example adding sifted flour to a creamed cake mixture or incorporating beaten egg whites – so that you don't knock out all the air you've carefully beaten or whisked in. Use the edge of a large spoon or plastic spatula to cut down cleanly through the centre of the mixture until you touch the bottom of the bowl, then turn the spoon right-way up and bring it up through the mixture to the top. Turn the spoon over so that the contents flop gently onto the rest of the mixture (*see photo, right*). Give the bowl a quarter turn so that you start from a different place, then cut down again through the mixture, lift it and flop it over again. Keep doing this folding action, using the least number of movements possible, until you can't see any more unmixed streaks.

HOW TO CHECK IF YOUR CAKE IS COOKED

The most reliable way to test delicate cakes and most sponges is the fingertip test: gently press the top of the sponge in the centre with your fingertip – the sponge is ready if it springs back into place and has started to shrink back from the sides of the tin. If a slight dent remains in the sponge after you press it (or it starts to sink), then keep it in the oven for a few more minutes.

The cocktail stick or skewer test is used for heavier and richer cakes, all fruitcakes and dense chocolate cakes. Simply stick a fine metal or wooden skewer or a wooden cocktail stick, into the centre of the cake. If it comes out clean rather than damp with cake mixture, the cake is ready, although in some exceptions (Brownies are a good example) the skewer should come out still slightly sticky.

HOW TO TURN OUT A SPONGE

Carefully run a round-bladed knife around the inside of the tin to loosen the sponge, and leave it to firm up for 30–60 seconds. To avoid your wire rack leaving marks on the top of a delicate sponge, cover a clean board with a clean, dry tea towel and turn the sponge out upside down onto it. Peel off the lining paper from the base, then set the wire rack on top and turn the whole thing over again. Carefully remove the board and towel. You can leave the sponge to cool, right-side up, on the wire rack.

HOW TO CUT A SPONGE INTO LAYERS

This sounds tricky but is actually very easy once you get the hang of it. First, make a small vertical nick or cut up the side of the sponge with the tip of a small knife – this will help you align the layers in the right place when sandwiching them back together. Gently but firmly press down on top of the sponge with the flat of your hand, and, using a long serrated knife (a bread knife is ideal), carefully saw the sponge horizontally in half to make two even layers (*see photo, above left*). If possible keep the cake still and let the knife do the work.

HOW TO MELT CHOCOLATE

Break the chocolate into even-sized pieces so that it all melts at the same rate. Put it into a heatproof bowl set over a saucepan of steaming hot (but not boiling) water; be careful not to let the base of the bowl touch the water as it will burn the chocolate (*see photo, left*). As the chocolate softens, stir it gently so it melts evenly. You can use the chocolate as soon as it is liquid and smooth, around 30°C (86°F). If it overheats and reaches 50°C (122°F) it will 'seize' and become grainy, hard and unusable, so you'll have to start again. White chocolate is very easy to overheat and the pieces may keep their shape, even when the chocolate is melted, because of the high fat and sugar content. Some recipes advise to take the bowl off the heat once the chocolate is half melted to continue melting off the heat.

HOW TO WHIP WHIPPING CREAM
Make sure the cream is thoroughly chilled before you start to stop it curdling (in warm weather, chill the whisk and bowl, too). Use a hand-held wire whisk (balloon is best), hand-held rotary whisk, electric whisk, or free-standing mixer. If you need the cream for folding in, whip on a medium speed to a **soft peak**. If you are going to pipe the cream, whisk for a couple more seconds again until it holds a **firm peak** (*see photo, above right*).

HOW TO FILL A PIPING BAG
Drop the piping nozzle, if you are using one, into the piping bag, then snip off enough of the tip so that the nozzle fits snugly and just peeps out. Twist the bag right above the nozzle (so the mixture doesn't ooze out while you're filling it), then put the bag in a tall glass or jug and fold the top of the bag over the rim (the container will support the bag so that it's easier to fill). Spoon the mixture into the bag until about two-thirds full (*see photo, right*). Unfold the bag from the rim and twist the top to push the icing down to the (still twisted) nozzle end, pushing out any air pockets, then twist it again to compact the mixture and prevent it escaping. Untwist the nozzle end and squeeze the bag so that the mixture fills the nozzle. Practise the flow and shape of the piping before you begin.

Help!

No matter how experienced you are in the kitchen, sometimes things just go wrong. Here are the most frequently encountered baking issues, and how to resolve them.

MY CREAMED MIXTURE HAS CURDLED!

Don't panic, all is not lost! First work out why this has happened, then try to fix it.

The butter and egg emulsion 'splits' (looks curdled) for three reasons: the butter was too cold and not sufficiently beaten (see page 31) before and after the sugar was added; the eggs were too cold, making it difficult to make the emulsion work; or the eggs were added too quickly, without enough beating in between each addition. As well as the tips on page 24, to prevent it happening next time, use a warmed mixing bowl if the kitchen is chilly, make sure the butter is properly softened and that the eggs are removed from the fridge in plenty of time. You can often save the mixture by beating in a tablespoon of hot water, or by adding another egg yolk.

If you haven't been able to rescue it, bake the cake anyway – it may be smaller and heavier, but it will taste just as good.

WHY DIDN'T MY CAKE RISE?

Check your ingredients: did you use the correct flour, or did you accidentally leave out the raising agents? If baking powder was an ingredient, check it's still active; if it's stale or damp it will have lost the power to produce carbon dioxide, which is what raises the sponge (see page 11).

It may be that your oven wasn't heated to the correct temperature before baking. Most cake mixtures need to be cooked fairly quickly, making the fat melt as the starch expands to form the structure of the cake. An oven thermometer will tell you if your oven's thermostat is out by a few degrees and you can also check if there are any hot or cool spots; if you're baking more than one cake at a time this can alter the bake, so they may need rotating around, or given a little longer in the oven.

WHY ARE THERE CRACKS ON THE TOP OF MY CAKE?

Cracks are usually caused by the oven being too hot, or the cake was on too high a shelf. The cake forms a crust too quickly but the rest of the mixture continues to rise, causing the surface to crack. Don't worry too much though, it will still taste good.

WHY DID MY CAKE SINK IN THE MIDDLE?

Your cake wasn't baked enough. This is another reason to get an oven thermometer to double-check your oven's thermostat. If the centre of the cake mixture doesn't get hot enough (around 100 °C/212 °F), then the structure won't become set and firm, and the middle of the cake will collapse as it cools. Use a kitchen timer and test for doneness (see page 33) before pulling the cake out of the oven to cool. And if you open the oven door too soon after putting the cake in it's likely to sink, so don't be tempted to take a peek until time is up.

Disguise the dip with whipped cream and fresh fruit, or cover it up with icing.

WHY IS THERE A PEAK IN THE MIDDLE OF MY CAKE?

A slight dome is usual, a mountain isn't! If your cake tin is too small for the mixture the outside edges will set and form a crust while the middle keeps on rising, as this is the last part of the cake mixture to get hot. Or maybe the cake mixture was too dry and stiff: check the egg sizes (see page 13) and always weigh everything accurately.

You can trim the cool dome away neatly and cover the surface of your cake with a dusting of icing sugar or a layer of icing.

WHY HAS MY CAKE OVERFLOWED ITS TIN?

It's likely that the tin was too small for the cake mixture, so check the recipe for the tin you need to use. Sandwich cakes usually call for 'deep' sandwich tins, with sides 4–5cm high – these are not the same as layer-cake tins or pans, which are shallower.

To make it ready for the table, just cut off the excess and neaten up the top and sides. Again, finish with a dusting of icing sugar or cover with a layer of icing.

WHY HAS MY SPONGE CAKE STUCK TO THE TIN?

Sponge cakes are sweet, sticky mixtures, so you really need to spend a little time on preparation. Non-stick tins are great but you still need to brush the inside and rim with melted butter and cover the base with baking paper (see page 29). To be really careful you can sprinkle a little extra of the flour used to make the cake into the tin, then shake and twirl the tin so that the flour coats the sides evenly before tapping out the excess. Rich fruitcakes and some deep sponge cake recipes call for the sides to be lined too, as this helps the edges set firm as well as helping the cake bake evenly – if you skip this step you risk the edges becoming scorched and dry, so it's worth taking the extra time to do it.

To get a truly stuck-in cake out of its tin, run a round-bladed knife all the way around

the inside of the tin and then leave it for about 5 minutes so it firms up a bit. It will shrink slightly as it cools, so hopefully you'll be able to get it out. If it's still not budging, leave it to cool and set firm, place a plate on top, then turn it upside down and shake it out onto the plate.

WHY IS MY CAKE RATHER TOUGH?

Over-working an all-in-one or rubbed-in cake mixture (see pages 24 and 22) can develop the gluten in the flour, causing a tougher, bread-like bake, so be careful not to over-beat these mixtures (see page 31): stop as soon as everything has been incorporated and looks smoothly amalgamated.

There's not a great deal you can do if your cake is heavy like this, but try using it in a trifle, moistened with sherry.

WHY IS MY CAKE SOGGY?

Check your recipe to see whether your cake needs to be turned out onto a wire rack as soon as it comes out of the oven (see page 34) – this will prevent the steam inside it condensing and making the base soggy. Sponge cakes are usually unmoulded fairly rapidly, but larger cakes and fruitcakes are best left to cool in the tin to prevent them cracking or splitting, and only unmoulded once they have firmed up. A few cakes, often those baked in springclip tins, need a few minutes for the edges to firm up before the sides of the tin are unclipped. Always run a round-bladed knife around the inside of the tin to loosen the cake unless the recipe says otherwise. And if you do have a cake with the dreaded soggy bottom, you can always just cut off the base.

BAKE IT
BETTER
Recipes

Spicy Scottish
Rock Cakes

A great starter cake for getting to grips with the **rubbed-in method**. No fancy equipment here – just grab a spoon and mixing bowl to make a batch of these little foolproof cakes.

225g self-raising flour
½ teaspoon ground mixed spice
¼ teaspoon ground ginger
couple of good pinches of salt
85g unsalted butter, chilled and diced
85g golden caster sugar

100g luxury dried fruit mix
1 medium egg
2 tablespoons milk
1 tablespoon demerara or coarse sugar crystals, for sprinkling

1. Preheat the oven to 200°C (180°C fan), 400°F, Gas 6. Line the baking sheet with baking paper.

2. **Sift** the flour, mixed spice, ground ginger and salt into a mixing bowl, then drop in the butter pieces, tossing them in the flour to lightly coat. Using the very tips of your fingers and thumbs, **rub in** the butter with the flour so that the mixture looks like fine crumbs. Add the golden caster sugar and luxury dried fruit mix to the bowl and stir with a wooden spoon or plastic spatula.

3. Break the egg into a separate bowl, pour in the milk and beat them with a fork until just combined.

4. Pour the mixture into the dry ingredients and stir together until you have a very firm, stiff dough. The dough should hold its shape, but if it refuses to come together or there are stray dry crumbs at the bottom of the bowl, stir in a little more milk, a teaspoon at a time. Don't add too much milk, though – if the mixture is too soft it will collapse into scones in the oven!

5. Use a spoon to scoop out the mixture and make 10 heaped peaky mounds on the prepared baking sheet. Space them well apart so that they can spread out. Sprinkle over the demerara or coarse sugar crystals.

6. Bake for 12–15 minutes until a good golden colour and firm to the touch. If they look like they are not baking evenly after the first 8 minutes, rotate the baking sheet.

7. Put the rock cakes on a wire rack and leave them to cool. Eat warm or at room temperature, preferably on the day you baked them.

Try Something Different

To make Fat Rascals (a richer version of Rock Cakes), replace the ground ginger with ½ tsp ground cinnamon and a few gratings of nutmeg, and use single cream instead of milk. Top each cake with 3 blanched almonds before you bake them. Eat warm, smothered with melting butter.

Easy does it

HANDS-ON TIME:
10 minutes

BAKING TIME:
12–15 minutes

MAKES:
10 cakes

SPECIAL EQUIPMENT:
None

METHOD USED:
Rubbed-in method, see page 22

STORAGE:
Eaten the day you make them

Vanilla Traybake

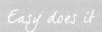

A traybake is a simple no-fuss **all-in-one** mixture. It uses just a few ingredients you probably already have, but can be glammed up with anything from glittery sprinkles to silver balls.

Easy does it

HANDS-ON TIME:
10 minutes

BAKING TIME:
20–25 minutes

MAKES:
20 squares

SPECIAL EQUIPMENT:
25.5 × 20.5 × 5cm traybake tin, cake tin or roasting tin

METHOD USED:
All-in-one method, see page 24

STORAGE:
Keep for up to 4 days in an airtight container

For the sponge
125g unsalted butter, softened
125g caster sugar
1 teaspoon vanilla extract
2 medium eggs, at room temperature, beaten
1 tablespoon milk, at room temperature
150g self-raising flour

For the frosting
250g icing sugar
100g unsalted butter, softened
1 tablespoon milk
½ teaspoon vanilla extract, or to taste
decorations of your choice

1. Preheat the oven to 180°C (160°C fan), 350°F, Gas 4. Grease and **line** the tin with butter and baking paper.

2. To make the sponge, put the butter, sugar, vanilla, eggs, milk and flour into a large bowl or the bowl of a food-mixer. **Beat** everything together with a spoon or the whisk attachment. Start slowly – otherwise the mixture could fly out of the bowl – and scrape down the sides of the bowl every now and then with a spatula. After about 2 minutes the mixture should look very smooth and light. If you are adding any flavourings or other ingredients, stir them in now.

3. Scrape the mixture into the prepared tin and spread it out right into the corners, making sure it is nice and even. Bake for 20–25 minutes until the top is a good golden brown. You can **check** your cake is cooked by gently pressing the centre lightly with your fingertips; it should spring back.

4. Stand the tin on a wire rack and run a round-bladed knife around the inside to loosen the sponge. Leave it to cool and firm up in the tin and it will be easier to remove. When cooled, **turn out** the cake and peel off the paper.

5. While the sponge is cooling, make the frosting. **Sift** the icing sugar into a mixing bowl then add the butter, milk and vanilla extract and beat well with a spoon or electric whisk on a low speed until the frosting is smooth and light.

6. Swirl the icing over the top of the cold sponge with a palette knife. If you want to decorate the top, do it now before the icing firms up. Cut into squares to serve.

Try Something Different

Stir in 5 tbsp chocolate chips, dried fruit, fudge chunks or chopped mixed nuts at the end of Step 3.

Lemon Drizzle Loaf Cake

This lemon drizzle is made using the same speedy **all-in-one** technique as the Vanilla Traybake (page 42), but with a little baking powder added to achieve a good rise.

Easy does it

HANDS-ON TIME:
15 minutes

BAKING TIME:
55–60 minutes

MAKES:
1 large cake

SPECIAL
EQUIPMENT:
900g loaf tin (about
26 × 12.5 × 7.5cm)

METHOD USED:
All-in-one method,
see page 24

STORAGE:
Keep for up to
5 days in an airtight
container

For the sponge
175g unsalted butter, very soft (but not runny)
250g caster sugar
3 medium eggs, at room temperature, beaten
100ml milk, at room temperature
finely grated zest of 2 medium unwaxed lemons
250g self-raising flour
½ teaspoon baking powder
good pinch of salt

For the drizzle
finely grated zest of 1 medium unwaxed lemon
juice of 2 medium lemons
100g caster sugar

1. Preheat the oven to 180°C (160°C fan), 350°F, Gas 4. Grease the tin with butter and **line** the base and two short sides with a long strip of baking paper.

2. To make the sponge, put the butter into a large mixing bowl or the bowl of a food-mixer. Add the sugar, eggs, milk and finely grated lemon zest and **sift** over the flour, baking powder and salt.

3. **Beat** everything together using a wooden spoon or the whisk attachment. Start slowly so that the ingredients don't fly out of the bowl, then increase the speed to medium–high and beat for a minute or so until you have a very smooth and creamy cake mixture.

4. Scrape the mixture out of the bowl and into the prepared tin and spread it out evenly, working the cake mixture into the corners and making sure the surface is level.

5. Bake for 55–60 minutes until the top is a good golden brown. **Check** your cake is cooked by inserting a cocktail stick or skewer into the centre; if it comes out clean, it is ready.

6. While the cake is in the oven, make the drizzle. Put the lemon zest, juice and sugar into a small bowl and stir well until you have a thick, sticky glaze.

7. Stand the cake in its tin on a wire rack and prick the sponge all over with a cocktail stick. Quickly pour the drizzle over the sponge, using the back of the spoon to push it into the holes.

8. Leave the cake, still in its tin, on the rack until cold. When it is cold, run a round-bladed knife around the inside of the tin to loosen the cake, then use the ends of the lining paper to lift it out. Peel off the lining paper, cut it into slices and serve.

Simple and Pretty Fruitcake

A moist and light 'cut and come again' fruitcake that uses the very forgiving **rubbed-in** method, which makes it a perfect fruitcake for first-time bakers to make.

HANDS-ON TIME:
20 minutes

BAKING TIME:
70 minutes

MAKES:
1 large cake

SPECIAL EQUIPMENT:
20.5cm springclip tin or deep round cake tin

METHOD USED:
Rubbed-in method, see page 22

STORAGE:
Keep for up to 4 days in an airtight container

For the sponge
350g self-raising flour
good pinch of salt
175g unsalted butter, cold and firm (but not hard), diced
175g golden caster sugar
finely grated zest of 1 unwaxed lemon
275g luxury mixed fruit
150g marzipan, cut into 1.5cm cubes

3 medium eggs, at room temperature
4 tablespoons milk, at room temperature

For the topping
2 tablespoons apricot jam, sifted, or apricot glaze
2 teaspoons boiling water
2 tablespoons toasted flaked almonds

1. Preheat the oven to 180°C (160°C fan), 350°F, Gas 4. Grease and **line** the tin with baking paper.

2. To make the sponge, **sift** the flour and salt into a mixing bowl, then add the butter, tossing it in the flour to coat. Using the very tips of your fingers, **rub in** the butter with the flour until the mixture looks like fine crumbs. Stir in the golden caster sugar and lemon zest with a wooden spoon. When everything is combined, add the mixed fruit and marzipan and mix well.

3. Beat the eggs and milk in a small bowl with a fork until just combined, then tip them into the mixing bowl and stir well with the spoon until you have a stiff mixture. Scrape the mixture into the prepared tin and spread evenly.

4. Bake for 70 minutes until golden brown. **Check** your cake is cooked by inserting a cocktail stick or skewer into the centre; if it comes out clean, it is ready. You might need to test in several places to avoid hitting a lump of marzipan, though!

5. Set the tin on a wire rack, run a round-bladed knife around the inside of the tin to loosen the cake, then unclip the sides. Leave on the rack until cold.

6. To make the topping, mix the jam with the boiling water until smooth, then brush it over the cake. Scatter over the almonds and leave to set for 20 minutes.

Try Something Different

Add a simple icing: sift 50g icing sugar into a bowl and stir in 2½ tsp lemon juice until smooth. Taste, adding a few more drops of lemon juice if needed, then pour the icing over the cake, letting it gently run down the sides.

Double Chocolate
Cupcakes

Using oil instead of butter in this super-speedy cupcake recipe means you don't have to cream the mixture.

For the sponge
140g self-raising flour
40g cocoa powder
good pinch of salt
200g caster sugar
125ml sunflower oil
1 medium egg, at room temperature
175ml milk, at room temperature
1 teaspoon vanilla extract
100g white chocolate chips (or milk or dark)

For the frosting
75g icing sugar
25g cocoa powder
50g unsalted butter
50g caster sugar
2 tablespoons milk
extra chocolate chips, to decorate

HANDS-ON TIME:
10 minutes

BAKING TIME:
18–20 minutes

MAKES:
12 cupcakes

SPECIAL EQUIPMENT:
12-hole cupcake tray, Food-processor (optional)

METHOD USED:
All-in-one method, see page 24

STORAGE:
Keep for up to 2 days in an airtight container

1. Preheat the oven to 190°C (170°C fan), 375°F, Gas 5. Line the cupcake tray with paper cases.

2. Put the flour, cocoa powder, salt and sugar into the food-processor bowl fitted with the blade. Pulse 3–4 times until everything is thoroughly mixed. (If you don't have a food-processor make the cakes using the traditional all-in-one whisking method, it just takes a bit longer.)

3. Pour the sunflower oil into a jug with the egg, milk and vanilla extract and beat everything together with a fork until the egg is broken up.

4. Switch on the food-processor and pour the liquid through the feed tube with the machine running. Once it is all added, stop the machine, scrape down the bowl, then pulse again for 10 seconds, until the mixture is smooth, creamy and streak-free.

5. Remove the blade and stir the chocolate chips into the cake mix using a plastic spatula. Divide the mixture between the paper cases so they are half filled.

6. Bake for 18–20 minutes. **Check** the cakes after 15 minutes and if they aren't cooking evenly, rotate the tray. They are ready when they spring back when you gently press them in the centre. Carefully remove from the tray and cool on a wire rack.

7. To make the frosting, **sift** the icing sugar and cocoa into a heatproof bowl. Melt the butter with the sugar and milk in a small pan over a low heat, stir well, then bring to the boil. Once boiling, immediately pour the mixture over the icing sugar and cocoa. Mix until smooth. Leave for 10 minutes until thickened and spreadable.

8. Swirl the frosting over the cooled cupcakes using a round-bladed knife or an offset palette knife. Scatter over the extra chocolate chips before the frosting sets.

Old-fashioned Banana Bread

Banana bread is another classic **all-in-one** cake. This one has a lovely dense texture, so there is no need for baking powder. Dark brown bananas add intense flavour and sweetness.

100g unsalted butter
250g self-raising flour
good pinch of salt
150g light muscovado sugar
100g pecan or walnut pieces
50g dark chocolate chips

250g peeled very ripe bananas
(2–3 depending on size)
2 medium eggs, at room
temperature, beaten

Easy does it

HANDS-ON TIME:
15 minutes

BAKING TIME:
55 minutes

MAKES:
1 large loaf cake

SPECIAL
EQUIPMENT:
900g loaf tin (about
26 × 12.5 × 7cm)

METHOD USED:
All-in-one method,
see page 24

STORAGE:
Eat the same day or
keep for up to
4 days in an airtight
container

1. Preheat the oven to 180°C (160°C fan), 350°F, Gas 4. Grease the tin with butter and **line** the base and two short sides with a long strip of baking paper.

2. Gently melt the butter in a small pan over a low heat, or in the microwave. Leave to cool.

3. Put the flour, salt, muscovado sugar, pecan or walnut pieces and dark chocolate chips into a mixing bowl and mix thoroughly with a wooden spoon, crushing any big lumps of sugar.

4. Put the peeled bananas onto a large plate and mash them fairly roughly with a fork — you don't want a smooth purée, it's better to have some small lumps left in to give a little texture to the cake. Add the bananas to the bowl with the eggs and the cooled melted butter, then mix together for 1 minute, using a wooden spoon, until everything is well combined.

5. Scrape the mixture into the prepared tin and spread it around evenly, getting it right into the corners of the tin.

6. Bake for 55 minutes until the top of the cake is a good golden brown. **Check** your cake is cooked by inserting a cocktail stick or skewer into the centre; if it comes out clean, it is ready. Test it in a few other spots, too, in case you hit a lump of molten chocolate.

7. Stand the tin on a wire rack and leave to cool for 5 minutes. Run a round-bladed knife around the inside of the tin to loosen the cake, then carefully lift it out using the ends of the lining paper. Leave it on the wire rack until it is completely cold, then peel off the lining paper.

Quick Berry Muffins

These lovely and easy fresh fruit muffins are a great way to practise making a **creamed** cake mixture. It's a simple way to add lightness to a bake that you'll use again and again.

60g unsalted butter, softened
150g caster sugar
1 medium unwaxed lemon
2 medium eggs, at room temperature
275g self-raising flour
½ teaspoon bicarbonate of soda

125ml natural yoghurt (unsweetened, not Greek-style)
200g fresh blueberries or raspberries
2 tablespoons coarse sugar crystals, for sprinkling

1. Preheat the oven to 200°C (180°C fan), 400°F, Gas 6. Line the cupcake tray with paper cases.

2. Put the butter into a mixing bowl or the bowl of a food-mixer and beat well with a wooden spoon or the whisk attachment until the mixture is creamy. Gradually beat in the caster sugar, scraping down the bowl every now and then. Finely grate the zest of the lemon into the bowl and beat it in.

3. Break the eggs into a separate bowl and beat them with a fork until broken up, then beat into the mixing bowl, a tablespoon at a time, until the mixture is quite soft.

4. **Sift** the flour and bicarbonate of soda into the bowl and stir in with a plastic spatula or wooden spoon.

5. Squeeze 1 tablespoon of juice from the lemon and stir this into the yoghurt, then stir this into the muffin mixture. As soon as it is thoroughly combined, add the fresh blueberries or raspberries to the bowl and gently **fold** in so the fruit does not break up.

6. Spoon the mixture into the paper cases, dividing it equally so that they are equally filled, then sprinkle over the coarse sugar crystals.

7. Bake for 20–25 minutes until golden brown. **Check** that the centres of the muffins feel firm when gently pressed. Set the tray on a wire rack and leave the muffins to cool in the tray for 5 minutes, then lift them out of the tray and onto the wire rack to cool. Eat warm or at room temperature on the day you baked them.

Try Something Different

For a slightly healthier bake, replace the white flour with wholemeal self-raising flour, add an extra tablespoon of yoghurt, and replace the sugar crystals with granola for a healthy crunchy topping.

HANDS-ON TIME:
10 minutes

BAKING TIME:
20–25 minutes

MAKES:
12 muffins

SPECIAL EQUIPMENT:
12-hole cupcake tray, deep-hole bun tray or muffin tray

METHOD USED:
Creamed method, see page 24

STORAGE:
Eaten the day you make them

Apple and Ginger Crumble Cake

This delicious apple and ginger cake shows you how to add flavour to a basic **all-in-one** mixture.

For the sponge
175g unsalted butter, softened
175g golden caster sugar
3 medium eggs, at room temperature, beaten
200g self-raising flour
good pinch of salt
2 teaspoons ground ginger
3 lumps (about 60g) stem ginger, drained and finely chopped
3 tablespoons ginger syrup from the stem ginger jar
3 Braeburn apples (about 400g)

For the topping
150g plain flour
½ teaspoon ground ginger
75g golden caster sugar
75g unsalted butter, chilled and diced
icing sugar, for dusting
whipped cream flavoured with sugar and vanilla, to serve

HANDS-ON TIME:
25 minutes

BAKING TIME:
60–70 minutes

MAKES:
1 large cake

SPECIAL EQUIPMENT:
20.5cm springclip tin or loose-based deep round cake tin

METHOD USED:
All-in-one method, see page 24

STORAGE:
Keep for up to 2 days in an airtight container

1. Preheat the oven to 180°C (160°C fan), 350°F, Gas 4. Grease and **line** the tin with butter and baking paper.

2. To make the sponge, put the butter, sugar and eggs into a large mixing bowl or the bowl of a mixer and **sift** over the flour, salt and ground ginger.

3. Add the ginger and syrup. Beat with a spoon or the whisk attachment (start slowly so you don't make a mess), for 2 minutes until you have a streak-free creamy batter. Spread the mixture into the prepared tin.

4. Peel, quarter and core the apples, then thinly slice them. Arrange in an even layer on top of the mixture.

5. To make the crumble topping, **sift** the flour and ground ginger into a small bowl. Stir in the sugar, then add the butter pieces and toss them in the flour until coated. Using the tips of your fingers, **rub** in the butter with the flour until the mixture looks like fine crumbs.

6. Sprinkle the topping evenly over the apples, then press down gently but firmly with the flat of your hand – this will make the cake easier to slice. Then dip your fingers in cold water and lightly flick it over the crumble topping to just dampen it – another clever trick to make slicing easier!

7. Bake for 60–70 minutes until well risen and lightly golden. **Check** your cake is cooked by inserting a cocktail stick or skewer into the centre; if it comes out clean, it is ready.

8. Set the tin on a wire rack and run a round-bladed knife around the inside of the tin to loosen the cake, then carefully unclip the sides or gently take the cake out of the tin. Leave until cold on the wire rack, then dust with icing sugar. Serve with whipped cream flavoured with a little sugar and vanilla extract.

Malted Tea-loaf Cake

You only need a pan, wooden spoon and tin for this **melted** mixture – the perfect method for making moisture-rich cakes. Here the tea-soaked fruit produces a deliciously sticky cake.

350g mixed dried vine fruits (raisins, sultanas and currants)
250ml strong 'builders'' tea (without milk)
100g unsalted butter, diced
100g dark muscovado sugar

2 teaspoons ground mixed spice
225g wholemeal plain flour
good pinch of salt
2½ teaspoons baking powder
2 medium eggs, at room temperature, lightly beaten

1. Put the dried fruits, tea, butter, dark muscovado sugar and ground mixed spice into a large pan. Set over a low heat and stir gently with a wooden spoon until the butter has melted, then increase the heat to medium and let the mixture simmer for 5 minutes, stirring now and then. Turn off the heat and leave to cool for 15–20 minutes until you can comfortably dip your little finger into it.

2. While the mixture is cooling down, preheat the oven to 180°C (160°C fan), 350°F, Gas 4. Grease the tin with butter and **line** the base and two short sides with a long strip of baking paper.

3. Set a sieve over the pan of barely warm mixture and **sift** in the wholemeal plain flour, salt and baking powder (tip any bran that's left in the sieve into the pan). Pour in the beaten eggs, then mix well with the wooden spoon until combined. Scrape the mixture into the prepared tin and spread it out evenly.

4. Bake for 45 minutes until the cake is well risen and a good golden brown. **Check** your cake is cooked by inserting a cocktail stick or skewer into the centre; if it comes out clean, it is ready.

5. Stand the tin on a wire rack and leave until cold, then use the ends of the lining paper to help lift the cake out.

Try Something Different

For a slightly different, citrusy flavour, use mixed fruit that includes chopped orange and lemon peel.

Easy does it

HANDS-ON TIME:
15 minutes

BAKING TIME:
45 minutes

MAKES:
1 medium loaf cake

SPECIAL EQUIPMENT:
450g loaf tin (about 19 × 12.5 × 7.5cm)

METHOD USED:
Melted method, see page 26

STORAGE:
Wrap in baking or greaseproof paper, store in an airtight container and eat within 4 days. The cake will deepen in flavour as it matures and is even better the day after baking

Blackberry Cinnamon Streusel

Tangy blackberries add a jam-like layer between the moist **all-in-one** sponge and crunchy streusel topping – a crumbly combination of flour, butter, sugar and, in this case, cinnamon.

For the sponge
125g self-raising flour
1½ teaspoons ground cinnamon
125g unsalted butter, softened
125g golden caster sugar
½ teaspoon vanilla essence
2 medium eggs, at room temperature
175g fresh blackberries

For the streusel topping
75g self-raising flour
50g demerara sugar
1 teaspoon ground cinnamon
50g unsalted butter, at room temperature, diced
2 tablespoons flaked almonds (optional), for sprinkling
icing sugar, for dusting

1. Preheat the oven to 180°C (160°C fan), 350°F, Gas 4. Grease and **line** the tin with butter and baking paper.

2. To make the sponge, **sift** the flour and ground cinnamon into a mixing bowl or the bowl of a food-mixer, then add the butter and golden caster sugar.

3. In a separate bowl, beat together the vanilla essence and eggs with a fork for a few seconds, just until the eggs are broken up. Tip these into the mixing bowl. Beat everything together with a wooden spoon or the whisk attachment, starting slowly so that the mixture doesn't fly out of the bowl. After a minute or so, scrape down the sides of the bowl, then beat again until the ingredients are combined and the mixture is very smooth and streak-free.

4. Scrape the mixture into the prepared tin and spread it out evenly, then scatter the fresh blackberries over the top nice and evenly.

5. To make the streusel topping, put the flour, demerara sugar and ground cinnamon into a bowl and mix them together. Add the butter and toss the pieces in the flour until coated. Using the tips of your fingers and thumbs, **rub** in the butter and flour until you have small pea-sized lumps.

6. Scatter the streusel topping over the fruit in an even layer, then top with the flaked almonds (if using). Bake for about 55 minutes, until the topping is a good golden brown. If you think the top is browning too quickly, cover it loosely with a sheet of baking or greaseproof paper. **Check** your cake is cooked by inserting a cocktail stick or skewer into the centre (avoiding the berries); if it comes out clean, it is ready.

7. Set the tin on a wire rack and run a round-bladed knife around the inside to loosen the cake, then leave for 5 minutes to firm up before unclipping the tin. Cool on the rack and dust with icing sugar to serve.

HANDS-ON TIME:
20 minutes

BAKING TIME:
55 minutes

MAKES:
1 medium cake

SPECIAL EQUIPMENT:
20.5cm springclip tin or loose-based deep round cake tin

METHOD USED:
All-in-one method, see page 24

STORAGE:
Store in an airtight container in the fridge and eat within 3 days. The cake tastes best if you let it come back up to room temperature or gently warm it before serving

Chocolate Brownies

Who can resist a brownie? This rich brownie is made from high-quality dark chocolate and uses a combination of the **melted** and **whisked methods** to create that signature squishy texture.

225g unsalted butter, diced
100g good-quality dark chocolate (about 70 per cent cocoa solids), broken or chopped into even-sized pieces
200g caster sugar

4 medium eggs, at room temperature
½ teaspoon vanilla extract
50g plain flour
50g cocoa powder
good pinch of salt
75g walnut pieces

1. Preheat the oven to 180°C (160°C fan), 350°F, Gas 4. Grease and **line** the tin with butter and baking paper.

2. Put the butter into a small heavy-based pan and drop in the chocolate pieces. Set the pan over the lowest possible heat and melt gently, stirring frequently with a wooden spoon, until the mixture is smooth but not hot. Remove the pan from the heat and leave the mixture to cool until needed.

3. Put the caster sugar, eggs and vanilla extract into a large mixing bowl and **whisk** with a wire hand whisk until thoroughly combined, then whisk in the cooled melted chocolate mixture.

4. **Sift** the plain flour, cocoa powder and salt into the bowl, and mix thoroughly with a wooden spoon or plastic spatula. Stir in the walnuts, then scrape the mixture into the prepared tin and spread it out evenly. Bang the tin on the worktop to settle the mixture and knock out any pockets of air.

5. Bake for about 25 minutes. **Check** your brownies are cooked by inserting a cocktail stick or skewer halfway between the sides of the tin and the centre of the cake. Don't test the centre because you want that to be slightly soft as the cake will continue cooking for few minutes once it's out of the oven, and you want a squidgy, not a dry, brownie! The mixture will puff up in the oven but don't be alarmed if it starts to sink as it cools – that's normal and will give you a dense, fudgy brownie.

6. Stand the tin on a wire rack then run a round-bladed knife around the inside of the tin to loosen the cake. Leave the cake to cool completely in the tin before cutting it into squares.

Easy does it

HANDS-ON TIME:
20 minutes

BAKING TIME:
25 minutes

MAKES:
36 squares

SPECIAL EQUIPMENT:
20.5cm square brownie tin or shallow cake tin

METHOD USED:
Melted method, see page 26
Whisked method, see page 26

STORAGE:
Keep in an airtight container and eat within a week

Sticky Gingerbread

Rich, moist gingerbread is made using the **melted** method. This one is made with spices and thick black treacle.

HANDS-ON TIME:
20 minutes

BAKING TIME:
45–50 minutes

MAKES:
1 large loaf cake

SPECIAL
EQUIPMENT:
900g loaf tin (about
26 × 12.5 × 7cm)

METHOD USED:
Melted method,
see page 26

STORAGE:
Wrap the loaf in
fresh baking paper
or foil and store in
an airtight container.
Keep for up to
1 week

225g self-raising flour
good pinch of salt
1 teaspoon bicarbonate of soda
1 tablespoon ground ginger
1 teaspoon ground cinnamon
1 teaspoon ground mixed spice
115g unsalted butter, chilled and diced
60g glacé ginger, or crystallised or drained stem ginger, very finely chopped

115g golden syrup
115g black treacle
115g dark brown muscovado sugar
275ml milk (not skimmed)
1 medium egg, at room temperature, beaten

1. Preheat the oven to 180°C (160°C fan), 350°F, Gas 4. Grease the tin with butter and **line** the base and two short sides with a long strip of baking paper.

2. **Sift** the flour, salt, bicarbonate of soda, ginger, cinnamon and mixed spice into a large mixing bowl then add the butter pieces, tossing them in the flour until coated. Using your fingertips, **rub** in the butter until the mixture looks like fine crumbs. Stir in the ginger with a wooden spoon, then put the bowl to one side.

3. Weigh the golden syrup and black treacle directly into a pan then warm gently until melted and runny but not hot. Leave to cool until barely warm.

4. Put the muscovado sugar and milk into a separate pan over a low heat and stir until the sugar dissolves. Let it cool until barely warm, then pour it into the flour mixture, followed by the melted syrup mixture and egg. Mix together with a wire hand whisk or a spoon until it's a thick but runny batter the consistency of double cream.

5. Transfer the mixture to the prepared tin and bake for 45–50 minutes until well risen and firm to the touch. **Check** your cake is cooked by inserting a cocktail stick or skewer into the centre; if it comes out clean, it is ready.

6. Stand the tin on a wire rack. Run a round-bladed knife around the inside of the tin to loosen the cake. Leave it in the tin until completely cold, then turn it out.

Try Something Different

Top the gingerbread with a simple zesty lemon glacé icing. Sift 65g icing sugar into a bowl and mix in 1 tbsp lemon juice until you have a smooth, runny, thin icing. Spoon this over the loaf while it is still in the tin, gently spreading it and letting it trickle down the sides. Scatter over small chunks of the same ginger you used for the cake. Leave to set before slicing.

Lemon-scented Madeleines

These little lemony cakes are made from a **whisked sponge,** giving them their light texture – no baking powder required. A fab way to get the hang of whisking eggs to the right consistency.

140g unsalted butter, diced
155g plain flour
good pinch of salt
finely grated zest of 1 large unwaxed lemon
4 medium eggs, at room temperature
140g caster sugar
icing sugar, for dusting

Easy does it

HANDS-ON TIME:
15 minutes

BAKING TIME:
10–12 minutes

MAKES:
12 Madeleines

SPECIAL EQUIPMENT:
Metal or silicone Madeleine mould tray

METHOD USED:
Whisked method, see page 26

STORAGE:
Keep for up to 5 days in an airtight container

1. Preheat the oven to 190°C (170°C fan), 375°F, Gas 5. Carefully brush the tray with melted butter, chill, then brush the tray a second time.

2. Gently melt the butter in a pan and leave to cool until needed.

3. **Sift** the flour and salt into a bowl and stir in the lemon zest until thoroughly combined. Put to one side for now.

4. Put the eggs and caster sugar into a large mixing bowl or the bowl of a mixer and **whisk** for about 4 minutes, until the mixture is very thick, pale and mousse-like. The whisk should leave a thick, ribbon-like trail when lifted.

5. Sprinkle the flour mixture over the egg and sugar mousse and gently but thoroughly **fold** it in using a large metal spoon. Drizzle the melted butter over the top and gently fold in in the same way until there are no more streaks.

6. Spoon about a heaped teaspoon of the mixture into each of the prepared moulds – they should be about two-thirds to three-quarters full, so adjust as needed as moulds vary. You will need to bake these in two batches.

7. Bake for 10–12 minutes until golden and the sponge springs back when gently pressed in the middle.

8. Stand the tray on a wire rack and leave to cool and firm up for a couple of minutes, then tip out the Madeleines onto the wire rack and leave to cool.

9. To bake the next batch, wipe out the warm moulds with kitchen paper, then brush them with melted butter, chill for 2 minutes, and re-butter as before. Then fill them again with more mixture.

10. Dust with icing sugar just before serving.

Try Something Different

Lemon zest adds a delicate flavour, but you could use 1 tsp orange flower water or the zest of 1 small orange instead, or 2 or 3 drops of almond extract.

Apple and Maple Syrup Traybake

Once you've mastered a basic traybake you can make this extra-light Bramley apple and maple syrup version using the **whisked** method.

HANDS-ON TIME:
25 minutes

BAKING TIME:
30–35 minutes

MAKES:
20 squares

SPECIAL
EQUIPMENT:
25.5 × 20.5 × 5cm
traybake tin, cake tin
or roasting tin

METHOD USED:
Whisked method,
see page 26

STORAGE:
Keep for up to
3 days in an airtight
container

For the sponge
400g Bramley cooking apples
¼ teaspoon ground cinnamon
2 tablespoons maple syrup
2 medium egg whites, at room temperature
2 medium whole eggs, at room temperature
finely grated zest of ½ unwaxed lemon
150g light brown muscovado sugar
125ml sunflower oil

50g pecan or walnut pieces, plus extra for topping (optional)
275g plain flour
(white or wholemeal)
½ teaspoon baking powder
1 teaspoon bicarbonate of soda

For the topping
3 tablespoons maple syrup
75g unsalted butter, softened
75g light brown muscovado sugar
175g full-fat cream cheese

1. Preheat the oven to 180°C (160°C fan), 350°F, Gas 4. Grease and **line** the tin with butter and baking paper.

2. Peel, quarter and core the apples, then cut into 1cm chunks. Put them in a bowl with the ground cinnamon and the maple syrup, and toss them together until thoroughly coated. Put the apple to one side until needed.

3. Put the egg whites into a mixing bowl or the bowl of a food-mixer and **whisk** with an electric hand whisk or the whisk attachment, until it makes stiff peaks when you lift the whisk. Put to one side or scoop into a smaller bowl.

4. Break the whole eggs into a separate larger mixing bowl or the empty food-mixer bowl. Add the lemon zest and the sugar, and whisk until the mixture is thick and very frothy and the whisk leaves a distinct ribbon-like trail when lifted. There's no need to wash the whisk before the next step.

Continued

5. Now keep whisking the mixture and at the same time pour in the 125ml sunflower oil in a thin, steady stream.

6. Stir in the apple mixture and the 50g pecan or walnut pieces (if using) with a large metal spoon or plastic spatula. **Sift** the 275g flour, ½ teaspoon baking powder and 1 teaspoon bicarbonate of soda into the bowl (if you are using wholemeal flour, tip the bran left in the sieve into the bowl), and gently stir in. **Fold in** the beaten egg whites in three batches.

7. Gently scrape the mixture into the prepared tin and spread it out evenly. Bake in the centre of the oven for 30–35 minutes until the sponge is golden brown. **Check** your cake is cooked by inserting a cocktail stick or skewer into the centre; if it comes out clean, it is ready.

8. Set the tin on a wire rack and run a round-bladed knife around the inside of the tin to loosen the sponge. Leave until cold then **turn out** the cake and peel off the baking paper.

9. While the sponge is cooling, make the topping. Put the 3 tablespoons maple syrup, 75g butter, 75g muscovado sugar and 175g cream cheese into a mixing bowl and **beat** with a wooden spoon or electric whisk until creamy and smooth.

10. Turn out the cooled sponge onto a serving plate or cake board and spread the topping over the traybake using a palette knife. If you want to add some chopped nuts, sprinkle them over now. Leave in a cool spot to firm up, then cut into 20 pieces.

Carrot Spice Cake

Always popular, this classic carrot cake is quick to put together using the **all-in-one method** and has a delightful combination of creamy and crunchy textures.

HANDS-ON TIME:
25 minutes

BAKING TIME:
25–30 minutes

MAKES:
1 large cake

SPECIAL EQUIPMENT:
2 × 20.5cm round sandwich cake tins

METHOD USED:
All-in-one method, see page 24

STORAGE:
Keep for up to 4 days in an airtight container

For the sponge

230g self-raising flour
1 teaspoon baking powder
1½ teaspoons ground cinnamon
¼ teaspoon grated nutmeg
½ teaspoon ground ginger
½ teaspoon ground mixed spice
200g caster sugar
100g walnut pieces, plus 1 tablespoon finely chopped for topping

3 medium eggs
150ml sunflower oil
500g carrots, coarsely grated

For the cream cheese frosting

1 large unwaxed lime
200g full-fat cream cheese
50g unsalted butter, softened
150g icing sugar

1. Preheat the oven to 180°C (160°C fan), 350°F, Gas 4. Grease and **line** the cake tins with butter and baking paper.

2. To make the sponge, **sift** the self-raising flour, baking powder, ground cinnamon, grated nutmeg, ground ginger and ground mixed spice into a large bowl. Add the caster sugar and walnuts and mix with a wooden spoon.

3. In a separate bowl, add the eggs to the sunflower oil and beat lightly with a fork. Pour into the dry ingredients, tip in the grated carrots and mix again. Divide the carrot mixture equally between the two prepared cake tins – if you want to be really accurate, use your scales, or you can just do it by eye.

4. Bake for 25–30 minutes until the cakes are golden and a cocktail stick or skewer inserted into the centre comes out clean. **Turn out** the cakes onto a wire rack, carefully peeling off the lining paper, then leave them until cold.

5. While the cakes are cooling, make the cream cheese frosting. Finely grate the lime zest into a mixing bowl, squeeze in 1½ teaspoons of the lime juice and add the cream cheese and butter. Sift the icing sugar over the top then beat everything together with an electric whisk until the mixture is smooth, creamy and spreadable. If it is a warm day or your kitchen is hot, cover the bowl at this point and pop it in the fridge to chill for a couple of minutes to thicken up the frosting.

6. Now put your cake together. Flip one sponge so it's top-side down on a serving plate and spread it with half the cream cheese frosting. Pop the second sponge on top and spread it with the rest of the frosting using a round-bladed or palette knife. You don't need to make a perfectly neat finish, but it looks nice with a little swirl made with a flick of the knife at the end. Finally, scatter the finely chopped walnuts around the edge.

Gluten-free Sticky Orange Polenta Cake

This rich gluten-free sponge is made from a standard **creamed** mixture, but polenta and almonds replace the flour.

HANDS-ON TIME:
20 minutes

BAKING TIME:
50 minutes

MAKES:
1 medium cake

SPECIAL EQUIPMENT:
20.5cm springclip tin

METHOD USED:
Creamed method,
see page 24

STORAGE:
Eat the same day
or the next day.
Keep in an airtight
container

For the sponge

250g unsalted butter, softened
1 large navel orange, organic/unwaxed
225g golden caster sugar
4 medium eggs, at room temperature
60g medium/fine polenta
250g ground almonds
2 teaspoons gluten-free baking powder

For the syrup

3 tablespoons golden caster sugar
1 tablespoon orange liqueur, orange brandy or orange juice

To decorate

2 tablespoons toasted flaked almonds
extra orange zest (optional)

1. Preheat the oven to 180°C (160°C fan), 350°F, Gas 4. Grease and **line** the cake tin with butter and baking paper.

2. To make the sponge, put the butter into a large mixing bowl or the bowl of a mixer. Finely grate the zest of the orange onto the butter (putting the fruit to one side for now) and **beat** well with a spoon or whisk attachment until creamy and mayonnaise-like. Gradually beat in the sugar, a couple of tablespoons at a time, scraping down the sides of the bowl every now and then. When all the sugar is in, beat for another couple of minutes, until the mixture is light and fluffy. Scrape down the sides of the bowl once more.

3. Break the eggs into a separate bowl and beat with a fork until broken up. Gradually beat the eggs into the butter mixture, a tablespoon at a time, beating well each time. Don't worry if the mixture looks slightly runny or curdled once the last bit of egg has been added.

4. **Sift** the polenta, almonds and baking powder into the bowl. Cut the orange in half, squeeze out the juice and add 1 tablespoon to the bowl, keeping the rest for the syrup. Gently **fold** everything together using a large metal spoon or plastic spatula until thoroughly combined and streak-free.

5. Scrape the mixture into the tin and spread it out evenly. Bake for about 50 minutes until the sponge is a good golden brown and **check** that the middle springs back when gently pressed.

6. While the sponge is baking, make the syrup. Put the sugar, 5 tablespoons of the reserved juice and the liqueur (if using) into a small pan and heat gently, stirring, until the sugar dissolves. Bring to the boil, simmer for 2–3 minutes to make a light syrup, then lower the heat to keep it warm.

7. As soon as the cake is ready, put it on a wire rack still in its tin. Prick the hot sponge all over with a cocktail stick, then quickly spoon the hot syrup over – the cake will sink in the middle. Scatter with the almonds and zest (if using), then cool completely.

8. Run a round-bladed knife around the inside of the tin to loosen the cake, then unclip the sides and take it out of the tin.

Victoria Sandwich Cake

The nation's favourite cake, traditionally the ingredients are calculated by weighing the eggs in their shells and using this weight for the butter, sugar and flour.

For the sponge
3 medium eggs, at room temperature
about 175g unsalted butter, softened
about 175g caster sugar
¾ teaspoon vanilla extract
about 175g self-raising flour
1 tablespoon water from the warm water tap

For the filling
6 tablespoons good-quality raspberry jam
150ml double or whipping cream, well chilled (optional)
icing sugar, for dusting

Easy does it

HANDS-ON TIME:
20–25 minutes

BAKING TIME:
20 minutes

MAKES:
1 medium cake

SPECIAL EQUIPMENT:
2 × 20.5cm round deep sandwich cake tins

METHOD USED:
Creamed method, see page 24

STORAGE:
Eat immediately or keep in an airtight container for a jam-filled cake, or in the fridge if you've used cream, for up to 3 days. (Take it out of the fridge 20 minutes before serving.)

1. Preheat the oven to 180°C (160°C fan), 350°F, Gas 4. Grease and **line** the cake tins with butter and baking paper.

2. Weigh the eggs – 3 medium eggs in their shells weigh around 175g – then use this same weight for the butter, sugar and flour.

3. Put the soft (but not oily) butter into a large mixing bowl or the bowl of a food-mixer and **beat** well with a wooden spoon or the whisk attachment until very creamy and mayonnaise-like. Scrape down any butter mixture from the sides of the bowl with a plastic spatula, then gradually beat in the sugar a couple of tablespoons at a time. Scrape the mixture off the sides of the bowl again and beat well for 1 minute or until the mixture looks very light and fluffy. Scrape down the mixture again.

4. Break the 3 eggs into a small jug, add the ¾ teaspoon vanilla extract and beat with a fork just until the eggs are broken up. Gradually add to the butter mixture a tablespoon at a time, beating well after each addition and scraping down the sides of the bowl from time to time. If the mixture looks like it might be 'splitting' or curdling, rather than appearing smooth and creamy, stir in a tablespoon of the flour with each of the last two additions of egg.

5. **Sift** the rest of the flour onto the mixture. Start to gently **fold** in the flour with a large metal spoon or plastic spatula and after two or three movements add the warm water. Keep folding in until the flour is well mixed in and there are no streaks.

Continued

6. Divide the mixture between the two prepared tins – if you want to be really precise, use your scales, or just do it by eye, then spread it evenly.

7. Bake for 20–25 minutes until the sponges are a light golden brown, starting to shrink back from the sides of the tin. Check the sponges after 15 minutes and if they aren't baking evenly, rotate the trays. **Check** that the sponge springs back when lightly pressed in the middle.

8. When cooked, take them out of the oven and run a round-bladed knife around the outside of each tin to loosen the sponge. Leave for a minute to firm up, then carefully **turn out** the cakes onto a wire rack. Leave until they are completely cooled.

9. If you are using cream, put a bowl and whisk (or whisk attachment) in the fridge to chill.

10. To assemble the cake, set one sponge crust-side down on a serving plate. Using the back of a tablespoon, evenly spread the sponge with the 6 tablespoons raspberry jam.

11. Pour the cream, if using, into the chilled bowl and **whip** with the chilled whisk or attachment until it thickens and soft peaks form when you lift out the whisk. Spoon the cream onto the cake and then gently smooth it evenly over the jam. Top with the second sponge, crust-side up, and dust with icing sugar.

Try Something Different

To make a Chocolate Sponge Cake, replace the flour in the recipe above with 150g self-raising flour and 30g cocoa powder, sifting the cocoa powder into the bowl with the flour and replacing the warm water with 2 tbsp room-temperature milk. Sandwich the sponges with chocolate spread instead of jam, or make up a classic chocolate butter icing using 125g soft unsalted butter creamed with 400g sifted icing sugar and flavoured with 3 tbsp cocoa powder plus 3–4 tbsp milk or cold coffee. This will make enough to fill and top the cake.

Rich Lemon Cupcakes

Take your cupcakes to the next level by adding buttermilk to the classic **creamed** mixture. The buttermilk is mildly acidic, giving your bake a lovely fine and light texture.

125g unsalted butter, softened
200g golden caster sugar
1 large unwaxed lemon
2 medium eggs, at room temperature
200g self-raising flour
100ml buttermilk

For the frosting
75g unsalted butter, softened
250g icing sugar, sifted
3 tablespoons Lemon Curd (see page 145)

1. Preheat the oven to 180°C (160°C fan), 350°F, Gas 4. Line the cupcake tray with paper cases.

2. Put the butter into a mixing bowl or the bowl of a food-mixer and **beat** for 1 minute with a wooden spoon or the whisk attachment until the mixture is very creamy and mayonnaise-like.

3. Sprinkle the caster sugar into the bowl. Finely grate the zest of the lemon on top, then halve the lemon, squeeze the juice into a small bowl and put to one side. Beat the mixture until very light and fluffy, scraping down any mixture that has splattered onto the sides of the bowl every now and then.

4. Break the eggs into a small bowl and beat with a fork until broken up. Gradually beat the eggs into the mixture a tablespoon at a time, beating well after every time and scraping down the sides of the bowl as before.

5. **Sift** one-third of the flour into the bowl and **fold** in with a large metal spoon, then add one-third of the buttermilk and fold in. Repeat twice more until all the flour and buttermilk have been well mixed in. Lastly, fold in 2 teaspoons of the reserved lemon juice (you won't need the rest).
Continued

Easy does it

HANDS-ON TIME:
15 minutes

BAKING TIME:
20–25 minutes

MAKES:
12 cupcakes

SPECIAL EQUIPMENT:
12-hole cupcake tray or deep-hole bun tray

METHOD USED:
Creamed method, see page 24

STORAGE:
Keep for up to 4 days in an airtight container

6. Spoon the mixture into the paper cases so that they are evenly filled. Bake for 20–25 minutes until the cakes are a light golden brown and feel just firm to the touch. Check the cakes after 15 minutes and if they aren't baking evenly, turn the tray around.

7. To make the frosting, beat the 75g butter in a large bowl or mixer bowl until creamy using a wooden spoon or electric whisk attachment.

8. At a slow speed, gradually beat in the 250g sifted icing sugar. Once all the sugar has been incorporated, beat in the 3 tablespoons lemon curd to make a smooth and light icing. The frosting should be firm enough to spread, pipe or swirl on top of your cupcakes; if it seems soft or slightly runny, cover the bowl and pop it in the fridge until it is firm but spreadable.

9. Remove the tray of cupcakes from the oven and set it on a wire rack. Leave to cool for 2 minutes, then transfer the cupcakes to the rack and leave until cold.

10. Swirl the frosting over the cooled cupcakes using a round-bladed knife or offset palette knife.

11. Sprinkle over any decorations immediately before the frosting firms up.

Try Something Different

For vanilla cupcakes, replace the lemon zest and juice with ¾ tsp vanilla extract. Top with vanilla frosting, replacing the lemon curd with 4 tbsp milk and 1 tsp vanilla extract. Or replace the lemon zest and juice with the finely grated zest of 1 medium orange and 2 tsp of the squeezed juice and flavour the topping with 2 tbsp orange juice instead of the lemon curd.

Fudgy Chocolate Birthday Cake

This is a really good chocolate cake that should be in everyone's repertoire, quickly made using the no-fuss **all-in-one** method. Its glossy frosting is easy to apply yet gives a great finish.

HANDS-ON TIME:
25 minutes

BAKING TIME:
60–70 minutes

MAKES:
1 large cake

SPECIAL
EQUIPMENT:
22cm springclip tin

METHOD USED:
All-in-one method,
see page 24

STORAGE:
Tastes best the day
after you make it.
Keep in an airtight
container for up to
5 days

For the sponge

200g self-raising flour
75g cocoa powder
large pinch of salt
1 teaspoon baking powder
250g light brown muscovado sugar
250g unsalted butter, very soft
6 medium eggs, at room temperature, beaten
100g chocolate chips (milk, dark or white chocolate)

For the frosting

100g good-quality milk chocolate
100g unsalted butter, at room temperature, diced
100g icing sugar
2 tablespoons cocoa powder

1. Preheat the oven to 180°C (160°C fan), 350°F, Gas 4. Grease and **line** the tin with butter and baking paper.

2. To make the sponge, **sift** the self-raising flour, cocoa powder, salt and baking powder into a large mixing bowl or the bowl of a food-mixer. Add the muscovado sugar, breaking up any lumps, the butter and the eggs.

3. Using a wooden spoon or the whisk attachment, start **beating** everything together, slowly at first so that the mixture doesn't fly everywhere, then a little faster until the mixture is silky smooth and light. Scrape down any mixture splattered on the sides of the bowl from time to time so that the batter is evenly mixed with no streaks.

4. Stir in the chocolate chips using the wooden spoon or a plastic spatula, then scrape the mixture into the prepared tin. Spread the mixture evenly, then make a saucer-like hollow in the middle of the cake, about 2cm deep. This helps the sponge rise evenly; you may still get a little cracked dome, but it will avoid a large volcanic peak.

5. Bake the cake in the centre of the oven for 50–60 minutes until the top is firm to the touch. **Check** your cake is cooked by inserting a cocktail stick or skewer into the centre; if it comes out clean, it is ready.

Continued

6. Set the tin on a wire rack, then run a round-bladed knife around the inside of the tin to loosen the sponge, and gently unclip the tin. Leave until completely cold before icing.

7. To make the frosting, break up the 100g milk chocolate into even-sized pieces and put it into a heatproof bowl large enough to hold all the frosting ingredients. Set the bowl over a pan of steaming hot but not boiling water – don't let the base of the bowl touch the water – and leave to **melt** gently, or melt the chocolate carefully in the microwave in 10-second bursts.

8. Remove the bowl from the pan and stir in the 100g unsalted butter. When melted and smooth, sift the 100g icing sugar and 2 tablespoons cocoa powder into the bowl and mix with a wooden spoon to make a smooth and thick but spreadable icing.

9. Spread and swirl the frosting evenly over the top of the cooled cake using a round-bladed knife or an off-set palette knife.

10. Now carefully spread and smooth the frosting evenly around the sides of the cake. If you want to add any decorations or candles pop them on now while the frosting is still soft.

Try Something Different

For a darker and less sweet topping, use 50g good-quality milk chocolate and 50g good-quality dark chocolate (around 70 per cent cocoa solids).

Double Marble Cake

This impressive cake cuts into pretty chocolate brown and golden marbled slices – made by dividing the **creamed** mixture and adding cocoa to half of it. Rich icing makes it even more special.

HANDS-ON TIME:
30 minutes

BAKING TIME:
60 minutes

MAKES:
1 large loaf cake

SPECIAL
EQUIPMENT:
900g loaf tin (about
26 × 12.5 × 7cm)

METHOD USED:
Creamed method,
see page 24

STORAGE:
Keep for up to
5 days in an airtight
container

For the sponge
225g unsalted butter, softened
225g caster sugar
½ teaspoon vanilla essence
4 medium eggs, at room temperature, beaten
225g self-raising flour
good pinch of salt
3 tablespoons cocoa powder
2 tablespoons milk, at room temperature

For the topping
100g good-quality dark chocolate (around 70 per cent cocoa solids), broken up
40g unsalted butter, very soft
50g white chocolate, broken up

1. Preheat the oven to 180°C (160°C fan), 350°F, Gas 4. Grease the tin with butter and **line** the base and two short sides with a long strip of baking paper.

2. To make the sponge, put the butter into a large mixing bowl or the bowl of a food-mixer and **beat** thoroughly with a wooden spoon or the whisk attachment until the mixture is creamy and mayonnaise-like.

3. Gradually beat in the caster sugar, scraping down any mixture on the sides of the bowl now and then, then beat until the mixture becomes light and fluffy. Beat in the vanilla essence.

4. Gradually beat in the eggs, a tablespoon at a time, beating well after each addition and adding a tablespoon of the flour with the last two additions.

5. **Sift** the rest of the flour and the salt onto the mixture and gently **fold in** using a large metal spoon.

6. Scoop out half the mixture into a separate bowl and put it to one side.

7. Sift the cocoa powder into one bowl of cake mixture, add the milk and very gently stir everything together using a large metal spoon or plastic spatula until all the cocoa is mixed in.

Continued

8

10. Set the tin on a wire rack, run a round-bladed knife around the inside of the tin to loosen the cake, then leave for 5 minutes to firm up. Gently remove the cake from the tin – use the ends of the lining paper to help lift the cake out – and leave on the wire rack until cold.

11. While the cake is cooling, make the topping. Put the 100g dark chocolate into a heatproof bowl and set over a pan of steaming hot but not boiling water – don't let the base of the bowl touch the water – and leave to **melt** gently, then remove the bowl from the pan (the chocolate can also be melted in the microwave in 10-second bursts). Stir until smooth, then stir in 25g of the soft butter.

8. Using soup spoons, put alternate spoonfuls of the two mixtures into the prepared tin until you have used them both up. Bang the tin on the worktop several times to level the contents and knock out any pockets of air. Now carefully swirl the two mixtures together with the handle of a teaspoon or a table knife.

9. Bake the cake for 60 minutes until the top is lightly golden and firm. **Check** your cake is cooked by inserting a cocktail stick or skewer into the centre; if it comes out clean, it is ready.

12. Melt the 50g white chocolate in a separate bowl in the same way, but take extra care as white chocolate melts at a lower temperature and can seize and become grainy. Stir the remaining butter into the melted white chocolate.

13. Spread the dark chocolate mixture over the top of the cake – don't worry if it drips down the sides. Using a teaspoon, drizzle the melted white chocolate mixture randomly over the top of the cake and marble and swirl it into the dark chocolate using the handle of a teaspoon or a table knife. Leave the cake until it is set and firm before you slice it.

13

Try Something Different

For an even more chocolatey indulgence, sprinkle 1½ tbsp dark chocolate chips over the sponge mixtures once they are in the loaf tin but before you bang the tin on the worktop (step 8).

Coffee and Walnut Cake

Just like the Victoria Sandwich Cake on page 74, this rich-tasting classic coffee sponge is made by the **creamed** method, but a little added baking powder keeps it nice and light.

Easy does it

HANDS-ON TIME:
30 minutes

BAKING TIME:
20–25 minutes

MAKES:
1 medium cake

SPECIAL EQUIPMENT:
2 × 20.5cm round deep sandwich cake tins

METHOD USED:
Creamed method, see page 24

STORAGE:
Keep for up to 4 days in an airtight container

For the sponge
100g walnut pieces
175g unsalted butter, softened
175g golden caster sugar
3 medium eggs, at room temperature
175g self-raising flour
good pinch of salt
½ teaspoon baking powder
1 tablespoon instant coffee powder or granules
1 tablespoon boiling water

For the filling and frosting
125g unsalted butter, diced
3 tablespoons instant coffee powder or granules
2 tablespoons boiling water
300g icing sugar
4 tablespoons single, double or whipping cream, at room temperature
walnut halves, to decorate

1. Preheat the oven to 180°C (160°C fan), 350°F, Gas 4. Grease and **line** the tins with butter and baking paper.

2. To make the sponge, first put the walnut pieces in an ovenproof dish or roasting tin and toast in the oven for 5–6 minutes until lightly coloured. Leave to cool, then chop slightly smaller. Set aside 25g for the filling and frosting and keep the remaining 75g for the sponge.

3. Put the butter into a mixing bowl or the bowl of a food-mixer and **beat** well with a wooden spoon or the whisk attachment until creamy and mayonnaise-like. Gradually beat in the golden caster sugar, a couple of spoonfuls at a time, scraping down the sides of the bowl from time to time where the mixture has splattered. Once all the sugar has been mixed in, scrape down the sides of the bowl again to get the last bits of mixture and beat for another minute or so until the mixture is light and fluffy.

4. Break the eggs into a separate bowl and beat with a fork just until broken up. Gradually beat the eggs into the butter mixture, beating well after each addition. Add 1 tablespoon of the flour with each of the last two bits of egg to help prevent the mixture curdling.

5. Add the reserved 75g toasted walnut pieces to the bowl, then **sift** the remaining flour, salt and baking powder into the bowl. In a small bowl, dissolve the instant coffee in the boiling water and pour it into the mixing bowl. Using a large metal spoon or plastic spatula, gently **fold** these ingredients into the butter mixture until thoroughly combined and there are no streaks.

6. Divide the mixture equally between the two prepared tins and spread evenly.
Continued

7. Bake for 20–25 minutes until the sponges are a good golden brown. **Check** your cake is cooked by lightly pressing in the centre – the cake will spring back if it's done. Run a round-bladed knife around the inside of the tins to loosen the sponges, leave them to firm up for 2 minutes, then **turn out** onto a wire rack and leave them to cool completely.

8. While the sponges are cooling, make the coffee filling and frosting. Put the 125g unsalted butter into a small pan and melt over a low heat.

9. In a small bowl, dissolve the 3 tablespoons instant coffee in the 2 tablespoons boiling water and sift the 300g icing sugar into a large heatproof mixing bowl.

10. Bring the butter to a boil, then quickly take it off the heat and pour it straight onto the icing sugar, swiftly followed by the warm coffee and the 4 tablespoons cream. Beat well with a wooden spoon until the mixture is very smooth, then stir in the reserved 25g toasted walnut pieces. Leave the frosting until it is thick enough to spread (you can speed this up by covering the bowl and putting it in the fridge for a few minutes).

11. To assemble the cake, set one sponge crust-side down on a serving plate and spread with half the coffee icing mixture. Cover with the second sponge, crust-side up, and spread and swirl the rest of the frosting on top.

12. To finish the cake, decorate with the walnut halves.

Classic Swiss Roll

Mastering the **whisked** sponge in this light and traditional Swiss roll is a great way to prepare for trying more challenging bakes like the Lemon Curd Layer Cake on page 144.

HANDS-ON TIME:
20 minutes

BAKING TIME:
9–10 minutes

MAKES:
1 medium cake

SPECIAL EQUIPMENT:
20 × 30cm Swiss roll tin

METHOD USED:
Whisked method, see page 26

STORAGE:
Eat the day you make it or the next day. Keep in an airtight container

For the sponge

3 medium eggs, at room temperature
75g caster sugar
75g plain flour
pinch of salt

For the filling

extra caster sugar, for sprinkling
200ml whipping or double cream
½ teaspoon vanilla extract
6 rounded tablespoons raspberry or strawberry jam

1. Preheat the oven to 220°C (200°C fan), 425°F, Gas 7. Grease and **line** the tin with butter and baking paper.

2. To make the sponge, break the eggs into a large mixing bowl or the bowl of a food-mixer and **whisk** with an electric mixer on a high speed for a few seconds just until slightly frothy.

3. Add the caster sugar to the bowl and whisk on a high speed for about 5 minutes or until the mixture becomes very thick, pale and mousse-like, and the whisk leaves a ribbon-like trail when it is lifted.

4. **Sift** the flour and pinch of salt onto a sheet of greaseproof paper, then sift half of it again, straight onto the egg mousse. Using a large metal spoon, very gently **fold** in the flour. Sift the rest of the flour onto the mixture and fold in until you can no longer see any streaks or specks. Gently scrape any splattered mixture down the sides of the bowl and check the bottom of the bowl for any unmixed flour, stirring it in as you go.

5. Carefully transfer the mixture to the prepared tin and spread it out evenly, making sure the corners are well filled. Bake for 9–10 minutes until the sponge is golden brown. You can **check** your cake is cooked by lightly pressing it in the centre – the cake will spring back when done.

Continued

Try Something Different

For a Gingerbread Swiss Roll, make the sponge as in the main recipe, replacing the caster sugar with light brown muscovado sugar and adding 1 tsp ground ginger and ½ tsp each ground cinnamon and ground mixed spice. Bake, roll and cool, spread with ginger conserve instead of the jam and then spread with 200ml whipping or double cream, whipped to **soft peaks,** as before and roll up.

6. While the sponge is baking, lay a sheet of baking paper (about 30 × 40cm) on the worktop and sprinkle it with a little caster sugar.

7. As soon as the sponge is ready, flip it out onto the baking paper and carefully lift off the tin. Peel off the lining paper.

8. Using a large, sharp knife make a shallow cut about 2cm in from one short end – this will help give the roll a neat spiral when it is sliced.

9. Now, starting from the end with the cut, gently roll up the warm sponge with the paper so that the paper is rolled inside it. Set the roll on a wire rack and leave until cold.

10. When ready to fill, **whip** 200ml whipping or double cream with ½ teaspoon vanilla extract to soft peaks. Gently unroll the sponge and trim off the edges with a sharp knife. Spread with about 6 rounded tablespoons raspberry or strawberry jam, leaving a 2.5cm border round the edge.

11. Cover the jam with a layer of the whipped cream, again leaving a 2.5cm border. Now gently re-roll the sponge, starting once more from the end with the cut. Sprinkle with a little more caster sugar to finish.

Three-layer
Banana Cake

Light muscovado sugar adds a lovely toffee flavour to a simple **all-in-one** vanilla sponge mixture.

For the sponge

175g unsalted butter, softened
150g golden caster sugar
25g light brown muscovado sugar
175g self-raising flour
3 medium eggs, at room temperature
½ teaspoon vanilla extract
1 tablespoon milk, at room temperature

For the filling and topping

300ml whipping cream, well chilled
100g vanilla fudge, cut into 5mm chunks
2 ripe medium bananas, thinly sliced

To decorate

extra fudge or grated chocolate (or both)

HANDS-ON TIME:
25 minutes

BAKING TIME:
18–20 minutes

MAKES:
1 large cake

SPECIAL EQUIPMENT:
3 × 20.5cm round sandwich cake tins

METHOD USED:
All-in-one method, see page 24

STORAGE:
Eat the day you make it or the next day. Keep in an airtight container

1. Preheat the oven to 180°C (160°C fan), 350°F, Gas 4. Grease and **line** the cake tins with butter and baking paper. If you only have one tin, bake in batches. Rinse the hot tin under cold running water to quickly cool it, then dry, re-grease and line it again.

2. To make the sponge, put the butter, two sugars and flour into a large bowl or the bowl of a food-mixer.

3. Break the eggs into a smaller bowl, add the vanilla extract and milk, and beat with a fork until just mixed. Pour into the larger bowl, then **beat** with a wooden spoon or the whisk attachment until smooth and blended with no streaks. Divide the mixture into the three tins, either by eye or, if you want to be more accurate, by weight. Spread the mixture evenly into the prepared tin or tins.

4. Bake for 18–20 minutes until the cakes are a light golden brown. Have a look at the cakes after 15 minutes, if they aren't baking evenly, turn the tins around. **Check** your cake is cooked by pressing it – it should be just firm to the touch.

5. Run a round-bladed knife around the inside of the tins to loosen the sponges, then **turn out** onto a wire rack. Leave until cold. Put a bowl and whisk (or whisk attachment) in the fridge to chill.

6. When you are ready to assemble the cake, make the filling and topping. Pour the cream into the chilled bowl and **whip** with the chilled whisk or attachment, until it thickens and soft peaks form when you lift the whisk out of the cream.

7. Set one layer of sponge crust-side down on a serving plate. Spread with one-third of the whipped cream, then scatter over half the fudge. Top with half the banana slices, then cover with a second sponge. Spread half the remaining cream over this sponge and scatter over the rest of the fudge, then cover with the remaining banana slices. Finally, set the last layer of sponge on top, crust-side uppermost, and spread with the remaining cream. Decorate with fudge, grated chocolate, or both.

Battenberg Cake

A step up from the free-and-easy Double Marble Cake (page 86), the Battenberg uses colouring in a more precise way to make a pretty cake that looks impressive.

For the sponge
125g unsalted butter, softened
125g caster sugar
2 medium eggs, at room temperature
2 drops of almond extract
100g self-raising flour
good pinch of salt
50g ground almonds
2 teaspoons milk
a few drops pink or red food colouring

To finish
icing sugar or cornflour, for dusting
350g white marzipan
8 tablespoons apricot jam

HANDS-ON TIME:
50 minutes

BAKING TIME:
20–25 minutes

MAKES:
1 medium cake

SPECIAL EQUIPMENT:
Battenberg tin (20 × 15cm) or 20cm square shallow cake tin

METHOD USED:
Creamed method

STORAGE:
Keep for up to 5 days in an airtight container

1. If you are using a special Battenberg tin that comes with dividers to make four strips of sponge, grease and **line** the tin with butter and baking paper. Alternatively, you can make your own: cut a 20 × 28cm rectangle of parchment-lined foil, set it paper-side up on the worktop and fold it in half widthways. Open it out, then push the centrefold upwards to make a pleat 4cm high. Crease it firmly, then press the parchment-lined foil sheet with the centre pleat onto the base of greased tin to line it – the pleat will run down the centre of the tin to divide it into two separate sections each 20 × 10cm. Preheat the oven to 180°C (160°C fan), 350°F, Gas 4.

2. To make the sponge, put the butter into a mixing bowl or the bowl of a food-mixer and **beat** until creamy with a wooden spoon or the whisk attachment. Scrape down the sides of the bowl and whisk in the sugar a couple of tablespoons at a time. Scrape down the sides of the bowl where any

mixture has splattered up, then beat really well for a couple of minutes until the mixture is light and fluffy. Scrape down the sides of the bowl again to get any stray mixture.

Continued

3. Break the 2 medium eggs into a separate bowl, add the 2 drops of almond extract and beat with a fork until broken up, then gradually beat into the butter mixture a tablespoon at a time, beating well after each addition. Add a tablespoon of the 100g self-raising flour with each of the last two additions of egg to prevent the mixture from curdling.

4. Sift the remaining flour, good pinch of salt and 50g ground almonds into the bowl. Add the 2 teaspoons milk and carefully **fold** everything together with a large metal spoon. Transfer half the mixture (for this cake it is best to do this by weight, to be really accurate) into a clean bowl. Add the pink or red food colouring a few drops at a time to one portion, mixing it in really well so that it turns pink with no coloured streaks.

5. Spoon the uncoloured pale yellow cake mixture into two sections of the prepared Battenberg tin or into one side of the prepared square tin (check the pleated divider is still straight and dead centre). Spoon the pink cake mixture into the other two sections of the Battenberg tin or the other section of the prepared square tin. Carefully spread each portion so that the surface is level and the corners are evenly filled.

6. Bake for 20–25 minutes until the sponges are well risen. **Check** the cake is cooked – it should be springy when gently pressed. Set the tin on a wire rack, run a round-bladed knife around the inside of the tin to loosen the sponges, then leave to cool until barely warm. Carefully **turn out** onto the wire rack and peel off the lining paper – the cakes will still be fragile so handle them gently – then leave until cold.

7. To assemble the cake, set the sponges on a chopping board. If you used the square tin to bake two sponges, cut each sponge in half lengthways using a serrated bread knife to make four strips of cake – two pink and two yellow. Trim all the strips so that the short sides are exactly square as they may have risen unevenly.

8. Make sure the worktop is spotlessly clean and crumb-free, then dust it lightly with icing sugar or cornflour. Knead the 350g white marzipan for a minute so that it is supple, then roll it out to a neat 20 × 30cm rectangle.

9. Heat the 8 tablespoons apricot jam with 1 tablespoon cold water, then push it through a sieve to make a smooth purée. Brush one long side of one pink strip very lightly with jam and set it jam-side down on one short side of the marzipan rectangle, lined up next to the edge. Brush the three other long sides of this piece of cake lightly with jam.

10. Now brush one long side of a yellow cake strip with jam and set it jam-side down onto the marzipan next to the pink strip and touching it. Brush the top of this strip with jam.

11. Set the other yellow strip on top of the pink strip that is in place on the marzipan.

12. Lightly brush one long side of the second pink strip with jam and set it next to this second yellow strip, so that it sits on top of the yellow strip that's on the marzipan. You will now have created the familiar checkerboard pattern.

13. To finish the cake, brush the top and long sides of the assembled cake with the apricot jam (gently re-warmed), then roll the cake or wrap the marzipan neatly around and over the cake, leaving the checkerboard ends visible. Press the marzipan join to seal it (it should be along one bottom edge). Gently smooth the marzipan with your hands to neaten it up and press out any air pockets, then trim the ends and crimp the top edges of marzipan by pinching it gently, at regular intervals, between your index finger and thumb.

14. Dust lightly with icing sugar just before serving.

Whisky
Dundee
Cake

This may seem straightforward, but the mark of a good Dundee cake is a rich and moist firm sponge mixture evenly peppered with fruit, carefully baked with a not-too-dark crust.

Needs a little skill

HANDS-ON TIME:
30 minutes

BAKING TIME:
2½ hours

MAKES:
1 large cake

SPECIAL EQUIPMENT:
20cm round deep cake tin

METHOD USED:
Creamed method
page 24

STORAGE:
Once matured, keep for up to 3 weeks in an airtight container

For the sponge
550g mixed dried fruit (raisins, sultanas, currants) and chopped mixed peel
5 tablespoons whisky
275g plain flour
couple of good pinches of salt
1 teaspoon baking powder
65g ground almonds
225g unsalted butter, softened

finely grated zest of 1 large unwaxed lemon
225g golden caster sugar
4 medium eggs, at room temperature
18 blanched almond halves

For the glaze
2 tablespoons milk
4 teaspoons golden caster sugar

1. To make the fruitcake, put the mixed dried fruit and mixed peel into a bowl, spoon over the whisky, stir well, then cover with clingfilm and leave to macerate overnight.

2. Next day, preheat the oven to 180°C (160°C fan), 350°F, Gas 4. Grease and **line** the tin with butter and baking paper.

3. **Sift** the flour, salt, baking powder and ground almonds into a bowl and put to one side until needed.

4. Put the butter and lemon zest into a large mixing bowl or the bowl of a food-mixer and **beat** with a wooden spoon or the whisk attachment until the mixture is very creamy and mayonnaise-like. Scrape down the sides of the bowl where any mixture has splattered and gradually beat in the golden caster sugar, scraping down the sides of the bowl from time to time.

5. Break the eggs into a separate bowl and beat with a fork until broken up, then beat into the butter mixture a tablespoon at a time, beating well each time. Add a tablespoon of the sifted flour mixture with each of the last two additions of egg to prevent the mixture from curdling.
Continued

Try Something Different

For a less traditional version, you could make this with a luxury bag of mixed fruit that contains glacé cherries, pineapple or chunks of dried apricot. It's well worth soaking the fruit in whisky for 12 hours for a really special flavour, but you could use brandy or dark rum instead.

6. Sift half the remaining flour mixture into the bowl – this helps to really lighten the cake mixture – and **fold in** with a large metal spoon or plastic spatula. Fold in half the fruit and whisky mixture into the bowl. Sift the rest of the flour on top, then fold in followed by the rest of the fruit and whisky. When everything is really well mixed, scrape the mixture into the prepared tin and spread evenly. Bang the tin on the worktop a couple of times to settle the mixture and knock out any air pockets.

7. Make a shallow hollow in the centre of the cake's surface so it will rise evenly. Dip your fingers in cold water and press them lightly over the top of the cake – this prevents the crust getting too hard.

8. Press the 18 blanched almond halves in a neat circle on top of the cake. Wrap a few sheets of newspaper around the outside of the tin and tie with kitchen string. Set the tin on top of another folded newspaper on the baking sheet. This stops the crust being over-baked.

9. Bake for 40 minutes, then reduce the oven temperature to 170°C (150°C fan), 325°F, Gas 3 and bake for a further 1¾ hours. **Check** your cake is cooked by inserting a cocktail stick or skewer into the centre; if it comes out clean, it is ready. Keep an eye on your cake from time to time while it is baking and if you think it is in danger of turning too dark on top, cover it with a sheet of baking paper or foil.

10. At the end of the baking time, make the glaze. Bring the 2 tablespoons milk to the boil and stir in the 4 teaspoons golden caster sugar until dissolved. When the cake is ready, pull it out of the oven and brush it with the hot glaze. Return the cake to the oven for another minute – this will add a nice sheen to the nuts. Set the cake, still in the tin, on a wire rack and leave overnight, then unmould the cake and peel off the lining paper. Wrap in fresh baking paper and foil and leave for 7–10 days before cutting.

Red Velvet Cupcakes

Eye-catching red velvet cupcakes seem to be everywhere these days. The perfect party piece, they are a great excuse, if you need one, to practise your piping skills.

Needs a little skill

HANDS-ON TIME:
50 minutes

BAKING TIME:
15–18 minutes

MAKES:
12 cupcakes

SPECIAL EQUIPMENT:
12-hole cupcake tray or deep-hole bun tray; piping bag fitted with a star piping nozzle; paper cupcake cases

METHOD USED:
Creaming method page 24

STORAGE:
Keep for up to 3 days in an airtight container in the fridge. (Take them out of the fridge 30 minutes before serving)

For the sponge
175ml milk
100g good-quality dark chocolate (about 70 per cent cocoa solids), broken up
125g caster sugar
60g unsalted butter, softened
1 medium egg, at room temperature
½ teaspoon vanilla extract
150g self-raising flour
good pinch of salt
1–2 teaspoons red gel food colour

For the topping
150g good-quality white chocolate (around 25 per cent cocoa solids), broken up
110g full-fat cream cheese, brought to room temperature at least an hour beforehand
75g unsalted butter, softened
½ teaspoon vanilla extract

To finish
freeze-dried raspberry pieces or red sugar sprinkles

1. Preheat the oven to 180°C (160°C fan), 350°F, Gas 4. Line the cupcake tray with paper cases.

2. To make the sponge, start by pouring the milk into a small pan. Add the dark chocolate pieces and 50g of the caster sugar. Stir gently with a wooden spoon over a very low heat until melted – be patient, and don't let the mixture get hot. Remove the pan from the heat and give it a quick whisk with a wire hand whisk to make sure the milk is blended and smooth. Leave it on one side to cool.

3. Put the butter into a mixing bowl and **beat** well with an electric whisk or the food-mixer whisk attachment until it goes pale and creamy. Stop the whisk and scrape any butter on the sides back down into the bowl. Add just a couple of tablespooons of the remaining 75g caster sugar, and beat again; repeat, stopping occasionally to add the rest of the sugar a couple of tablespoons at a time, and scrape down the sides as you go. When all the sugar has been added, beat well for another minute until all the ingredients are thoroughly combined and your mixture is nice and fluffy.
Continued

4. Break the egg into a small bowl, add the ½ teaspoon vanilla extract and lightly beat with a fork. Now turn on your whisk again and gradually pour the eggs into the butter mixture, whisking all the time. When all the egg has been added, keep whisking for another minute until light and airy.

5. Now you're ready to add the chocolatey milk. Mix the 150g self-raising flour and a pinch of salt together. Take one-third of this mix and **fold** it into the butter mixture with a large metal spoon or plastic spatula. Now fold in a third of the chocolate milk in the same way.

6. Add half of the remaining flour and fold that in, followed by half of the remaining chocolate milk. Continue to fold in alternate spoonfuls of flour and milk until all the ingredients are thoroughly combined and silky smooth.

7. Add your red gel food colouring. It's best to start with 1 teaspoon and fold it in until you can see no streaks. You can always add more after mixing, depending on how vibrant you'd like your red velvet sponge to be.

8. Using a metal spoon, half fill each paper case with the mixture, then bake for 15–18 minutes until the cakes spring back when you lightly press their tops. Stand the tray on a wire rack and cool for 2 minutes, then take them off the tray and pop them back on the wire rack to cool completely.

9. When the cakes are cold, make the frosting. First **melt** the chocolate. Break the 150g white chocolate into pieces and put them in a small heatproof bowl set over a pan of steaming hot (but not boiling) water. Be careful not to let the base of the bowl touch the hot water. White chocolate is quite sensitive and melts at a lower temperature than dark chocolate, so remove the bowl from the pan while there are still a few unmelted lumps and stir to melt, until smooth.

10. Put the 110g cream cheese in a separate bowl and beat until smooth and creamy, then beat in the 75g butter a little at a time. Add the melted chocolate and ½ teaspoon vanilla extract, and beat again for 2 minutes until the mixture is like whipped cream.

11. Now it's time to pipe the frosting on to your cakes. Fill a **piping** bag fitted with a star piping nozzle by spooning in the frosting mixture. Twist the bag, sealing the end to stop the buttercream from escaping.

12. Holding the top of the piping bag, firmly squeeze down the filling as you start to pipe the frosting on to your cake. Start at the edge, keeping the tip of the nozzle just above the cake.

13. Keep squeezing the icing out in a steady spiral until you have covered the whole cake.

14. To get the perfect finish, end the frosting in the very centre and just lift up the nozzle a fraction so that the icing forms a gentle swirling peak.

15. Decorate with the dried raspberry pieces or red sugar sprinkles and leave to firm up for at least half an hour on the worktop before serving.

Cherry Cake

This lovely **creamed** sponge's slightly firmer texture helps keep the delicious ruby red cherries in every mouthful of cake, rather than sunk down at the bottom.

200g glacé cherries
200g self-raising flour
good pinch of salt
good pinch of baking powder
175g unsalted butter, softened
175g caster sugar
3 medium eggs, at room temperature
50g ground almonds

1. Halve the glacé cherries, then put them into a colander or sieve and rinse well under hot running water to remove the sticky sugar coating. Drain thoroughly, then pat dry on kitchen paper and leave to dry on a fresh sheet of kitchen paper. It's really important to rinse and dry the cherries otherwise their sticky syrup drags them down through the cake mix to sit at the bottom; but you want the cherries scattered evenly throughout the cake.

2. Preheat the oven to 170°C (150°C fan), 325°F, Gas 3. Grease and **line** the tin with butter and baking paper.

3. **Sift** the flour, salt and baking powder onto a sheet of greaseproof paper and put to one side until needed.

4. Put the butter into a mixing bowl or the bowl of a mixer and **beat** with a wooden spoon or the whisk attachment until the mixture is creamy and mayonnaise-like.

5. Gradually beat in the sugar, a couple of tablespoons at a time. Once all the sugar has been added, scrape down the sides of the bowl, then beat the butter mixture well for 2 minutes until very light and fluffy.
Continued

Try Something Different

Madeira cake is made in the same way as Cherry Cake but flavoured with lemon zest and juice rather than cherries and ground almonds. Follow the recipe above, adding the finely grated zest and squeezed juice of ½ unwaxed large lemon after the eggs have been beaten in, and mix thoroughly before folding in 250g plain flour, sifted, with a pinch of salt and 1 teaspoon baking powder.

Needs a little skill

HANDS-ON TIME:
25 minutes

BAKING TIME:
60–70 minutes

MAKES:
1 medium cake

SPECIAL EQUIPMENT:
20.5cm round deep cake tin (preferably loose-based) or a springclip tin

METHOD USED:
Creamed method page 24

STORAGE:
Keep for up to 6 days in an airtight container

6. Scrape down the sides of the bowl again. Break the 3 medium eggs into a small bowl and beat with a fork until just broken up. Gradually beat the eggs into the butter mixture a tablespoon at a time, beating well each time and scraping down the sides of the bowl from time to time.

7. Sprinkle the 50g ground almonds over the mixture and **fold** in with a large metal spoon.

8. Tip the cherries into a separate bowl, sprinkle over a tablespoon of the sifted flour and toss gently – the flour stops the cherries sinking to the bottom of the cake.

9. Sift the rest of the flour, once more, onto the cake mixture and gently but thoroughly fold in using a large metal spoon. Scatter the cherries and any remaining flour in the bowl on top and fold in until evenly distributed.

10

10. Scrape the mixture into the prepared tin and spread evenly. Make a slight hollow (about 1cm deep) in the centre of the mixture so that the cake rises evenly.

11. Bake for 60–70 minutes until the cake is a good golden brown and firm to the touch. **Check** your cake is cooked by inserting a cocktail stick or skewer into the centre; if it comes out clean, it is ready.

12. Set the tin on a wire rack and leave until completely cold. Carefully take out of the tin and peel off the lining paper.

Devil's Food Cake

This American cake is made with soured cream, giving the sponge a moist, light crumb. The marshmallow frosting requires patient whisking, but the luxurious result is worth the effort.

For the sponge
100g good-quality dark chocolate (around 70 per cent cocoa solids), chopped
175ml boiling water
4 tablespoons cocoa powder
1 teaspoon bicarbonate of soda
125g unsalted butter, softened
350g caster sugar
2 medium eggs, at room temperature
1 teaspoon vanilla essence
300g plain flour, sifted
125ml soured cream, at room temperature

For the filling and frosting
2 medium egg whites, at room temperature
350g white caster sugar
1½ tablespoons maple syrup
1 teaspoon vanilla extract
good pinch of salt

HANDS-ON TIME:
80 minutes

BAKING TIME:
20 minutes

MAKES:
1 large cake

SPECIAL EQUIPMENT:
3 × 20.5cm round deep sandwich cake tins

METHOD USED:
Creamed method page 24

STORAGE:
Keep for up to 2 days in an airtight container

1. Preheat the oven to 180°C (160°C fan), 350°F, Gas 4. Grease and **line** the tins with butter and baking paper.

2. To make the sponge, put the chocolate into a heatproof medium bowl set over a pan of steaming hot but not boiling water – don't let the base of the bowl touch the water – and **melt** the chocolate gently. Remove the pan from the heat.

3. While the chocolate is melting, pour the boiling water over the cocoa powder in a heatproof bowl and mix or whisk with a hand wire whisk thoroughly until smooth and lump-free. Stir in the bicarbonate of soda and leave to cool for 5 minutes. Pour this onto the melted chocolate and stir or whisk until smooth and thoroughly combined. Put to one side until needed.

4. Put the butter into a large mixing bowl or the bowl of a food-mixer and **beat** well with an electric whisk or the whisk attachment until creamy. Gradually add the sugar to the bowl and beat well, scraping down the mixture that has splattered on the sides of the bowl from time to time, until thoroughly combined. Don't worry about the texture of the mixture at this point, it won't be light and fluffy like a regular creamed mixture, and that's fine.

5. Beat the eggs with the vanilla, using a fork, just until broken up. Gradually beat into the butter mixture, beating well after each addition and scraping down the sides of the bowl as before.
Continued

6. On a very slow speed, mix in the 300g sifted plain flour in five batches, alternating with the 125ml soured cream. When thoroughly combined, scrape down the sides of the bowl and mix in the lukewarm chocolate mixture, again on a very slow speed. When thoroughly combined with no streaks, divide the mixture equally among the three prepared tins (for this cake it is best to do this by weight for accuracy), and then spread out evenly.

7. Bake for about 20 minutes. **Check** your cakes are cooked by inserting a cocktail stick or skewer into the centre of each cake; if it comes out clean, they are ready. After 14 minutes, check the cakes are cooking evenly, if they aren't, turn the tins round. Run a round-bladed knife around the inside of each tin to loosen the sponges, then leave for a minute to firm up. **Turn out** onto a wire rack and leave to cool completely.

8. Now make the marshmallow fluff filling and frosting. Put the 2 egg whites and 350g caster sugar into a large heatproof bowl and set it over a pan of gently simmering water – don't let the base of the bowl touch the water.

9. Using an electric hand whisk or a rotary hand whisk, **whisk** for 10 seconds, then whisk in the 1½ tablespoons maple syrup, 1 teaspoon vanilla extract, 125ml cold water and a good pinch of salt. Keep whisking on full speed for 12 minutes until the mixture has become thick and glossy and holds a soft peak when the whisk is lifted. Remove the bowl from the pan and whisk again for about 15 minutes until the mixture has cooled to room temperature and has turned snowy white, very thick and holds a stiff peak when the whisk is lifted. Spoon out one-third of the mixture for the filling, then cover the bowl as the rest will be used for the frosting (this stops dark cake crumbs sneaking in).

10. Before the mixture starts to set, quickly assemble the cake: set one sponge crust-side down on a serving platter and spread with half the filling mixture, top with a second sponge, crust-side up, and spread with the rest of the filling. Cover with the third sponge, crust-side up.

11. Divide the reserved frosting in half, then, using a clean round-bladed knife or off-set palette knife, quickly cover the top and sides of the cake with one portion of the frosting. This is the 'crumb catcher' layer – it doesn't need to be neat. Clean the knife and swirl the remaining frosting over the top of the first layer so that it is evenly covered. Leave uncovered on the worktop to firm up for about 5 hours or overnight before serving.

Try Something Different

To make **chocolate curls**, gently melt 50g dark chocolate (70 per cent cocoa solids) and pour it onto a clean marble slab (or onto your worktop). Quickly spread with a metal spatula to a rectangle 20 × 30cm and 2mm thick. When it is set and has turned dull, use a sharp knife at a 45-degree angle away from you to shave off delicate curls of chocolate.

Black Forest
Gateau

The ultimate celebratory cake. Layers of light **whisked** sponge made with cocoa instead of flour will hone your whisking skills. You can really go to town on the decadent decorations.

For the sponge
9 medium eggs, at room temperature
large pinch of salt
200g caster sugar
90g cocoa powder

For the filling and topping
850g jar black cherries in light syrup flavoured with Kirsch (700g drained weight), or 850g jar black cherries in light syrup and 3 tablespoons Kirsch
475ml double or whipping cream, well chilled

1 teaspoon vanilla extract
3 tablespoons caster sugar
50g dark chocolate, grated

HANDS-ON TIME:
55 minutes

BAKING TIME:
20–25 minutes

MAKES:
1 large cake

SPECIAL EQUIPMENT:
3 × 20.5cm round deep sandwich cake tins; Piping bag fitted with a star nozzle

METHOD USED:
Whisked method page 26

STORAGE:
Keep in the fridge in an airtight container and eat within 48 hours

1. Preheat the oven to 180°C (160°C fan), 350°F, Gas 4. Grease and **line** the tins with butter and baking paper.

2. To make the sponge, separate the eggs, putting the whites into a large and spotlessly clean grease-free bowl or the bowl of a food-mixer, and the yolks into a separate bowl. Add the salt to the egg whites and **whisk** with an electric whisk or the whisk attachment until stiff peaks form when you lift the whisk from the bowl. Put the bowl to one side for now (or transfer the whites to another bowl if you need to reuse the mixer bowl).

3. Add the sugar to the egg yolks and whisk for 4–5 minutes until the mixture is very thick and mousse-like and the whisk leaves a thick ribbon-like trail when lifted out of the mixture.

4. Very gently **fold** the egg whites into the yolk mixture in three batches using a large metal spoon. **Sift** the cocoa powder over the mixture and fold in really well so that there are no streaks.

5. Divide the mixture equally among the three prepared tins (for this cake it is best to do this by weight for accuracy) and gently spread evenly.

6. Bake the sponges for 20–25 minutes. **Check** the cakes are cooked – they should spring back when gently pressed in the centre and start to shrink away from the sides of the tins. After 16 minutes, check the cakes, if they are not baking evenly, turn the tins round.

7. Set the tins on a wire rack and leave to cool completely then **turn out** and peel off the lining paper.
Continued

8. Thoroughly drain the 850g cherries, reserving the syrup. Pat the cherries dry on kitchen paper and reserve 12 for decoration.

9. To make the filling and topping, **whip** the 475ml well-chilled double or whipping cream with the 1 teaspoon vanilla until soft peaks form when you lift the whisk (the cream whips best if the bowl and whisk first have been chilled too). Sprinkle the 3 tablespoons caster sugar over the cream and whip for no more than 10 seconds until combined and the cream is slightly thicker. Set aside one-quarter of the cream for piping, covering and chilling it until needed. Divide the rest of the cream in half: one portion to sandwich the sponges, the other to cover the cake.

10. To assemble the cake, set one sponge crust-side down onto a serving plate. Brush well with some of the reserved cherry syrup (add the 3 tablespoons Kirsch to the 75ml light syrup). Leave to soak for 5 minutes, then spread half the 'sandwiching' cream. Arrange half the remaining cherries over the cream. Set another sponge on top, crust-side down, and brush with more of the syrup. Leave for 5 minutes, then gently spread with the rest of the 'sandwiching' cream and add the cherries as before. Set the last sponge on top, crust-side up, and lightly brush with cherry syrup.

11. Put the cake into the fridge (very lightly covered with a sheet of clingfilm) for about 30 minutes, just to firm it up, then cover the top and sides of the cake with the 'covering' cream using a clean round-bladed knife or offset palette knife. Chill as before for 15 minutes.

12. To decorate the cake, spoon the cream reserved for piping into the **piping** bag fitted with the star nozzle and decorate the top of the cake with piped swirls or a neat rope, then add the reserved cherries and the 50g grated dark chocolate. Cover lightly and chill until ready to serve. Store in the fridge in an airtight container and eat within 48 hours.

Flourless Rich Chocolate Torte

This dark chocolate delight is perfect for gluten-free bakers, because the rich sponge is made using ground almonds instead of flour. Get creative with chocolate shards for a dramatic look.

For the sponge
125g good-quality dark chocolate (about 70 per cent cocoa solids)
125g unsalted butter, at room temperature
100g caster sugar
4 medium eggs, at room temperature, separated
good pinch of salt
125g ground almonds

For the covering
125g good-quality dark chocolate (about 70 per cent cocoa solids)
1 tablespoon Amaretto liqueur
100g unsalted butter, softened

To decorate
75g good-quality dark chocolate (around 70 per cent cocoa solids)
cocoa powder or edible gold dust, for dusting (optional)

HANDS-ON TIME:
55 minutes

BAKING TIME:
25 minutes

MAKES:
1 medium cake

SPECIAL EQUIPMENT:
20.5cm round springclip tin or deep round cake tin; Non-stick baking paper, or a re-usable silicone sheet, or a printed-chocolate transfer sheet

METHOD USED:
Creamed method page 24

STORAGE:
Bake the day before. Keep for up to 5 days in an airtight container

1. Preheat the oven to 180°C (160°C fan), 350°F, Gas 4. Grease and **line** the tin with butter and baking paper.

2. To make the sponge, break up the dark chocolate into even-sized pieces and put it into a heatproof bowl set over a pan of steaming but not boiling water – don't let the base of the bowl touch the water – and **melt** the chocolate very gently, stirring occasionally until smooth. Remove the bowl from the pan and leave to cool until needed.

3. Put the butter into a large mixing bowl or the bowl of a mixer and **beat** well with an electric whisk or the whisk attachment until the butter is creamy and mayonnaise-like. Set aside 1 tablespoon of the caster sugar for the egg whites, then gradually beat the remaining sugar into the butter, a couple of tablespoons at a time. Scrape down any mixture from the sides of the bowl and beat for a further 2 minutes until the mixture looks pale and fluffy. Scrape

down any mixture again and beat in the 4 egg yolks, one at a time, beating well after each addition.

4. Put the 4 egg whites and salt into a separate large spotlessly clean and grease-free mixing bowl and **whisk** until soft peaks form when the whisk is lifted. Sprinkle the reserved sugar over the whites, then whisk for another minute or so until the mixture stands in stiff peaks.

5. Pour the cooled melted chocolate into the butter and sugar mixture and gently **fold** in using a large metal spoon or plastic spatula. Add the ground almonds and fold in. Add one-quarter of the egg whites and gently stir them in with the metal spoon or spatula to loosen the mixture. Then very gently fold in the rest of the whites in three batches, taking care not to knock out all the air.
Continued

6. Spoon the mixture into the prepared tin and spread evenly then bake for about 25 minutes. **Check** your cake is cooked by inserting a cocktail stick or skewer halfway between the side of the tin and the centre; it should come out clean but the centre should still be moist – it's important not to overcook this cake.

7. Set the tin on the worktop on top of a wet tea towel, run a round-bladed knife around the inside of the tin to loosen the sponge and leave to cool.

8. When the cake is cold, **turn out** from the tin and make the covering. Break up the 125g dark chocolate, put it into a heatproof bowl with the 1 tablespoon Amaretto liqueur and **melt** gently as in Step 2. Remove the bowl from the pan and stir in the 100g soft unsalted butter a little at a time to make a smooth, glossy icing. If the chocolate starts to firm up before all the butter has been mixed in, set the bowl back over the hot water and stir until smooth again.

9. Set the cake upside down on a wire rack set over a baking sheet (to catch the drips) and, as soon as the chocolate is pourable but starting to thicken, pour it over the cake and let it drizzle down the sides. Use a round-bladed knife or an off-set palette knife to neaten up and smooth the sides. Transfer to a cake board or serving plate and leave until almost set.

10. Meanwhile, make the chocolate decoration. Break up the 75g dark chocolate and melt as in Step 2. Remove the bowl from the pan and stir until smooth, then pour the chocolate onto the baking paper/silicone sheet/chocolate transfer sheet and spread it out fairly thinly. Work quickly so that the chocolate doesn't firm up before it is evenly spread. Leave in a cool spot but not in the fridge (otherwise beads of moisture will appear on it when it is back at room temperature).

11. Once set, carefully break the chocolate into shards and arrange them on top of the cake – you can either gently press them into the icing so that the shards stand vertically like sails, or scatter them over the icing to cover.

12. Dust with cocoa powder or edible gold dust, using a clean brush to gently scatter it evenly.

13. To cut neat slices, use a warmed large, sharp knife. Keep a jug of very hot water and kitchen paper at hand so you can dip the knife in the water and wipe it clean and dry between each slice.

Try Something Different

For an even more intense flavour, replace the ground almonds with ground whole unblanched almonds. Lightly toast 125g whole unblanched almonds for 7 minutes in an oven preheated to 180°C (160°C fan), Gas 4. Leave to cool, then grind to a fine powder in a food processor.

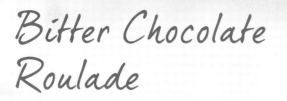
Bitter Chocolate Roulade

This elaborate roulade is made from a butterless and flourless mix which produces a rich yet delicate sponge. It requires a little more skill than the Classic Swiss Roll (page 94).

For the sponge

175g good-quality dark chocolate (around 75 per cent cocoa solids), chopped or broken up
6 medium eggs, at room temperature
good pinch of salt
175g caster sugar
1 tablespoon cocoa powder
icing sugar, for dusting

For the filling

225ml double cream, well chilled
250g tin crème de marrons (sweetened chestnut purée)
1 teaspoon dark rum or brandy

To decorate

marrons glacés (candied chestnuts) or grated chocolate or chocolate shavings or curls

1. Preheat the oven to 180°C (160°C fan), 350°F, Gas 4. Grease and **line** the Swiss roll tin with butter and baking paper.

2. To make the sponge, put the dark chocolate into a heatproof bowl and set over a pan of steaming but not boiling water – don't let the base of the bowl touch the water – and **melt** the chocolate, gently stirring now and then.

3. Remove the bowl from the pan, stir gently until smooth, then leave to cool until needed.

4. Separate the eggs, putting the whites in one large mixing bowl or the bowl of a food-mixer, and the yolks into another.

5. Add the salt to the egg whites and **whisk** using an electric whisk or the whisk attachment, until the whites stand in stiff peaks when the whisk is lifted. Put to one side for now (or spoon the whites into another clean bowl if you need the mixer bowl).

6. Add the caster sugar to the egg yolks and whisk for about 3 minutes with the electric whisk or the whisk attachment (there's no need to wash the whisk) until the mixture is very thick and mousse-like and the mixture leaves a distinct ribbon-like trail when the whisk is lifted.

7. Pour the cooled melted chocolate into the bowl and gently **fold** in with a plastic spatula or large metal spoon. Add one-quarter of the stiffly whisked egg whites to the bowl and stir in to loosen the mixture, then gently fold in the rest of the whites in three batches using the plastic spatula or large metal spoon.

8. **Sift** the cocoa powder into the bowl and fold in very carefully – you want to make sure it is well mixed in with no streaks, but you don't want to knock out all the air.
Continued

Needs a little skill

HANDS-ON TIME:
45 minutes

BAKING TIME:
20–25 minutes

MAKES:
1 large cake

SPECIAL EQUIPMENT:
23 × 33cm Swiss roll tin or 27 × 37cm sheet parchment-lined foil folded into a 23 × 33 × 2cm case and set foil-side down on a baking sheet

METHOD USED:
Whisked method page 26

STORAGE:
Cover lightly and keep cool stored in an airtight container in the fridge. Serve the same or next day

12

9. Carefully scrape the mixture into the prepared tin (or the case set on a baking sheet) and gently ease it into the corners so that the tin is evenly filled (otherwise the sponge will rise unevenly). Again, be really careful not to knock out the air.

10. Bake for 20–25 minutes until well risen. **Check** the cake is cooked – the top should feel just firm when lightly pressed with a finger. Remove the tin

or case from the oven and set it on a wire rack (carefully slide the case onto the rack from the baking tray). Leave the sponge to cool in the tin or case – don't worry if it sinks, that's fine.

11. While the sponge is cooling, make the filling. **Whip** the 225ml chilled double cream to soft peaks. Stir together the 250g chestnut purée and 1 teaspoon dark rum or brandy until smooth, then lightly stir through the whipped cream to give a marbled or swirled effect. Cover the bowl and chill until needed.

12. Set a large sheet of baking paper on the worktop and dust it lightly with icing sugar. **Turn out** the sponge onto the paper, then remove the tin and lining paper. Using a large sharp knife, make a shallow cut along one short edge about 2cm in. Gently spread the marbled chestnut cream over the sponge to within 2cm of all the edges.

13. Gently fold the cut edge over and then roll up the sponge fairly tightly using the sugar-dusted baking paper to help you pull the roll into a neat shape. Finish with the join at the far end underneath. The roll WILL crack so don't worry if it does, it's all part of the look! Trim off the ends with a large sharp knife to neaten.

14. Gently lift the roll onto a serving plate and decorate with marrons glacés or chocolate shavings or curls (see page 127 and above for how to do this).

Try Something Different

If you prefer a chocolate-based filling, omit the chestnut purée and fold in 50g finely grated dark chocolate (around 70 per cent cocoa solids) instead. Decorate with shavings of white chocolate or dust with cocoa powder or icing sugar, rather than use the marrons glacés as decoration.

Sachertorte

The legendary rich **whisked** chocolate cake from the Hotel Sacher in Vienna is a great way to practise writing with icing – have a few trial runs first before you go for it on the cake!

Needs a little skill

HANDS-ON TIME:
50 minutes

BAKING TIME:
35 minutes

MAKES:
1 medium cake

SPECIAL EQUIPMENT:
22cm springclip tin or deep round cake tin; Small disposable piping bag

METHOD USED:
Whisked method page 26

STORAGE:
Keep for up to 1 week in an airtight container

For the sponge
175g good-quality dark chocolate (about 70 per cent cocoa solids), chopped or broken up
100g unsalted butter, diced
4 medium eggs, plus 1 egg white, at room temperature
good pinch of salt
115g caster sugar
½ teaspoon vanilla paste
100g plain flour

For the apricot glaze
175g apricot conserve
1 teaspoon lemon juice

For the chocolate glaze
165g good-quality dark chocolate (about 70 per cent cocoa solids)
20g unsalted butter
4 tablespoons double cream

To finish
30g dark chocolate, melted, for piping

1. Preheat the oven to 170°C (150°C fan), 325°F, Gas 3. Grease and **line** the tin with butter and baking paper.

2. To make the sponge, put the chocolate pieces into a heatproof bowl with the butter. Set the bowl over a pan of steaming hot but not boiling water – don't let the base of the bowl touch the water – and **melt** the chocolate gently, stirring occasionally. When melted and smooth, remove the bowl from the pan and put to one side until needed.

3. Separate the 4 eggs; put the 5 egg whites into a large mixing bowl or the bowl of a food-mixer, and the 4 yolks into a separate large bowl. Add the salt to the egg whites, then **whisk** with an electric whisk or the whisk attachment until soft peaks form. Whisk in 85g of the caster sugar a tablespoon at a time, then whisk briefly until it makes stiff peaks. Put the meringue to one side (or spoon it carefully into another bowl if you need the mixer bowl) until required.

4. Add the remaining 30g caster sugar and the vanilla paste to the yolks and whisk (there's no need to wash the whisk) on a high speed until very thick and the whisk leaves a ribbon-like trail when lifted from the mixture. Gently stir in the chocolate mixture with a plastic spatula – the mixture will be very thick.

5. **Sift** the flour onto the mixture, add one-quarter of the meringue and stir in with a large metal spoon or spatula – take your time as the mixture will be quite stiff. **Fold** in the rest of the meringue in three batches. When thoroughly combined – check there are no lumps of meringue or streaks of chocolate – transfer the mixture to the prepared tin and spread evenly.
Continued

6. Bake for about 35 minutes until just firm when gently pressed in the centre. **Check** your cake is cooked by inserting a cocktail stick or skewer into the centre; if it comes out clean, it is ready. Set the tin on a wire rack, run a round-bladed knife around the inside of the tin to loosen, then leave to cool for 5 minutes before unmoulding. Leave on the wire rack until completely cool.

7. To make the apricot glaze, gently heat the 175g apricot conserve in a small pan with the 1 tablespoon cold water and 1 teaspoon lemon juice, then press the mixture through a sieve into a small bowl. Set the cake upside down on a board and slice into two **layers** horizontally. Return the bottom layer (this was the top of the cake) to the wire rack, cut side up, and spread with half the apricot mixture. Cover with the second sponge layer so that the cut sides are together. Gently reheat the glaze and brush it over the top and sides of the sponge. Leave to cool on the wire rack.

8. To make the chocolate glaze, break or chop up the 165g dark chocolate and put it into a heatproof bowl with the 20g unsalted butter and gently melt, as in Step 2. Remove the bowl from the heat and stir until smooth. Heat the 4 tablespoons double cream until almost, but not quite, boiling, then leave for a minute before pouring onto the chocolate mixture in a thin, steady stream while stirring constantly. Pour the glaze over the cake and gently ease it down the sides to completely and

smoothly coat the cake, using a palette knife to carefully smooth it if needed. Leave to set in a cool spot but not in the fridge. Transfer the cake to a board.

9. To decorate, melt the remaining 30g dark chocolate as in Step 2 and pour it into a small **piping** bag.

10. Snip off the tip of the piping bag to give a fine writing tip about 3mm across. Pipe 'Sacher' across the top of the cake for a professional finish!

Gâteau le Progrès

Circles of hazelnut meringue sandwiched between layers of rich buttercream and topped with praline. A mouthwatering tower of sweetness and a bake that is worth the challenge.

For the hazelnut meringue
6 medium egg whites, at room temperature
good pinch of salt
300g golden caster sugar
150g ground roasted hazelnuts
2 tablespoons cornflour

For the praline
200g whole unblanched hazelnuts
200g caster sugar

For the buttercream
100g caster sugar
4 medium egg yolks, at room temperature
250g slightly salted butter, softened (but not oily)

HANDS-ON TIME:
2 hours

BAKING TIME:
1½ hours

MAKES:
1 medium cake

SPECIAL EQUIPMENT:
2 × baking sheets;
Large piping bag fitted with a 1.5cm plain nozzle; Sugar thermometer

METHOD USED:
None of the traditional methods. This is a meringue cake.

STORAGE:
The cake is at its best after a night in the fridge, but will keep for up to 5 days in an airtight container in the fridge. (Remove 30 minutes before serving)

1. Preheat the oven to 120°C (100°C fan), 250°F, Gas ½. Lightly grease the baking sheets with butter.

2. Using a cake tin or plate as a guide, draw three 20cm circles on a sheet of baking paper the same size as your baking sheets. Don't cut out the circles but press the baking paper, marked-side down, onto the baking sheets.

3. To make the meringue, put the egg whites and salt into a large, spotlessly clean mixing bowl or the bowl of a food-mixer. Spoon 6 tablespoons of the golden caster sugar into a separate small bowl and put to one side until needed. Put the remaining sugar into a separate bowl and thoroughly mix in the ground roasted hazelnuts and cornflour.

4. Using an electric whisk or the whisk attachment, **whisk** the egg whites on a slow speed until frothy, then increase the speed until the mixture stands in soft peaks when the whisk is lifted out of the bowl. Sprinkle the reserved 6 tablespoons sugar over the whites, then whisk for a further 30 seconds until the mixture stands in stiff peaks and is glossy and smooth.

5. Sprinkle one-third of the hazelnut mixture over the meringue and fold in lightly using a large metal spoon. **Fold** in the rest of the hazelnut mixture in two batches.
Continued

6. Transfer the mixture to the **piping** bag fitted with the plain nozzle. Pipe three discs of meringue inside the drawn circles. Start in the centre and pipe spirals. If necessary, go back and fill in any gaps. The trick with this cake is to pipe the meringue discs in neat rings and then get them nice and crisp in the oven.

7. Bake the meringue rounds for about 1½ hours until a very light golden colour and crisp and dry. For more even baking, you can rotate the sheets after 50 minutes. Remove the baking sheets from the oven, set them on a wire rack and leave the rounds until cold before peeling them off the lining paper.

8. While the rounds are cooling make the praline. Have an oiled baking sheet ready. Put the 200g whole unblanched hazelnuts and 200g caster sugar into a medium-sized heavy-based pan and set over a low heat. Stir occasionally with a metal spoon (the caramel will stick like glue to a wooden spoon) until the sugar starts to melt. When all the sugar has melted, increase the heat to medium and cook until the sugar starts to colour, giving the pan a gentle shake from time to time. Then stir gently to make sure the nuts brown evenly, taking care as the mixture will be very hot, until the mixture turns a chestnut-brown colour. Immediately – and carefully, it's very hot! – pour it onto the baking sheet and leave until cool and set.

9. Now make the buttercream. Put the 100g caster sugar and 6 tablespoons cold water into a small heavy-based pan and heat gently, without boiling, until the sugar dissolves. Bring to the boil until the syrup reaches 110°C/230°F on a sugar thermometer – this will take about 5 minutes. Watch the syrup carefully as you don't want it to start to caramelise.

10. Meanwhile, put the 4 medium egg yolks into a heatproof bowl set on top of a damp cloth (to prevent the bowl slipping) and whisk for a few seconds with an electric whisk just to combine. When the syrup reaches the required temperature, pour it into the bowl in a thin steady stream, whisking constantly at high speed. You can also do this in a large free-standing mixer but aim for the yolks and not the sides of the bowl or the syrup will harden before it is whisked in. Continue whisking for about 5 minutes until the mixture becomes very thick and mousse-like, pale in colour and the whisk leaves a distinct ribbon-like trail when lifted. Whisk until the mixture is completely cold, then gradually whisk in the 250g soft butter.

11. Break the praline into pieces, then put half into a food-processor and pulse to a fine powder. Don't overwork the praline, though, or it will turn oily and paste-like. Stir the powder into the buttercream, cover and, if necessary, chill until firm enough to spread.

12. Select 12 large chunks of the reserved hazelnut praline for the decoration and set aside, then roughly chop the rest using a large sharp knife. Keep this for decorating the sides of the cake.

13. Set one meringue round on a serving plate. Divide the buttercream into four equal portions. Spread one portion over the meringue round and set a second one on top. Spread with a second portion of buttercream and add the last round of meringue. Spread one portion of buttercream over the top and use the remaining portion to cover the sides. Press the chopped praline around the bottom of the sides of the cake. Pop it in the fridge to chill for 15 minutes.

14. Decorate the top of the cake with the reserved praline chunks before serving.

Try Something Different

Swap the hazelnuts for unblanched almonds for a different flavour. Lightly toast 150g whole unblanched almonds for 7 minutes in an oven preheated to 180°C (160°C fan), Gas 4. Leave to cool then grind to a fine powder in a food processor and use for the meringue layers instead of the ground almonds. You could also replace the hazelnuts in the praline with 200g whole unblanched almonds.

Sea Salt
Caramel Cake

This classic **creamed** sponge has a lovely light crumb thanks to the addition of buttermilk. The feathered salty caramel topping looks simple but takes a little skill to master.

Needs a little skill

HANDS-ON TIME:
90 minutes

BAKING TIME:
22–25 minutes

MAKES:
1 large cake

SPECIAL
EQUIPMENT:
3 × 20.5cm deep
round sandwich
cake tins; Small
disposable piping
bag

METHOD USED:
Creamed method
page 24

STORAGE:
Bake the the day
before. Keep for
up to 4 days in an
airtight container

For the sponge
250g unsalted butter, softened
300g golden caster sugar
4 medium eggs
1 teaspoon vanilla extract
300g self-raising flour
good pinch of salt
5 tablespoons buttermilk, at room temperature

For the filling and frosting
250g unsalted butter, softened
450g dark muscovado sugar
175ml double cream
250g icing sugar, sifted
1/4–1/2 teaspoon sea salt flakes
100g good-quality dark chocolate (about 70 per cent cocoa solids), chopped

1. Preheat the oven to 180°C (160°C fan), 350°F, Gas 4. Grease and **line** the tins with butter and baking paper.

2. To make the sponge, put the butter into a large mixing bowl or the bowl of a food-mixer and **beat** well with a wooden spoon or the whisk attachment until creamy. Gradually beat in the caster sugar, then scrape down any mixture that has splattered on the sides of the bowl and beat for a further 2 minutes or until well mixed and lighter (although the mixture will be slightly heavier than the usual creamed mixture).

3. Break the eggs into a separate bowl, add the vanilla extract and beat with a fork just until broken up, then gradually beat into the butter mixture, beating well after each addition and scraping down the sides of the bowl every now and then. Add a tablespoon of the self-raising flour with each of the last two additions of egg to prevent the mixture curdling.

4. **Sift** the remaining flour and salt into the bowl, then add the buttermilk and gently **fold** everything together using a large metal spoon.

5. When thoroughly combined, divide the mixture among the three prepared tins (for this cake it is best to do this by weight for accuracy) and spread evenly.

6. Bake the cakes for 22–25 minutes until the sponges are golden. **Check** the cakes are cooked – they should start to shrink away from the sides of the tins and spring back when lightly pressed in the centre with your fingertip. Check the cakes after 20 minutes, and if they are not baking evenly, turn the tins round.

Continued

7. Run a round-bladed knife around the inside of the tins to loosen the sponges, then leave to cool and firm up for 2 minutes before **turning out** onto a wire rack. Leave to cool completely.

8. While the sponges are cooling, make the filling and frosting. Put 175g of the butter into a heavy-based medium pan with the 450g dark muscovado sugar and 175ml double cream. Set over a low heat and stir with a wooden spoon until melted and lump-free, then increase the heat and bring the mixture to a boil, stirring constantly. As soon as it boils, turn down the heat so that the mixture gently simmers and simmer for 5 minutes, stirring frequently so that it doesn't catch around the base.

9. Carefully pour into a heatproof mixing bowl and immediately start to whisk with an electric whisk while slowly and gradually beating in the 250g sifted icing sugar. Continue beating after all the sugar has been added, for about 8 minutes or until the mixture is fluffy and cooled to barely warm. Gradually beat in the remaining 75g soft butter (if the mixture is too warm it will melt the butter and the mixture will be too soft). Sprinkle over ¼ teaspoon of the salt and stir in with a plastic spatula – taste and add a little more salt as needed. Cover the bowl and leave on the worktop for now.

10. Put the 100g chopped dark chocolate into a small heatproof bowl and set over a pan of steaming hot but not boiling water – don't let the base of the bowl touch the water – and **melt** the chocolate gently, stirring occasionally. Remove the bowl from the heat, stir again, then spoon two-thirds into a separate bowl and fold in 250g of the caramel mixture. Put to one side until spreadable. Cover the rest of the caramel and put to one side for now.

11. Set one sponge crust-side down on a serving platter and spread with half the chocolate caramel. Cover with another sponge, crust-side up, and spread with the rest of the chocolate caramel. Using a round-bladed knife or off-set palette knife, thickly spread the caramel over the top and sides of the cake to completely cover. If the caramel mixture has become too firm to spread easily, gently warm it over a pan of steaming water, and stir it until it is smooth and spreadable.

12. If necessary, gently reheat the remaining chocolate in the same way until smooth and melted. Spoon into the **piping** bag and snip off the end to make a writing tip about 3mm across. Hold the bag vertically over the cake and, starting in the centre, pipe a spiral of chocolate on top of the cake without touching the surface with the tip. Try to keep an even, gentle pressure on the bag so that the lines are equally thick and evenly spaced. Work quickly and don't worry about the odd wobble.

13. Now gently draw the tip of a cocktail stick through both the chocolate spiral and the caramel frosting, from the centre to the sides at six regular intervals, to make the first set of lines. Then draw the cocktail stick back from the edge towards the centre to make lines in between the first set of lines.

14. Leave the cake to firm up for about 3 hours, uncovered on the worktop, before serving.

Lemon Curd
Layer Cake

Home-made lemon curd is well worth the effort as nothing tastes as sublime, and it really lifts the flavours of this light genoise sponge. Layers of luscious loveliness.

For the sponge
6 medium eggs, at room temperature
175g caster sugar
175g plain flour
good pinch of salt
extra caster sugar, for sprinkling

For the syrup
100g caster sugar
2 large unwaxed lemons

For the lemon curd
125g unsalted butter, diced
225g caster sugar
finely grated zest and juice of 3 medium unwaxed lemons
3 medium eggs, at room temperature

For the filling and covering
250ml double cream, well chilled
2 tablespoons caster sugar
400g ricotta
2–3 teaspoons limoncello, to taste (optional)

To decorate
50g toasted flaked almonds
3 tablespoons Lemon Curd

HANDS-ON TIME:
2 hours

BAKING TIME:
20 minutes

MAKES:
1 large cake

SPECIAL EQUIPMENT:
2 × 20.5cm round deep sandwich cake tins; Small disposable piping bag

METHOD USED:
Whisked method page 26

STORAGE:
Put the cake into an airtight container and chill in the fridge for 8 hours or overnight so that it can firm up and the flavours can develop. (Remove from the fridge 30 minutes before serving.) Keep for up to 3 days in the fridge

1. Preheat the oven to 180°C (160°C fan), 350°F, Gas 4. Grease and **line** the tins with butter and baking paper.

2. To make the sponge, put the eggs into a large mixing bowl or the bowl of a food-mixer and **whisk** using an electric whisk or the whisk attachment for 30 seconds so that the eggs are frothy, then whisk in the caster sugar. Whisk on a high speed for about 5 minutes until the mixture is very thick, pale and mousse-like, and the whisk leaves a thick ribbon-like trail when lifted.

3. **Sift** the flour and salt into a separate bowl, then sift one-third of that again onto the whisked mixture. Very gently **fold** in using a large metal spoon,

taking care not to knock out the air you've whisked in. Repeat, adding the remaining flour in two batches. When there are no streaks of flour visible, divide the mixture equally between the two prepared tins (for this cake it is best to do this by weight for better accuracy) and gently spread evenly.

4. Bake the sponges for 20 minutes, or until well risen and a light golden brown. **Check** the cakes are cooked – they should be just firm when gently pressed in the centre. Run a round-bladed knife around the inside of the tins to loosen the sponges. Cover a wire rack with a sheet of baking paper and sprinkle with a little caster sugar. Carefully **turn out** the sponges onto the paper and leave to cool completely.
Continued

5. While the sponges are cooling, make the syrup. Put 100ml cold water and the 100g caster sugar into a small pan. Finely grate the zest from one of the large unwaxed lemons and add it to the pan. Using a vegetable peeler, carefully peel the zest from the second lemon into strips, then cut the strips into fine needle-like shreds. Wrap tightly in clingfilm and set aside for decoration. Halve both lemons and squeeze the juice – add 3 tablespoons to the pan. Set the pan over a low heat and stir until the sugar has dissolved, then bring to the boil and simmer for 2 minutes to make a light syrup. Leave to cool.

6. To make the lemon curd, put the 125g butter, 225g caster sugar and the finely grated zest and juice of the 3 lemons into a large heatproof bowl. Set the bowl over a pan of just simmering water, not letting the base of the bowl touch the water, and stir gently with a wooden spoon until the mixture is completely melted and smooth.

7. Carefully remove the bowl from the pan and strain the 3 eggs into the mixture (this will make sure there are no fragments of shell or tough white membranes). Stir well, then set the bowl over the pan of simmering water again and stir constantly until the mixture becomes thick and opaque. Take your time – don't be tempted to turn up the heat as the eggs will scramble if the

mixture gets anywhere near boiling. The lemon curd is ready when you can draw a finger through the mixture on the wooden spoon and leave a clear path.

8. Immediately lift the bowl from the pan and transfer the lemon curd to a clean bowl or a screw-topped jar. When cold, cover tightly and chill. The lemon curd can be kept in the fridge for up to 2 weeks.

9. To make the filling and covering, pour the 250ml chilled double cream into a mixing bowl and **whip** with an electric whisk or the whisk attachment (the cream whips better if the bowl and whisk have been chilled) until thick. Sprinkle the 2 tablespoons caster sugar into the bowl and whisk briefly just until soft peaks form when the whisk is lifted. Stir the lemon curd until smooth (reserving 3 tablespoons for the decoration), then mix 175g of it into the 400g ricotta, then gently fold this into the whipped cream. Cover and chill for about 30 minutes. Transfer slightly less than half the lemon cream to a separate bowl (this will be used to cover the cake), cover and return to the fridge.

10. Stir the 2–3 teaspoons limoncello (if using) into the rest of the lemon curd, cover and chill until needed.

11. When you're ready to assemble the cake, gently slice each sponge into two **layers** horizontally using a serrated bread knife.

12. Set one of the four sponge layers cut-side up on a serving plate. Brush with one-quarter of the syrup and leave for 2–3 minutes to soak in. Using a round-bladed knife or offset palette knife, spread one-third of the remaining lemon cream over the soaked sponge. Using another knife, carefully spread one-third of the limoncello lemon curd on top.

13. Set a second sponge layer, cut-side up, on top and repeat the process: brushing with syrup, spreading with lemon cream, then lemon curd. Set a third sponge layer on top, cut-side up, and repeat. Finally, top with the last sponge layer, this time with the crust-side uppermost, and brush with the last of the lemon syrup. Cover and chill for 30 minutes.

14. Spread the reserved lemon cream over the top and sides of the cake to cover.

15. Press the 50g flaked almonds around the sides of the cake using your hands.

16. Spoon the reserved lemon curd for the decoration into the **piping** bag, snip off the tip and pipe in a zigzag pattern over the top of the cake. Scatter the shreds of reserved lemon zest over the top of the cake.

Chess Cake

Appearances can be deceptive. An understated cake from the outside, but one that will impress when you cut the first slice to reveal a chequerboard of chocolate and vanilla squares inside.

Up for a challenge

HANDS-ON TIME:
60 minutes

BAKING TIME:
22–25 minutes

MAKES:
1 large cake

SPECIAL
EQUIPMENT:
3 × 20.5cm round
deep sandwich cake
tins (or springclip
tins); 2 piping bags,
each fitted with a
1.5cm plain piping
nozzle

METHOD USED:
Creamed method,
see page 24

STORAGE:
Bake the day before.
Keep for up to
4 days in an airtight
container

For the sponge
350g unsalted butter, softened
350g caster sugar
6 medium eggs, at room temperature
1 teaspoon vanilla extract
350g self-raising flour
couple of pinches of salt
50g cocoa powder
4 tablespoons milk, at room
temperature

For the white chocolate ganache filling
175g good-quality white chocolate
(around 25 per cent cocoa solids), very
finely chopped

125ml whipping cream
50g unsalted butter
pinch of salt

For the dark chocolate ganache covering
300g good-quality dark chocolate
(around 70 per cent cocoa solids),
very finely chopped
300ml whipping cream

For the decoration
Dark and white chocolate shavings
and curls, or grated chocolate, or
ready-made chocolate decorations

1. Preheat the oven to 180°C (160°C fan), 350°F, Gas 4. Grease and **line** the tins with butter and baking paper. Use springclip tins if you already have them, but sandwich tins are best as its easier to pipe the mixture into shallow tins.

To make the sponge
2. To make the whisked sponge, put the butter into a large mixing bowl or the bowl of a food-mixer and **beat** with a wooden spoon or the whisk attachment until very creamy and mayonnaise-like. Scrape down any mixture splattered on the sides of the bowl, then beat in the caster sugar a couple of tablespoons at a time, scraping down the sides of the bowl from every now and then. Beat well for a couple of minutes until the mixture is light and fluffy.

3. Break the eggs into a separate bowl, add the vanilla extract and beat with a fork until broken up. Gradually beat the eggs into the butter mixture, a tablespoon at a time, beating well after each addition and scraping down the sides of the bowl occasionally. Add 1 tablespoon of the flour with the last two additions of egg to stop the mixture curdling.

4. **Sift** the remaining flour and the salt into the bowl and carefully **fold** in using a large metal spoon. Divide the mixture in half – using scales or by eye. Put one half into a separate mixing bowl. Sift the cocoa powder into one portion with 2 tablespoons of the milk and fold in until combined and streak-free. Fill one of the **piping** bags fitted with the plain nozzle with the chocolate mixture.
Continued

5. Add the remaining 2 tablespoons milk to the other bowl of cake mixture and fold in with a clean spoon. Transfer this mixture to the other piping bag.

6. Set all three prepared tins in front of you. Take the bag containing the dark chocolate mixture and pipe a ring around the inside edge of one tin.

7. Now pick up the bag containing the vanilla mixture and pipe a ring inside the chocolate ring.

8. Repeat, piping alternate dark chocolate and white vanilla rings, until the tin is filled. Do not spread or even out the mixtures – leave them looking just like a bull's-eye. Do exactly the same thing with the second tin, starting with a dark chocolate ring around the edge.

9. For the third tin, do the reverse – use the vanilla mixture to pipe the outermost ring, then pipe a chocolate ring inside and repeat until it is filled.

10. Bake the sponges for 22–25 minutes. To **check** if the cakes are cooked, lightly press them in the centre – they should be well risen and springy. Run a round-bladed knife around the inside of the tins to loosen the sponges, leave to firm up for 1 minute, then **turn out** onto wire racks and leave to cool completely

To make the white chocolate ganache filling

11. Once the sponges are cold, make the white chocolate ganache filling. Put the 175g chocolate into a heatproof bowl. Put the 125ml cream, 50g butter and salt (this helps it set well) into a small pan and heat until steaming hot but not quite boiling. Pour onto the chocolate in a thin, steady stream. Leave to stand for a minute, then stir gently until melted and smooth. Leave for a few minutes until it is thick enough to spread easily.

12. Set one sponge with a dark chocolate outer ring crust- or top-side down on a cake board or serving platter. Spread with half the white chocolate ganache, then set the sponge with the white vanilla outer ring on top. Spread with the remaining white ganache and top with the last sponge, crust- or top-side uppermost. Leave to firm up while you make the other ganache.

To make the dark chocolate ganache filling

13. Make the dark chocolate ganache in the same way as the white: put the 300g chocolate into a heatproof bowl. Heat the 300ml cream until very hot but not boiling and pour over the chocolate. Leave to melt for a couple of minutes, then stir gently. Leave until thick enough to spread.

14. When it is nice and thick, using a knife, spread the ganache evenly over the top and sides of the cake. Leave in a cool spot – but not the fridge – until set and firm, then decorate with the grated chocolate, chocolate shavings and curls or decorations.

Porter
Fruitcake

A rich fruit cake hides beneath a light and pretty exterior. Perfect for celebrating the arrival of spring, the beautiful blooms look impressive but are surprisingly easy to make.

For the cake mixture
500g mixed dried fruit
100ml Porter or stout
175g unsalted butter, softened
175g dark muscovado sugar
3 medium eggs, at room temperature
200g plain flour
good pinch of salt
50g ground almonds
1½ teaspoons ground mixed spice
85g walnut pieces or chopped toasted almonds

For covering the cake
2 tablespoons apricot jam, warmed
800g ready-made marzipan
icing sugar, for dusting
500g royal icing made with 500g royal icing sugar (or use ready-to-roll or rolled white or ivory icing or ready-coloured icing/sugarpaste)

For the rose decoration
250g ready-to-roll white icing/sugarpaste
yellow edible gel, liquid or paste food colour
cornflour, for dusting
edible lustre, for spraying (optional)

Up for a challenge

HANDS-ON TIME:
2½ hours

BAKING TIME:
2 hours

MAKES:
1 large cake

SPECIAL EQUIPMENT:
20.5cm springclip tin or deep round cake tin; Newspaper; Cake board; Tape measure; Food preparation gloves (optional); Small silicone flower mould

METHOD USED:
Creamed method, see page 24

STORAGE:
Ready after 5 days, keep for up to 2 months in an airtight container

To make the cake mixture

1. Tip the dried fruit into a mixing bowl, pour over the Porter or stout and mix well. Cover tightly and leave to soak overnight on the worktop.

2. Next day, preheat the oven to 170°C (150°C fan), 325°F, Gas 3. Grease and **line** the tin with butter and baking paper.

3. Put the butter into a large mixing bowl or the bowl of a food-mixer and **beat** well with a wooden spoon or the whisk attachment for a couple of minutes until creamy and mayonnaise-like. Scrape down any mixture that has splattered on the sides of the bowl. Press out any lumps in the muscovado sugar (you may need to sieve it), then gradually beat it into the butter, scraping down the sides of the bowl from time to time. Once all the sugar has been added, beat well for 4 minutes or until the mixture turns much lighter in colour and texture. Scrape down the sides of the bowl again.

4. Break the eggs into a separate bowl and beat with a fork just until broken up, then gradually beat into the butter mixture, a tablespoon at a time, beating well after each addition. If it looks like it is about to curdle or separate, add a tablespoon of the plain flour with the last addition of egg.

5. **Sift** the remaining flour, salt, almonds and mixed spice into the bowl and gently but thoroughly **fold** in using a large metal spoon or plastic spatula. Add the soaked fruit (plus any liquid) and gently fold in.
Continued

6. When thoroughly combined, add the 85g walnut pieces or chopped toasted almonds and fold in. Spoon the mixture into the prepared tin and spread it out evenly. Bang the tin down on the worktop to knock out any pockets of air, then make a shallow hollow in the centre of the cake so that it will rise evenly.

7. Line the baking sheet with several sheets of newspaper and set the tin on top. Fold several more sheets into strips slightly wider than the depth of the tin and long enough to go all around it.

8. Wrap these strips around the tin and secure them with paper clips or string.

9. Bake the cake for about 2 hours. **Check** your cake is cooked by inserting a cocktail stick or skewer into the centre; if it comes out clean, it is ready. If necessary, turn the cake a couple of times during baking so that it cooks evenly and cover the top with a sheet of baking paper or foil if you think it is getting too brown. Remove the tin from the oven, set it on a wire rack and leave until cold, then turn out. Remove the lining paper, then wrap in baking paper then foil, and leave in a cool dry spot for at least a week – 4 weeks if possible – before finishing the cake.

To cover the cake
10. Unwrap the cake and set it upside down on the cake board or serving plate – the flat base of the cake is easier to cover. Brush all over with the 2 tablespoons warm apricot jam.

11. Knead the 800g ready-made marzipan until smooth and pliable. If there is a gap between the cake and the board, make a thin roll of marzipan and press it around the base of the cake so the sides are flat and straight. Lightly dust the worktop with icing sugar and roll out the remaining marzipan to a round large enough to cover the top and sides of the cake – for accuracy use a tape measure.

12. Carefully wrap the marzipan around the rolling pin and lift it over the cake. Gently unroll it so the edge just touches the board, then continue unrolling it over the cake so that it covers the cake evenly and touches the board all around.

13. Now remove all rings/bangles/watch – you don't want to make dents in the marzipan – then use the palms of your hands to smooth the marzipan on the top and down the sides of the cake, to make sure it is well fixed and there are no pockets of air. Find a sharp knife and roughly trim away the excess marzipan.

14. Lift the cake and board and trim the marzipan with a sharp knife until flush with the board. Leave the cake very loosely covered in a cool dry spot for a day or two to let the marzipan firm up.

15. Brush the marzipan very lightly with a little cooled boiled water (or brandy).

16. To make your own royal icing, mix the 500g royal icing with 80ml water, using an electric whisk or mixer, until smooth and thick. It should stand in soft peaks and leave a solid trail.

Continued

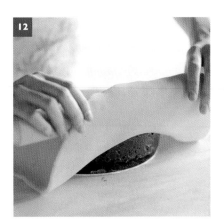

17. Spread the royal icing over the top and sides, smoothing it with a palette knife. (If you are using bought icing, knead and roll it out as for the marzipan then smooth and polish it with the palms of your hands.) Use the back of a small knife to tuck the icing under the edge of the cake (press the edge of the rolled icing under) so that it looks neat.

To make the decorations

18. When you are ready to decorate the cake, knead the 250g ready-to-roll white icing/sugarpaste until smooth, then tint it to the palest shade you want, kneading in just 1–2 drops of yellow edible gel, liquid or paste food colour for 4–5 minutes until even, with no streaks (it's a good idea to wear food preparation gloves to avoid staining your hands).

19. Divide the paste into four equal portions: wrap one portion tightly with clingfilm or put it in a snap-close plastic food bag, then add another 1–2 drops of food colour to the next portion, knead in and wrap tightly. Continue with all the remaining two portions of the sugarpaste.

20. When you are happy with the range of shades, you can make the shaped decorations. Make sure the silicone mould is clean and dry, then dust it lightly with cornflour. Mould a little of the lightest-coloured icing into a ball the size of a marble (or follow the instructions on the pack) and press it into the mould, making sure the icing is level with the edge of the mould.

21. Turn the mould over and gently release the blossom onto a sheet of baking paper. Repeat with the rest of the icing in this colour, then with the other three shades. Once dry and firm, the blossoms can be sprayed with edible lustre, if you like. You can paint the centres of the flowers darker yellow with edible cake decorating powder colour mixed with a couple of drops of vodka. For a little added sparkle, you could delicately brush dry powder lustre onto the petals – very fine brushes are available from specialist shops.

22. While the blossoms dry, if you like, cut a ribbon long enough to go around the cake and tie it around it tightly and neatly, securing it at the back.

23. When the blossoms are set and dry, start to attach them to the cake using small dabs of royal icing applied with a cotton bud.

24. Dot the flowers over the top of the cake, mixing up the colours for a pretty effect. Leave uncovered to firm up, then store carefully in a covered container in a cool and dry place.

Mile-high
Chocolate Cake

An out-of-this-world chocolate cake; three very light layers of **creamed** sponge mixture filled and frosted with a whipped mousse-like chocolate ganache buttercream.

For the sponge
60g cocoa powder
250ml boiling water
30g good-quality dark chocolate (around 70 per cent cocoa solids), finely chopped
225g unsalted butter, softened
325g light muscovado sugar
2 medium eggs plus 2 yolks, at room temperature
½ teaspoon vanilla extract
4 tablespoons crème fraîche
225g self-raising flour
good pinch of salt

For the filling and frosting
200g good-quality dark chocolate (about 70 per cent cocoa solids), finely chopped
125ml double cream
2 medium egg whites, at room temperature
200g caster sugar
450g unsalted butter, softened and diced

To finish
75g chocolate-coated popping candy
100g good-quality dark chocolate (around 70 per cent cocoa solids), finely chopped

Up for a challenge

HANDS-ON TIME:
2¼ hours

BAKING TIME:
25–28 minutes

MAKES:
1 large cake

SPECIAL EQUIPMENT:
3 × 20.5cm deep round sandwich cake tins; 20.5cm cake board; Strip of acetate or chocolate transfer sheet

METHOD USED:
Creamed method, see page 24

STORAGE:
Bake the day before. Keep for up to 4 days in an airtight container in the fridge

1. Preheat the oven to 180°C (160°C fan), 350°F, Gas 4. Grease and **line** the tins with butter and baking paper.

To make the sponge
2. **Sift** the cocoa powder into a heatproof bowl and pour on the boiling water while whisking constantly with a hand wire whisk. When the mixture is smooth and lump-free, whisk the chocolate in until melted and perfectly smooth. Put to one side to cool.

3. Meanwhile, put the butter into a large mixing bowl or the bowl of a food-mixer and **beat** with a wooden spoon or the whisk attachment until creamy and mayonnaise-like. Scrape down any mixture splattered on the sides of the bowl. You need lump-free muscovado sugar for this, so push it through a sieve,

then gradually beat the fine sugar into the butter on a low speed. Once all the sugar has been added, scrape down the sides of the bowl, then beat on a high speed for about 4 minutes until the mixture looks fluffy and paler in colour. Stop and scrape down the sides of the bowl from time to time.

4. Put the 2 whole eggs and 2 yolks into a separate bowl (reserve the whites for the filling and frosting). Add the vanilla extract and beat with a fork just until combined. Gradually beat the egg mixture into the butter mixture a tablespoon at a time, beating well after each addition and scraping down the sides of the bowl as before.
Continued

5. Add the cooled chocolate mixture to the bowl and start to **fold** in, using just two movements, with a large metal spoon or plastic spatula. Spoon the 4 tablespoons crème fraîche into the bowl, sift over the 225g flour and a good pinch of salt, and gently fold everything together.

6. When thoroughly combined, divide the mixture equally among the three prepared tins – if you want to be more precise use your scales, or you can just judge it by eye – and spread evenly. Bake for 25–28 minutes. Check the cakes after 20 minutes and if they are not baking evenly, turn the tins round. **Check** if the cakes are cooked – they're ready when the sponges are well risen, starting to shrink away from the sides of the tins and the centre of each sponge springs back when gently pressed in the centre.

7. Run a round-bladed knife around the inside of each tin to loosen the sponges, then **turn out** onto a wire rack and leave to cool.

To make the filling and frosting
8. While the sponges are cooling, make the filling and frosting. Put the chocolate into a heatproof bowl. Heat the 125ml double cream until just boiling, then pour it over the chocolate and stir until melted and smooth. Leave to cool to room temperature, stirring occasionally.

9. Meanwhile, put the 2 egg whites into a large heatproof mixing bowl. Add the 200g caster sugar and **whisk** with a hand-held electric mixer for a minute

until slightly frothy. Set the bowl over a pan of simmering water – don't let the base of the bowl touch the water – and whisk for 7 minutes to make a very thick, white, glossy stiff meringue. Lift the bowl off the pan and set it on a damp cloth (so it doesn't slip) and whisk for about 8 minutes until the mixture is completely cold.

10. Once the mixture is cold, gradually whisk in the 450g butter – if the meringue is even slightly warm the butter will melt and the mixture will turn oily. Once all the butter has been mixed in, whisk for 2 minutes to make a very light and fluffy buttercream.

11. Now whisk in the cooled but fluid ganache and whisk for 2 minutes to make a thick and light mousse-like mixture that stands in soft peaks when the whisk is lifted. It needs to be firm enough to spread easily and hold its shape so that it doesn't slide off the cake so, if necessary, cover the bowl and chill for a few minutes. Check the consistency every 5 minutes – you don't want it to become too firm to spread.

12. To assemble the cake, spoon half the buttercream into a separate bowl and reserve for covering the cake. Stir the 75g chocolate-coated popping candy into the remaining buttercream – this will be the filling. Set one sponge crust-side down on a cake board and spread with half the buttercream filling. Set a second sponge on top and cover with the rest of the buttercream filling. Place the third sponge on top, crust-side up.

13. Use half the reserved buttercream to cover the sides of the cake – use an off-set palette knife to spread the frosting evenly and neatly, making sure the sides are smooth and straight. Spoon the rest of the frosting on top of the cake and spread and swirl it to thickly cover. Leave in the fridge to firm up.

14. When the frosting feels just firm, gently **melt** the 100g chocolate in a heatproof bowl set over a pan of steaming hot but not boiling water – don't allow the base of the bowl to touch the water.

15. Cut the acetate or chocolate transfer sheet into a strip as wide as the cake is high and long enough to go around the cake – about 8 × 65cm. Set on a piece of baking paper, then spread with the melted chocolate so that the strip is completely and evenly covered.

16. Pick up the strip and press it around the sides of the cake so that the melted chocolate sticks to the frosting. If necessary trim the strip with small scissors so that it just meets. Leave in a cool spot until the chocolate has set. Carefully peel off the acetate, leaving the shiny chocolate band around the cake. Decorate with candles or more popping candy and leave in a cool spot until ready to serve.

Fraisier

An extravagant summery combination of orange biscuit sponge, Grand Marnier and strawberries. Crème diplomate is crème pâtissière lightened with the addition of whipped cream.

For the sponge
4 medium eggs, at room temperature
pinch of salt
125g caster sugar
1 large navel orange
125g plain flour
50g unsalted butter, melted
and cooled

For the crème diplomate
500ml creamy milk
6 medium egg yolks, at room
temperature
100g caster sugar
50g cornflour
1 tablespoon orange liqueur
50g unsalted butter, softened
150ml double cream, well chilled

For the syrup
75g caster sugar
4 tablespoons orange juice
1 tablespoon orange liqueur

To assemble
225g white marzipan
icing sugar, for dusting
pink or green gel/paste food colour
(optional)
about 600g medium-sized
strawberries (avoid 'giant' out-of-
season ones)
50g dark chocolate, chopped

HANDS-ON TIME:
3 hours

BAKING TIME:
25 minutes

MAKES:
1 large cake

SPECIAL
EQUIPMENT:
22cm springclip tin;
Strip of acetate to fit
inside the tin; Large
piping bag fitted
with a 1.5cm plain
piping nozzle; Small
disposable piping
bag

METHOD USED:
Whisked method,
see page 26

STORAGE:
Keep for up to
2 days in an airtight
container

1. Preheat the oven to 180°C (160°C fan), 350°F, Gas 4. Grease the tin with butter, dust it with flour and **line** the base with baking paper.

To make the sponge
2. Separate the eggs, putting the whites into a large, spotlessly clean, grease-free mixing bowl or food-mixer bowl, and put the yolks into a separate bowl. Add the pinch of salt to the whites and **whisk** with an electric whisk or the whisk attachment until they stand in soft peaks. Gradually whisk in half the sugar, a couple of tablespoons at a time, to make a light meringue. Put the bowl aside for now. If you need to re-use the bowl, carefully spoon the whites into another bowl or dish. There's no need to wash the whisk.

3. Add the remaining caster sugar to the yolks. Finely grate the zest from the navel orange into a small bowl. Halve the orange and squeeze out the juice. Add half the zest to the yolks and 2 tablespoons of the juice – keep the rest of the juice and zest for later. Start on a low speed, then gradually increase the speed and whisk for 4–5 minutes until the mixture becomes very thick, light and mousse-like, and falls in thick ribbons from the whisk when it is lifted. *Continued*

4. Fold in the meringue in three batches, using a large metal spoon or plastic spatula. **Sift** the 125g plain flour on top and very gently but thoroughly fold it into the mixture. Drizzle the 50g cooled melted butter into the bowl and delicately fold it in. Take your time, it is essential that all the ingredients are really well combined and there are no blobs of meringue or streaks of flour.

5. Pour the mixture into the prepared tin and spread it evenly. Bake for about 25 minutes until pale golden brown. To **check** if the cake is cooked, gently press the centres – they should spring back and the sides should have shrunk away from the sides of the tin. Set the tin on a wire rack, run a round-bladed knife around the inside to loosen the sponge, then leave for 5 minutes to firm up (be patient here, the sponge is fragile). **Turn out** onto the wire rack and leave until cold. Wash and thoroughly dry the tin so that it is ready to assemble the cake.

To make the crème diplomate
6. This is a crème pâtissière with added butter, whipped cream and liqueur. Heat the 500ml creamy milk in a heavy-based pan until steaming hot but not boiling, then remove from the heat.

7. Put the 6 egg yolks, 100g caster sugar, 50g cornflour and the reserved orange zest into a heatproof bowl and **beat** well with a wooden spoon for a couple of minutes until it is lighter in colour and very smooth. Stir in the hot milk, then when it is well combined, pour the mixture back into the pan.

8. Set the pan over a medium heat and stir constantly until the mixture boils and thickens to make a very thick, smooth custard. Remove the pan from the heat and stir in the 1 tablespoon orange liqueur and the 50g butter. Wash and dry the heatproof bowl, then pour the hot crème pâtissière into it. Press a piece of clingfilm or dampened baking paper onto the surface of the custard (to prevent a skin forming), leave to cool, then chill thoroughly for at least 4 hours.

9. Whip the 150ml chilled double cream until it is thick and stands in soft peaks when the whisk is lifted. Whisk the cold crème pâtissière until smooth, then fold in the whipped cream using a large metal spoon or plastic spatula. Cover and chill for up to 1 hour.

To make the syrup
10. Put the 75g caster sugar and 4 tablespoons orange juice into a small pan, add 2 tablespoons cold water and heat gently, stirring until the sugar has dissolved, then bring to the boil. Simmer for 3 minutes to make a light syrup. Remove from the heat and stir in the 1 tablespoon orange liqueur. Put to one side for now.

11. Set the 225g white marzipan onto the worktop lightly dusted with icing sugar and knead for 1 minute until smooth and pliable. If you want, you can colour the marzipan with a couple of drops of pink or green food colour, add it very gradually using the tip of a cocktail stick dipped in the pot and gently knead in until evenly coloured.

12. Roll out the marzipan to a 3mm thickness and cut out a neat 22cm disc using the cake tin as a guide. Cover with clingfilm.

13. To assemble the cake, cut the cold sponge into two **layers** horizontally to make two thin and even discs of sponge. Place the acetate strip inside the clean tin so that it lines the edge. Instead of acetate sheets for the assembly, you can use clingfilm or parchment-lined foil.

14. Set one sponge disc cut-side up in the tin and brush liberally with the reheated syrup. With the back of a spoon, very gently press down the edges of the (slightly swollen) sponge cake so that it touches the acetate, with no gaps or holes.

15. Pick 12 strawberries of the same height and hull them to remove the green stalks. Cut them in half vertically, then arrange them, pointed-end up, on top of the sponge layer with the cut side pressed against the acetate – squidge them up so they fit together closely (depending on their size, you may need one more or one less). Set aside 5 good-looking strawberries as decoration, then hull and thinly slice the rest. *Continued*

16. Set aside 1 tablespoon of the crème diplomate for the top of the cake and spoon the rest into a large **piping** bag fitted with a plain piping nozzle (depending on how big your bag is you may have to fill it in two batches). Pipe a spiral of the crème over the base.

17. Pipe vertically between the berries to fill in the gaps. Arrange the rest of the berries over the crème in an even layer to fill the space inside the strawberry ring, then pipe another spiral on top and spread it out so that it's level and the strawberries are completely covered.

18. Set the second sponge disc on top, cut-side down. Gently press the sponge with the palm of your hand so that there are no gaps between the crème and the sponge, and between the sponge edges and the acetate around the tin. Brush with the rest of the hot syrup, then leave to soak in for 10 minutes. Spread the reserved crème over the sponge using an offset palette knife to thinly cover. Press the marzipan disc on top, then cover and chill for at least 6 hours but preferably overnight.

19. Meanwhile, make the decorations. **Melt** the chocolate by putting the 50g chopped dark chocolate into a small heatproof bowl and set over a pan of steaming hot but not boiling water – don't let the base of the bowl touch the water. Dip the reserved strawberries in the chocolate so that they are half covered vertically – leaving half of each berry exposed. Leave to set on baking paper or acetate.

20. Spoon the rest of the chocolate into a small piping bag, snip off the tip and pipe small freehand flowers onto baking paper or acetate set on a baking tray. Leave to set.

21. To serve, unclip the tin and remove the acetate-wrapped cake. Set on a board or serving plate and gently peel off the acetate. Decorate the top with the dipped berries and the piped decorations. Once assembled, the cake needs plenty of time to firm up in the fridge – ideally overnight. Serve chilled and use a large sharp knife for cutting the slices.

Gâteau de
L'Opéra

Three layers of joconde – a **whisked** biscuit sponge flavoured with almonds – make up this glorious gâteau with layers of coffee buttercream, ganache and chocolate glaze.

For the sponge
6 medium egg whites, at room temperature
good pinch of cream of tartar
30g caster sugar
45g unsalted butter
6 medium eggs, at room temperature
225g icing sugar, sifted
225g ground almonds
75g plain flour

For the coffee syrup
60g caster sugar
1½ tablespoons instant coffee granules or powder
1 tablespoon brandy

For the coffee buttercream
85g caster sugar
2 medium egg yolks, at room temperature
150g unsalted butter, at room temperature, diced
1½ tablespoons instant coffee granules or powder
1 tablespoon boiling water

For the ganache
200g good-quality dark chocolate (about 70 per cent cocoa solids), finely chopped
175ml whipping cream
25g unsalted butter, at room temperature

For the under-glaze
50g good-quality dark chocolate (about 70 per cent cocoa solids), finely chopped

For the top glaze
100g unsalted butter
150g good-quality dark chocolate (about 70 percent cocoa solids), finely chopped

For the decoration
50g dark chocolate, finely chopped
edible gold leaf

Up for a challenge

HANDS-ON TIME:
3½ hours

BAKING TIME:
8–10 minutes

MAKES:
1 large cake

SPECIAL EQUIPMENT:
3 sheets of parchment-lined foil; 25cm square cake board; 1–3 baking sheets (see Step 2); Sugar thermometer; Small disposable piping bag; Tweezers

METHOD USED:
Whisked method, see page 26

STORAGE:
Keep for up to 3 days in an airtight container in the fridge. (Remove about 30 minutes before serving.)

1. Preheat the oven to 220°C (200°C fan), 425°F, Gas 7.

2. Cut each sheet of parchment-lined foil to a 26cm square, then set parchment-side up and fold in the sides by 1cm to make a shallow square cake case with 25cm sides – use the cake board to help form the shape. Set each case on a baking sheet – bake in batches if necessary.
Continued

Try Something Different

For a change of flavour, replace the ground almonds in the sponge layers with finely ground hazelnuts, and replace the brandy in the brushing liquid with rum, coffee liqueur or hazelnut liqueur.

To make the sponge

3. Put the 6 medium egg whites and a good pinch of cream of tartar into a large, spotlessly clean mixing bowl or the bowl of a food-mixer and **whisk** with an electric hand whisk or the whisk attachment until the whites stand in soft peaks when the whisk is lifted. Sprinkle the 30g caster sugar over the whites and whisk again for a few seconds until the whites form stiff peaks and the mixture is smooth and glossy. Put the bowl to one side until needed. If you need to re-use the bowl, gently spoon the meringue into a separate bowl or dish. There's no need to wash the beaters/whisk.

4. Melt the 45g butter and leave to cool while you whisk the eggs. Put the 6 whole eggs into a second large bowl or the bowl of the food-mixer, and add the 225g sifted icing sugar and the 225g ground almonds. Whisk on a high speed with the electric whisk or whisk attachment for about 4 minutes until the mixture is very thick and has massively increased in volume. **Sift** the 75g plain flour on top and very gently but thoroughly **fold** it in.

5. Once all the flour has been combined, fold in the meringue in three batches. Trickle the cool but still runny butter over the top and fold in until just combined. Divide the mixture equally – if you want to be precise use your scales – among the three cake cases set on baking sheets and spread it carefully to fill the corners. The surface should be flat and level and each sponge layer should be the same depth all over so that the cake is level when assembled.

6. Bake for 8–10 minutes. Check the cakes after 5 minutes to see if they are baking evenly, if not, turn the tins round. To **check** if the cakes are cooked, gently press the centres – the sponges should be lightly browned and spring back.

7. While the sponges are baking, cover two wire racks with baking paper. Remove the baked sponges from the oven and flip the baking sheets over onto the racks to **turn out** the sponges in their cake cases. Gently peel off the cake cases and leave the sponges to cool, loosely covered with a clean, dry tea towel or a sheet of baking paper. Once cold, the sponges can be wrapped individually in baking paper then clingfilm and left at room temperature overnight.

To make the coffee syrup

8. While the sponges cool, make the coffee syrup. Put the 60g caster sugar, 1½ tablespoons instant coffee granules or powder and 125ml cold water into a small pan and heat gently, stirring, until dissolved. Bring to the boil, then simmer for 1 minute to make a light syrup. Remove from the heat and stir in the 1 tablespoon brandy. Once cold, the syrup can be covered and left at room temperature for a couple of hours or stored in a jam jar in the fridge overnight (remove 1 hour before using).

To make the coffee buttercream

9. Next make the coffee buttercream filling. Put the 85g caster sugar and 4 tablespoons cold water into a small heavy-based pan and set over a low heat. Stir gently until the sugar has dissolved, then bring to the boil. Boil rapidly for 4–5 minutes until the syrup reaches 115°C (240°F) on a sugar thermometer. While the sugar is boiling, put the 2 medium egg yolks into a heatproof mixing bowl (put it on a damp cloth so it doesn't slip) or the bowl of a food-mixer and whisk for a few seconds until frothy with an electric whisk or the whisk attachment.

10. As soon as the syrup is up to temperature, pour it onto the yolks in a thin, steady stream while whisking at top speed – aim directly for the yolks or the syrup will harden before it is whisked in. Continue whisking for about 5 minutes until the mixture is very thick, pale and mousse-like. Then whisk for 1–2 minutes more until the mixture is completely cold. Whisk in the 150g soft (but not runny or oily) unsalted butter a few pieces at a time to make a smooth, thick buttercream. Dissolve the 1½ tablespoons instant coffee granules or powder in 1 tablespoon boiling water, leave to cool, then whisk in. Cover and chill until just spreadable (check the consistency every 5 minutes).

To make the ganache

11. To **melt** the chocolate, put the 200g dark chocolate into a heatproof bowl, heat the 175ml whipping cream until steaming hot but not quite boiling, and pour it over the chocolate. Leave it for a minute, then gently stir until melted and smooth. Gradually stir in the 25g butter to make a smooth and glossy ganache. Cover and chill until spreadable (check the consistency every 5 minutes).

Continued

To make the under-glaze

12. Now to assemble the cake: first **melt** the chocolate. Put the 50g finely chopped dark chocolate into a small heatproof bowl and set over a pan of steaming hot but not boiling water – don't let the base of the bowl touch the water – and leave to melt gently.

13. Set one sponge crust-side up on a sheet of baking paper and brush on a thin, even layer of melted chocolate. Leave until set hard (if you're impatient, chill it for 5 minutes), then gently flip the sponge over on the baking paper so that the chocolate-coated side is underneath. Carefully brush the upper surface with one-third of the coffee syrup so that the whole surface is evenly moistened. Leave for 10 minutes to soak in.

14. Spoon three-quarters of the coffee buttercream onto the sponge and spread evenly – take your time assembling the cake as it's vital it has a flat surface and doesn't tilt at the sides. Set a second sponge crust-side down on top, making sure all the sides are aligned. Brush this sponge with half the remaining coffee syrup as before and leave to soak in for 10 minutes.

15. Spoon all the ganache on top and spread evenly. Set the third sponge crust-side down on top and gently press it in place, checking again that the surface is level. Brush with the remaining coffee syrup and leave to soak in, as before. Spoon the remaining buttercream onto the sponge and spread to make a thin, even layer. Chill for 15 minutes until firm.

To make the top glaze

16. Meanwhile, make the top glaze. Melt the 100g unsalted butter in a small pan, then skim off the foam that rises to the top. Pour the clear butter into a small bowl, leaving behind and discarding the milky residue in the pan. Reheat the clarified butter until bubbling. Put 125g of the finely chopped dark chocolate into a heatproof bowl (reserve the remaining 25g for now) and pour in the very hot butter. Stir gently until the chocolate has melted, then stir in the reserved 25g chocolate to make a smooth, glossy glaze.

17. Take the cake out of the fridge and set it on the worktop still on its sheet of baking paper. Now, quickly pour the top glaze over the top of the cake to completely cover it – if necessary use a warmed offset palette knife to help spread the glaze evenly. Let the glaze drip down the sides of the cake (the sides will be trimmed off later). Return the cake to the fridge and chill until the glaze is firm and set.

18. Lastly, put the finishing touches to the cake. Dip a large, sharp knife in hot water, dry it quickly, then carefully trim the sides to make a perfectly square, neat cake. Transfer to the cake board.

19. Melt the chocolate for the decoration (as in Step 10) and spoon it into a small **piping** bag. Snip off the end to make a writing tip and pipe 'Opéra' diagonally across the cake, and decorate the edges with flourishes and musical notes. Finish with gold leaf – use tweezers or the tip of a small knife to very gently set the fragile flakes in place. Chill until firm, preferably overnight.

20. To cut neat slices, use a large sharp knife dipped in hot water, and keep kitchen paper at hand for cleaning the knife between slices.

What cake shall I bake?

Conversion table

WEIGHT	
Metric	**Imperial**
25g	1oz
50g	2oz
75g	2½oz
85g	3oz
100g	4oz
125g	4½oz
140g	5oz
175g	6oz
200g	7oz
225g	8oz
250g	9oz
280g	10oz
300g	11oz
350g	12oz
375g	13oz
400g	14oz
425g	15oz
450g	1lb
500g	1lb 2oz
550g	1lb 4oz
600g	1lb 5oz
650g	1lb 7oz
700g	1lb 9oz
750g	1lb 10oz
800g	1lb 12oz
850g	1lb 14oz
900g	2lb
950g	2lb 2oz
1kg	2lb 4oz

VOLUME	
Metric	**Imperial**
30ml	1fl oz
50ml	2fl oz
75ml	3fl oz
125ml	4fl oz
150ml	¼ pint
175ml	6fl oz
200ml	7fl oz
225ml	8fl oz
300ml	½ pint
350ml	12fl oz
400ml	14fl oz
450ml	¾ pint
500ml	18fl oz
600ml	1 pint
725ml	1¼ pints
1 litre	1¾ pints

SPOON MEASURES	
Metric	**Imperial**
5ml	1 teaspoon
10ml	2 teaspoons
15ml	1 tablespoon
30ml	2 tablespoons
45ml	3 tablespoons
60ml	4 tablespoons
75ml	5 tablespoons

LINEAR	
Metric	**Imperial**
2.5cm	1in
3cm	1¼in
4cm	1½in
5cm	2in
5.5cm	2¼in
6cm	2½in
7cm	2¾in
7.5cm	3in
8cm	3¼in
9cm	3½in
9.5cm	3¾in
10cm	4in
11cm	4¼in
12cm	4½in
13cm	5in
14cm	5½in
15cm	6in
16cm	6½in
17cm	6½in
18cm	7in
19cm	7½in
20cm	8in
22cm	8½in
23cm	9in
24cm	9½in
25cm	10in

Index

Acknowledgements

Hodder & Stoughton and Love Productions would like to thank the following people for their contribution to this book:

Linda Collister, Laura Herring, Caroline McArthur, Sam Binnie, Helena Caldon, Alasdair Oliver, Kate Brunt, Laura Del Vescovo, Joanna Seaton, Sarah Christie, Anna Heath, Damian Horner, Auriol Bishop, Anna Beattie, Rupert Frisby, Jane Treasure, Sharon Powers.

The author would also like to thank Alan Hertz, Barbara Levy and Simon Silverwood.

First published in Great Britain in 2015
by Hodder & Stoughton
An Hachette UK company

1

Copyright © Love Productions Limited 2015
Photography Copyright © David Munns 2015

The right of Linda Collister to be identified as the Author of the Work has been asserted by her in accordance with the Copyright, Designs and Patents Act 1988.

BBC and the BBC logo are trademarks of the British Broadcasting Corporation and are used under licence. BBC logo © BBC 1996.

A CIP catalogue record for this title is available from the British Library

Hardback ISBN 978 1 473 61525 0
Ebook ISBN 978 1 473 61526 7

Editorial Director: Nicky Ross
Editor: Sarah Hammond
Project Editor: Laura Herring
Series Editor: Linda Collister
Art Director: James Edgar
Layouts: Nicky Barneby
Photographer: David Munns
Food Stylist: Natalie Thomson, Lizzie Harris
Props Stylist: Victoria Allen

Typeset in Dear Joe, Mostra, Kings Caslon and Gill Sans
Printed and bound in Italy by L.E.G.O. Spa

Hodder & Stoughton policy is to use papers that are natural, renewable and recyclable products and made from wood grown in sustainable forests. The logging and manufacturing processes are expected to conform to the environmental regulations of the country of origin.

Hodder & Stoughton Ltd
Carmelite House
50 Victoria Embankment
London EC4Y 0DZ

www.hodder.co.uk

Continue on your journey to star baker with the other titles in *The Great British Bake Off: Bake It Better* series, the 'go to' baking books which give you all the recipes and baking know-how you'll ever need.

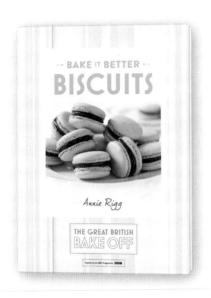

DON'T JUST BAKE. BAKE IT BETTER.